# Introductory Mathematics for Industry, Science, and Technology

# Introductory Mathematics for Industry, Science, and Technology

**Keith Roberts**
**Leo Michels**

Milwaukee Area Technical College

**Brooks/Cole Publishing Company**

I(T)P® An International Thomson Publishing Company

Pacific Grove • Albany • Belmont • Bonn • Boston • Cincinnati • Detroit
Johannesburg • London • Madrid • Melbourne • Mexico City • New York • Paris
Singapore • Tokyo • Toronto • Washington

**Brooks/Cole Publishing Company**
A Division of International Thomson Publishing Inc.


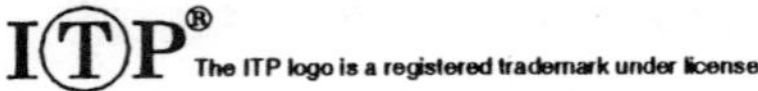




Printed in the United States of America
10 9 8

**Library of Congress Cataloging in Publication Data**

Roberts, Keith J.
Introductory mathematics for industry, science, and technology.

Includes index.
1. Mathematics—1961- —Programmed instruction.
I. Michels, Leo. II. Title.
QA39.2.R614 1985 510′.7′7 85-7714
ISBN 0-534-05148-0

Sponsoring Editor: *Craig Barth*
Editorial Assistant: *Eileen Galligan*
Production Editor: *Penelope Sky*
Production Assistants: *Dorothy Bell* and *Dan K. McQueen*
Manuscript Editor: *Cece Munson*
Permissions Editor: *Carline Haga*
Interior and Cover Design: *Katherine Minerva*
Art Coordinator: *Judith L. Macdonald*
Interior Illustration: *Carl Brown*
Typesetting: *Omegatype Typography, Inc., Champaign, Illinois*
Printing and Binding: *Malloy Lithographing, Inc., Ann Arbor, Michigan*

# PREFACE

## To the Instructor

This text is intended for introductory courses in basic mathematics for students in trade or technical areas, and can also serve as the foundation for more advanced study. Students who master the topics we present will be able to do practical work involving architectural and mechanical drafting, electronics, welding, air conditioning, aviation and automotive mechanics, machining and construction, and can acquire many other skills.

We have organized the material carefully to provide flexibility: the text may be used in a lecture class, in a laboratory setting, or for self-paced instruction; it is designed to be student-interactive and is highly readable.

Exercise sets appear regularly and frequently in each chapter. These sets are paired: answers to the regular exercises are in the back of the book, while answers to the supplementary sets are in the instructor's manual. Each chapter concludes with a self-test accompanied by answers. In all, students can practice over 3,000 problems, many enhanced by diagrams. Further, whenever necessary we show worked-out solutions as well as the answers.

Additional important features include the presentation of problem-solving techniques, instructions on using the calculator, explanation of the metric system, and how to estimate answers. We also discuss significant digits and dimensional analysis, and offer applied problems relating to actual situations in trades and technologies.

## To the Student

This text will help you master the skills needed in technical courses, such as basic math, the use of measurement systems, and the strategies of problem solving. Each chapter is divided into frames that present the individual concepts on which the major topics are based. Every concept is first explained and then illustrated with an example. Questions about the material test your understanding of the concept. The answers, on the right side of the page, are staggered, so that the answer to frame 1 is next to frame 2, and so on.

Exercise sets appear at intervals throughout the chapters, so you can check your understanding of the topics under discussion. The answers to the exercise sets are in the back of the book (no answers are given for the supplementary exercise sets). A self-test at the end of each chapter provides further opportunity for mastering the material; the answers to these are also in the back of the book.

As you work the problems and answer the questions in the frames, incorrect answers will show you what area needs more studying, and you should reread those frames that explain the subject. You will then be able to work the exercise sets successfully.

## Acknowledgments

We would like to thank our reviewers: Dorothy Crepin, Lower Columbia College, Longview, Washington; Molly Fails, Terra Technical College, Fremont, Ohio; Edward Laughbaum, Columbus Technical Institute, Columbus, Ohio; Don

Osborne, Sumter Area Tech, Sumter, South Carolina; Charles Rich, Southeast Community College, Milford, Nebraska; Anthony Vavra, West Virginia Northern Community College; and Roger Wege, Fox Valley Technical Institute, Appleton, Wisconsin.

*Keith Roberts*
*Leo Michels*

# CONTENTS

## Chapter 1 Basic Mathematics 1



## Chapter 2 Fractions 48



## Chapter 3 Basic Concepts of Algebra 75

## Chapter 7 The Metric System of Measurement 179



## Chapter 8 The Powers and Roots of Numbers 208



## Chapter 9 Geometry 223



## Chapter 10 Graphing 254

# Introductory Mathematics for Industry, Science, and Technology

# CHAPTER 1

# Basic Mathematics

Success in technical occupations requires a mastery of basic mathematical skills, because most information needed in technical fields is in quantitative form. Typically, first some measurements are obtained, and then calculations must be done using these measurements. This chapter will help you to increase your skills in working with whole numbers and decimals and in using a calculator.

## Section 1-1 Place Names of Whole Numbers and Decimal Numbers

**1**

The Hindu-Arabic number system is used throughout the world. It is a base-ten system and uses ten digits (0, 1, 2, 3, 4, 5, 6, 7, 8, 9) to express the values of numbers. The location of a digit determines the value of that digit in a number.

**Example** In the number 46, the digit 4 means four tens, or 40.

**a.** In 57, the digit 5 means five tens, or ______.
**b.** In 32, the digit 3 means ______ tens, or ______.

**2**

In the preceding frame, each of the digits in question was in the "tens" place. Therefore, the 4 had a value of 40, the 5 had a value of 50, and the 3 had a value of 30. The concept of determining the value of a digit by its location in the number is known as *place value*. Figure 1-1 shows the value of each place in a whole number.

**1. a.** 50
**b.** three; 30

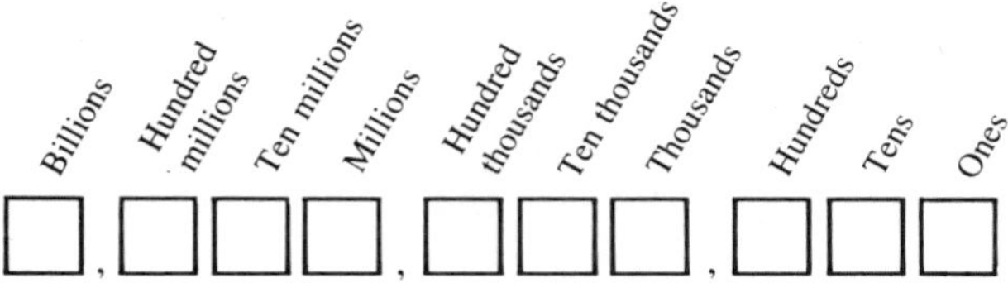


**Figure 1-1** Place values for whole numbers

**Example** The number 4,876 is a four-digit number.
The 4 is in the thousands place and means 4,000.
The 8 is in the hundreds place and means 800.
The 7 is in the tens place and means 70.
The 6 is in the ones place and means 6.

For the number 65,832, state which digit is in the:

**a.** ten-thousands place ______.
**b.** hundreds place ______.
**c.** ones place ______.

## 3

For the number 9,254,870, state which digit is in the:

**a.** millions place ______.
**b.** ten-thousands place ______.
**c.** tens place ______.

2. **a.** 6
**b.** 8
**c.** 2

## 4

A decimal number is a number whose value falls between that of two whole numbers.

> **Example** 3.8 is a decimal number. It represents a number whose value is greater than 3 and less than 4.

A decimal number with a value between 0 and 1 is represented by a 0 to the left of the decimal point.

> **Example** 0.4 is a decimal number.

Therefore, 4.5, 0.6, and 534.07 are all decimal numbers.

Which of the following are decimal numbers? (Circle your selections.)

**a.** 243 **b.** 0.07 **c.** 11,047 **d.** 2.015

3. **a.** 9
**b.** 5
**c.** 7

## 5

Each decimal number has a definite number of decimal places, which is determined by the number of digits to the right of the decimal point.

> **Example** 5.793 has three decimal places; 17.01 has two decimal places; and 0.0506 has four decimal places.

How many decimal places are in each of these numbers?

**a.** 1.785 ______ **b.** 0.501 ______ **c.** 9.0007 ______

4. **b** and **d**

## 6

The names of the places to the right of the decimal point are shown in Figure 1-2. Notice that all of these names end in *ths.*

5. **a.** 3
**b.** 3
**c.** 4

Decimal point
0.
Tenths | Hundredths | Thousandths | Ten thousandths | Hundred thousandths | Millionths

**Figure 1-2** Place values for decimal numbers

**Example** The number 58.279 has three decimal places.
The 2 is in the tenths place and means 0.2.
The 7 is in the hundredths place and means 0.07.
The 9 is in the thousandths place and means 0.009.
*Note:* The 5 and the 8 are whole-number digits.
The 5 is in the tens place and means 50.
The 8 is in the ones place and means 8.

For the decimal number 287.0543, state which digit is in the:

**a.** tenths place ______.
**b.** thousandths place ______.
**c.** ten-thousandths place ______.
**d.** hundreds place ______.

**7**

For the decimal number 0.000975, state which digit is in the:

**a.** thousandths place ______.
**b.** ten-thousandths place ______.
**c.** hundred-thousandths place ______.
**d.** millionths place ______.

**6. a.** 0
**b.** 4
**c.** 3
**d.** 2

**8**

Answers to frame 7.

**a.** 0 **b.** 9 **c.** 7 **d.** 5

# Section 1-2 Naming Whole Numbers and Decimal Numbers

**1**

Commas are placed in numbers to make them easier to read. Starting from the right and counting to the left, a comma is placed to the left of each third digit. In the example below, a comma is placed to the left of the 7, the 5, and the 1.

**Example** 6,142,583,709

Notice that this system means a comma is used after the billions place, the millions place, and the thousands place. The number is read as: six billion, one hundred forty-two million, five hundred eighty-three thousand, seven hundred nine (or 6 billion, 142 million, 583 thousand, 709).

The number 7,247,982,542 is read as:
________ billion, ________ million, ________ thousand, ________.

**2**

The number four billion, twenty-six million, two hundred thousand, one hundred seventeen is written:

4,026,200,117

**1.** 7
247
982
542

Notice that zeros are used as place holders in this number so that each digit has the correct place value.

**Example** Two million, seventy-six thousand, fifteen is written as 2,076,015.

**a.** Write the number seven million, twenty-five thousand, ten.

____________________

**b.** Write the number five billion, ninety-two million, two thousand, forty-five.

____________________

### 3

The number five million, twenty-seven is written:

5,000,027

Notice that, even though thousands are not named in the number, three zeros are used as place holders for the thousands digits. (This number could be thought of as five million, zero thousand, twenty-seven.)

**a.** Write the number seven million, eighteen.

____________________

**b.** Write the number nine billion, forty-seven thousand.

____________________

**2. a.** 7,025,010
**b.** 5,092,002,045

### 4

A decimal number smaller than 1 is read as if it were a whole number with the place name of the last digit added.

**Examples** The number 0.9 is read as nine-*tenths*.
The number 0.072 is read as seventy-two *thousandths*.
The number 0.006 is read as six-*thousandths*.
The number 0.0548 is read as five hundred forty-eight *ten-thousandths*.

Name the following decimal numbers.

**a.** 0.6 is read as ____________________.
**b.** 0.85 is read as ____________________.
**c.** 0.009 is read as ____________________.
**d.** 0.0783 is read as ____________________.

**3. a.** 7,000,018
**b.** 9,000,047,000

### 5

In the following examples, the word name is converted to a decimal number.

**Examples** Eight-tenths is written as 0.8.
Fifty-seven thousandths is written as 0.057.

**a.** Write the number seven-tenths.

____________________

**b.** Write the number thirteen-millionths.

____________________

**c.** Write the number forty-nine thousandths.

____________________

**d.** Write the number two hundred seven ten-thousandths.

____________________

**4. a.** six-tenths
**b.** eighty-five hundredths
**c.** nine-thousandths
**d.** seven hundred eighty-three ten-thousandths

**6**

A decimal number larger than 1 is read as a whole number, followed by the word "and," followed by the decimal part of the number.

**Examples** The number 25.4 is read as twenty-five and four-tenths. The number 147.08 is read as one hundred forty-seven and eight-hundredths.

**a.** Write the number sixty and seven-tenths.

______________

**b.** Write the number five and seventeen-hundredths.

______________

**c.** Write the number four hundred and eight-thousandths.

______________

**d.** Write the number two thousand, fifteen and six-tenths.

______________

**5. a.** 0.7
**b.** 0.000013
**c.** 0.049
**d.** 0.0207

**7**

Answers to frame 6.

**a.** 60.7  **b.** 5.17  **c.** 400.008  **d.** 2,015.6

# Exercise Set, Sections 1-1–1-2

## Place Names of Whole Numbers and Decimal Numbers

What is the place name of the 5 in each of these numbers?

**1.** 547 ______________
**2.** 3,562,481 ______________
**3.** 5,278,000,021 ______________
**4.** 26.57 ______________
**5.** 0.005 ______________
**6.** 15.074 ______________

## Naming Whole Numbers and Decimal Numbers

Write the following numbers.

**7.** four million, twenty-five thousand, twelve ______________

**8.** six billion, eighty-three million, two thousand, sixty-five ______________

**9.** seven thousand, twenty-one ______________

**10.** six hundred, two ______________

**11.** four thousand, three ______________

**12.** fifteen thousand, thirty-seven ______________

**13.** nine-tenths ______________

**14.** three hundred eight ten-thousandths ______________

**15.** seven and twenty-one thousandths ______________

**16.** thirty-seven hundredths ______________

**17.** sixteen ten-thousandths ______________

**18.** four and nine-hundredths ______________

**19.** twelve and nine-tenths ______________

**20.** five hundred seven and one hundred twelve ten-thousandths ______________

# Supplementary Exercise Set, Sections 1-1–1-2

Write the following numbers.

**1.** seven million, forty-two thousand, eighty-five

________________

**2.** fourteen billion, twenty-four million, nine thousand, fifty-four

________________

**3.** seventy-five thousand, two hundred, thirty-five

________________

**4.** eight hundred, three

________________

**5.** four thousand, seven

________________

**6.** seven-tenths

________________

**7.** ninety-five and seven-hundredths

________________

**8.** fifty-seven hundredths

________________

**9.** nineteen ten-thousandths

________________

**10.** four hundred six and two hundred fourteen ten-thousandths

________________

# Section 1-3 Addition of Whole Numbers and Decimal Numbers

## 1

The addition operation can be written either horizontally or vertically.

**Examples** *Horizontal Addition* *Vertical Addition*

$$3 + 4 = 7 \qquad \begin{array}{r} 3 \\ +4 \\ \hline 7 \end{array}$$

In every addition operation, the numbers to be added are called *addends,* and the answer is called the *sum.*

**a.** In the problem $3 + 4 = 7$, the 7 is called the ________.

**b.** In the problem $\begin{array}{r} 3 \\ +4 \\ \hline 7 \end{array}$ the 3 and 4 are called the ________.

## 2

Additions also can be written in any order.

**Example** $5 + 2 = 7$
$2 + 5 = 7$

This property is called the *commutative property of addition.* It can be illustrated as follows:

$$5 + 4 = 4 + 5$$

Complete the following examples of the commutative property of addition.

**a.** $3 + 2 = 2 +$ ______ **b.** $5 + 8 =$ ______ $+$ ______

**1. a.** sum
**b.** addends

**3**

Additions without regrouping (carrying) can be performed by adding the digits in each column. In the example below, the digits 0, 2, and 4 are in the ones place. Therefore, that column is called the ones column. The digits 3, 5, and 1 are in the tens place. That column is called the tens column.

2. **a.** 3
**b.** 8 + 5

$$\begin{array}{r} 30 \\ 52 \\ +14 \\ \hline 96 \end{array}$$

These numbers can be added without carrying because the sum of each of the columns is less than 10.

Add the following numbers.

**a.** $\begin{array}{r} 25 \\ +31 \\ \hline \end{array}$ **b.** $\begin{array}{r} 15 \\ 60 \\ +24 \\ \hline \end{array}$ **c.** $\begin{array}{r} 54 \\ +31 \\ \hline \end{array}$ **d.** $\begin{array}{r} 251 \\ 102 \\ +423 \\ \hline \end{array}$

**4**

When the sum in one of the columns of an addition is 10 or greater, it is necessary to regroup (carry). For example, in the problem

3. **a.** 56
**b.** 99
**c.** 85
**d.** 776

$$\begin{array}{r} 25 \\ +89 \\ \hline \end{array}$$

the right-hand column sum is 14. The 4 is written at the bottom of the ones column and the 1 is carried to the top of the tens column.

$$\begin{array}{r} 1\phantom{0} \\ 25 \\ +89 \\ \hline 4 \end{array}$$

This addition is then completed as follows.

$$\begin{array}{r} 1\phantom{0} \\ 25 \\ +89 \\ \hline 114 \end{array}$$

Add the following numbers.

**a.** $\begin{array}{r} 59 \\ +84 \\ \hline \end{array}$ **b.** $\begin{array}{r} 17 \\ +65 \\ \hline \end{array}$ **c.** $\begin{array}{r} 64 \\ +86 \\ \hline \end{array}$ **d.** $\begin{array}{r} 19 \\ +18 \\ \hline \end{array}$

**5**

Carrying can be performed for more than one column in an addition. For example, three steps must be completed in order to add the following numbers.

4. **a.** 143
**b.** 82
**c.** 150
**d.** 37

**Example** $\begin{array}{r} 569 \\ +437 \\ \hline \end{array}$

*Step 1.* Add the digits in the ones column. The sum is 16. The 6 is written below the line, and the 1 is carried to the tens column.

$$\begin{array}{r} {\scriptstyle 1} \\ 569 \\ +437 \\ \hline 6 \end{array}$$

*Step 2.* Add the digits in the tens column (including the carried 1). The sum is 10. The 0 is written below the line in the tens place; the 1 is carried to the hundreds column.

$$\begin{array}{r} {\scriptstyle 11} \\ 569 \\ +437 \\ \hline 06 \end{array}$$

*Step 3.* Add the digits in the hundreds column (including the carried 1). The sum is 10, and the 10 is written below the line. The sum is 1,006.

$$\begin{array}{r} {\scriptstyle 11} \\ 569 \\ +437 \\ \hline 1{,}006 \end{array}$$

Add the following numbers.

**a.** $\begin{array}{r} 789 \\ +428 \\ \hline \end{array}$ **b.** $\begin{array}{r} 972 \\ +859 \\ \hline \end{array}$ **c.** $\begin{array}{r} 1{,}942 \\ +3{,}566 \\ \hline \end{array}$

## 6

The same addition principles apply when adding decimal numbers. The concept of "columns" is important in all addition, but it is particularly important in adding decimals.

**5. a.** 1,217
**b.** 1,831
**c.** 5,508

Just as whole numbers can be added both horizontally and vertically, so can decimal numbers. When adding vertically, make sure that the decimal points are written above each other. This will ensure that the digits are in the proper columns.

**Examples** 3.56 + 2.89 can be written $\begin{array}{r} 3.56 \\ +2.89 \\ \hline \end{array}$

4.07 + 25.8 can be written $\begin{array}{r} 4.07 \\ +25.8\phantom{0} \\ \hline \end{array}$

Write the following horizontal additions as vertical additions. (You do not need to perform the addition.)

**a.** 50.7 + 194.85 **b.** 3.02 + 12

**c.** 25 + 1.8 + 0.5 **d.** 100 + 0.5 + 90.4

## 7

A zero written to the right of a decimal number helps in performing certain additions. For example, one of the additions from the previous frame can be performed as follows.

**6. a.** $\begin{array}{r} 50.7\phantom{0} \\ +194.85 \\ \hline \end{array}$

**b.** $\begin{array}{r} 3.02 \\ +12\phantom{.00} \\ \hline \end{array}$

**c.** $\begin{array}{r} 25\phantom{.0} \\ 1.8 \\ +\ 0.5 \\ \hline \end{array}$

**d.** $\begin{array}{r} 100\phantom{.0} \\ 0.5 \\ +\ 90.4 \\ \hline \end{array}$

**Example** $\begin{array}{r} 50.7\phantom{0} \\ +194.85 \\ \hline \end{array}$

Write a zero to the right of the shorter number.

```
  50.70
+194.85
```

Add the right column (hundredths).

```
  50.70
+194.85
      5
```

Add the tenths column and carry.

```
    1
  50.70
+194.85
     55
```

Bring down the decimal point.

```
    1
  50.70
+194.85
    .55
```

Add the ones column.

```
    1
  50.70
+194.85
   5.55
```

Add the tens column.

```
  1 1
  50.70
+194.85
  45.55
```

Add the hundreds column. The answer, then, is 245.55.

```
  1 1
  50.70
+194.85
 245.55
```

Do these additions.

**a.**
```
  3.02
+12
```

**b.**
```
  25
   1.8
+  0.5
```

**c.**
```
  100
    0.5
+  90.4
```

**8**

Answers to frame 7.

**a.** 15.02 **b.** 27.3 **c.** 190.9

# Section 1-4 The Calculator

**1**

The calculator will be used to solve addition problems. The keys needed for performing addition on a calculator are:

[C] key clears all numbers from the calculator

[CE] key clears only the last number entered

[+] key means add

[=] key means show the answer

*Note:* Some calculators do not have a [CE] key, only a [C] key. On those calculators, to clear the last number entered, press the [C] key once. To clear all numbers, press the [C] key twice. Read the calculator booklet to learn how to operate your calculator.

The Enter-Press-Display format for explaining how to use the calculator is illustrated.

Enter—number to be entered
Press—operation key to be used
Display—what is on the display after the operation key is pressed

The numbers 5 and 18 can be added on the calculator by following these steps.

| *Enter* | *Press* | *Display* |
|---|---|---|
| | [C] | 0 |
| 5 | [+] | 5 |
| 18 | [=] | 23 |

The [C] key was pressed to make sure there were no numbers in the calculator before starting a new problem. This operation should become automatic for you, and we will not continue to show it.

The problem above reads: 5 plus 18 equals 23, or 5 [+] 18 [=] 23. Enter the numbers and press the keys in the same sequence as they are read.

**2**

The addition below can be accomplished by using a calculator as follows:

$$\begin{array}{r} 5{,}287 \\ +1{,}958 \\ \hline 7{,}245 \end{array}$$

| *Enter* | *Press* | *Display* |
|---|---|---|
| 5287 | [+] | 5287 |
| 1958 | [=] | 7245 |

Thus, 5,287 + 1,958 = 7,245.

Use a calculator to perform these additions.

**a.** $\begin{array}{r} 2{,}957 \\ +9{,}854 \\ \hline \end{array}$

**b.** 5,784 + 187 = ______

**3**

Addition of decimal numbers also can be performed on a calculator. The addition

2. **a.** 12,811
**b.** 5,971

$$8.65 + 19.70 + 5.43 =$$

can be performed as follows:

| *Enter* | *Press* | *Display* |
|---|---|---|
| 8.65 | + | 8.65 |
| 19.70 | + | 28.35 |
| 5.43 | = | 33.78 |

The answer is 33.78.

Use a calculator to perform these decimal additions.

**a.** $\begin{array}{r} 15.09 \\ 4.70 \\ +35.90 \\ \hline \end{array}$

**b.** $29.7 + 0.8 + 107.2 =$ _______

**4**

Answers to frame 3.

**a.** 55.69

**b.** 137.7

## Section 1-5 Subtraction of Whole Numbers and Decimal Numbers

**1**

The subtraction operation can be written either horizontally or vertically.

| *Horizontal Subtraction* | *Vertical Subtraction* |
|---|---|
| $7 - 3 = 4$ | $\begin{array}{r} 7 \\ -3 \\ \hline 4 \end{array}$ |

In every subtraction the first number, or number on top, is called the *minuend.* The number that is being subtracted is called the *subtrahend,* and the answer is called the *remainder.* In the problem $4 - 3 = 1$, the 4 is the minuend, the 3 is the subtrahend, and the 1 is the remainder.

**a.** In $7 - 3 = 4$, the 7 is called the ____________.

**b.** In $7 - 3 = 4$, the 4 is called the ____________.

**c.** In $\begin{array}{r} 7 \\ -3 \\ \hline 4 \end{array}$, the 3 is called the ____________.

**2**

Subtraction is *not commutative;* therefore:

3 − 2 does *not* equal 2 − 3

Subtractions cannot be written in any order.

**1. a.** minuend
**b.** remainder
**c.** subtrahend

## 3

Subtractions without borrowing can be performed whenever the digit in the minuend (top) is greater than the digit in the subtrahend (bottom) in each column. The subtractions can then be performed by subtracting the digits in each column.

$$\begin{array}{r} 57 \\ -42 \\ \hline 15 \end{array}$$

Subtract the following numbers.

**a.** $\begin{array}{r} 287 \\ -\ 74 \\ \hline \end{array}$ **b.** $\begin{array}{r} 948 \\ -835 \\ \hline \end{array}$ **c.** $\begin{array}{r} 95 \\ -94 \\ \hline \end{array}$

## 4

When the digit in one column of the minuend is smaller than the digit in that column of the subtrahend, it is necessary to borrow.

3. **a.** 213 **b.** 113 **c.** 1

**Example** The subtraction at the right cannot be performed without borrowing, because 9 cannot be subtracted from 5 using positive numbers.

$$\begin{array}{r} 65 \\ -39 \\ \hline \end{array}$$

A 1 must therefore be borrowed from the tens column so that the subtraction 15 − 9 can be performed. This borrowing reduces the 6 to 5.

$$\begin{array}{r} {\scriptstyle 5\,1} \\ \not{6}5 \\ -39 \\ \hline \end{array}$$

The subtraction 15 − 9 = 6 is performed and the answer written in the ones column.

$$\begin{array}{r} {\scriptstyle 5\,1} \\ \not{6}5 \\ -39 \\ \hline 6 \end{array}$$

The subtraction 5 − 3 = 2 is performed and the answer written in the tens column. The answer is 26.

$$\begin{array}{r} {\scriptstyle 5\,1} \\ \not{6}5 \\ -39 \\ \hline 26 \end{array}$$

Subtract the following numbers.

**a.** $\begin{array}{r} 53 \\ -19 \\ \hline \end{array}$ **b.** $\begin{array}{r} 70 \\ -37 \\ \hline \end{array}$

## 5

In many subtractions it is necessary to borrow more than once.

4. **a.** 34 **b.** 33

**Example** $\begin{array}{r} 49{,}270 \\ -39{,}382 \\ \hline \end{array}$

A 1 must be borrowed from the 7 in the tens column so that the subtraction 10 − 2 can be performed.

$$\begin{array}{r} {\scriptstyle 6\,1} \\ 49{,}2\not{7}0 \\ -39{,}382 \\ \hline 8 \end{array}$$

A 1 must be borrowed from the 2 in the hundreds column so that the subtraction 16 − 8 can be performed.

$$\begin{array}{r} {\scriptstyle 1\;\;\;\;} \\ {\scriptstyle 1\,6\,1} \\ 49{,}\not{2}\not{7}0 \\ -39{,}382 \\ \hline 88 \end{array}$$

A 1 must be borrowed from the 9 in the thousands column so that the subtraction 11 − 3 can be performed.

```
    11
  8 161
  49,270
− 39,382
     888
```

A 1 must be borrowed from the 4 in the ten-thousands column so that the subtraction 18 − 9 can be performed. Since 3 − 3 = 0 (in the ten-thousands column), the answer is 9,888.

```
  1 11
 38 161
  49,270
− 39,382
   9,888
```

Subtract the following numbers.

**a.**
```
  5,280
− 3,591
```

**b.**
```
  47,908
− 19,479
```

**6**

---

When subtracting decimal numbers, the concept of columns is important. The numbers must be lined up with the decimal points above each other, just as in addition of decimals.

**5. a.** 1,689
**b.** 28,429

Subtract the following decimal numbers.

**a.**
```
  53.09
− 44.55
```

**b.** 15.87 − 5.90 = ________

**7**

---

It is helpful to add a zero to the right of a decimal when performing certain subtractions. A sample subtraction would be performed as follows.

**6. a.** 8.54
**b.** 9.97

**Example**
```
  23.1
−  5.89
```

Write a zero to the right of the shorter number.

```
  23.10
−  5.89
```

Borrow from the tenths column and subtract.

```
     01
  23.10
−  5.89
      1
```

Borrow from the ones column and subtract.

```
      1
   2 01
  23.10
−  5.89
    .21
```

Borrow from the tens column and subtract.

```
   1  1
  12.01
  23.10
−  5.89
   7.21
```

Subtract 1 − 0 in the tens column. The answer is 17.21.

```
   1 1
  12 01
  23.10
−  5.89
  17.21
```

Do these subtractions.

**a.** 45.9 − 7.61

**b.** 5.98 − 0.752

## 8

The subtraction below can be checked by using the calculator.

7. **a.** 38.29 **b.** 5.228

```
  25.90
−  1.56
  24.34
```

| *Enter* | *Press* | *Display* |
|---|---|---|
| 25.90 | [−] | 25.90 |
| 1.56 | [=] | 24.34 |

The answer is 24.34.

Use a calculator to perform these subtractions.

**a.** 5,285 − 1,760

**b.** 98.75 − 4.895

**c.** 7.008 − 2.017

## 9

Answers to frame 8.

**a.** 3525 **b.** 93.855 **c.** 4.991

# Exercise Set, Sections 1-3–1-5

## Addition of Whole Numbers and Decimal Numbers

Perform the following additions, and check your answers using a calculator.

**1.** 15 + 28

**2.** 25 + 37 + 15

**3.** 4,571 + 3,281

**4.** 97 + 95 + 81 + 65

**5.** 28,291 + 27,595

**6.** 5.08 + 0.17

**7.** 25.09 + 18.70

**8.** 97.07 + 0.09

**9.** 15 + 0.5 = ____________

**10.** 17 + 0.052 + 15.8 = ____________

**11.** 21,896 + 11,297 = ____________

**12.** 6.02 + 15.97 = ____________

**13.** 196 + 0.187 = ____________

**14.** 83.67 + 21.7 = ____________

**15.** 67.04 + 0.03 = ____________

**16.** 11.6 + 14.08 + 16.01 = ____________

**17.** 0.08 + 0.16 + 1.19 = ____________

**18.** 1.98 + 2.37 + 4.1 = ____________

**19.** 21.07 + 0.07 + 16.9 = ____________

## Subtraction of Whole Numbers and Decimal Numbers

Perform the following subtractions, and check your answers using a calculator.

**20.** $\begin{array}{r} 25 \\ -14 \\ \hline \end{array}$ **21.** $\begin{array}{r} 498 \\ -315 \\ \hline \end{array}$ **22.** $\begin{array}{r} 742 \\ -698 \\ \hline \end{array}$ **23.** $\begin{array}{r} 9{,}285 \\ -\ \ 796 \\ \hline \end{array}$ **24.** $\begin{array}{r} 15{,}010 \\ -14{,}985 \\ \hline \end{array}$ **25.** $\begin{array}{r} 95.2 \\ -\ \ 1.65 \\ \hline \end{array}$ **26.** $\begin{array}{r} 3.5 \\ -1.4 \\ \hline \end{array}$ **27.** $\begin{array}{r} 95.8 \\ -\ \ 5.9 \\ \hline \end{array}$

**28.** 76.2 − 0.07 = ____________

**29.** 45.9 − 42.89 = ____________

**30.** 24 − 7.98 = ____________

**31.** 83.02 − 24.61 = ____________

**32.** 53 − 0.987 = ____________

**33.** 16.9 − 8.11 = ____________

**34.** 0.819 − 0.112 = ____________

**35.** 1.01 − 0.009 = ____________

**36.** 0.973 − 0.013 = ____________

# Supplementary Exercise Set, Sections 1-3–1-5

Perform the following additions and subtractions.

**1.** 16 + 45 =

**2.** 31 + 24 + 18 =

**3.** 4,951 + 3,287 =

**4.** 34 + 91 + 87 + 81 + 92 =

**5.** 29,471 + 35,289 =

**6.** 5.12 + 0.11 =

**7.** 35.07 + 19.1 =

**8.** 88.08 + 0.02 =

**9.** 18 + 0.042 + 16.3 =

**10.** 1.28 + 31.05 + 4.7 =

**11.** 25 − 11 =

**12.** 785 − 219 =

**13.** 4,295 − 942 =

**14.** 18,071 − 17,401 =

**15.** 85.2 − 1.764 =

**16.** 0.5 − 0.3 =

**17.** 0.951 − 0.112 =

**18.** 35 − 0.01 =

**19.** 0.97 − 0.012 =

**20.** 4.5 − 4.05 =

# Section 1-6 Multiplication of Whole Numbers and Decimal Numbers

## 1

Multiplication problems can be written horizontally or vertically. Therefore:

$3 \times 8 = 24$ can be written as $\begin{array}{r} 8 \\ \times 3 \\ \hline 24 \end{array}$

In both forms of this multiplication, the 3 and 8 are called *factors* and the 24 is called the *product*.

**a.** In the problem $4 \times 5 = 20$, the 4 is called the ____________.

**b.** In the problem $\begin{array}{r} 9 \\ \times 2 \\ \hline 18 \end{array}$, the 18 is called the ____________.

**c.** In the problem $\begin{array}{r} 7 \\ \times 5 \\ \hline 35 \end{array}$, the 7 is called the ____________.

## 2

The multiplication $6 \times 8 = 48$ can also be written as $(6)(8) = 48$. Rewrite the multiplications below.

**a.** $5 \times 4 = 20$ can be written as ____________.
**b.** $(3)(6) = 18$ can be written as ____________.
**c.** $(9)(5) = 45$ can be written as ____________.
**d.** $6 \times 8 = 48$ can be written as ____________.

1. **a.** factor
   **b.** product
   **c.** factor

## 3

Multiplication is commutative. This means that multiplication problems can be written in any order. For example:

$3 \times 5 = 15$ can be written $5 \times 3 = 15$

This commutative property of multiplication can be illustrated as follows:

$$3 \times 5 = 5 \times 3$$

Complete the following examples of the commutative property of multiplication.

**a.** $9 \times 5 = 5 \times$ ______  **b.** $6 \times 8 =$ ______ $\times$ ______

2. **a.** $(5)(4) = 20$
   **b.** $3 \times 6 = 18$
   **c.** $9 \times 5 = 45$
   **d.** $(6)(8) = 48$

## 4

Any multiplication in which one factor is a one-digit number is called a short multiplication. These are short multiplications:

$\begin{array}{r} 64 \\ \times\ 2 \\ \hline \end{array}$ $\begin{array}{r} 8 \\ \times 7 \\ \hline \end{array}$ $3 \times 45 =$ $5 \times 7 =$

3. **a.** 9
   **b.** $8 \times 6$

Which of the following are short multiplications? (Circle your selections.)

**a.** $57 \times 2$  **b.** $35 \times 11$  **c.** $3 \times 95 =$  **d.** $25 \times 27 =$

## 5

Short multiplication without carrying is done as follows:

4. a and c

**Example** $43 \times 2$

The 3 in the ones place is multiplied by 2, and the product is written below the ones column.

$$\begin{array}{r} 43 \\ \times\ 2 \\ \hline 6 \end{array}$$

The 4 in the tens place is multiplied by 2, and the product is written below the tens column. The answer is 86.

$$\begin{array}{r} 43 \\ \times\ 2 \\ \hline 86 \end{array}$$

Multiply the following numbers.

**a.** $22 \times 4$  **b.** $53 \times 2$  **c.** $91 \times 7$  **d.** $13 \times 3$

## 6

Most short multiplications require carrying. This is similar to the carrying that is done in addition.

5. a. 88
b. 106
c. 637
d. 39

**Example** $173 \times 4$

The multiplication $3 \times 4 = 12$ is performed, and the 2 is written below the line in the ones column. The 1 is carried to the top of the tens column.

$$\begin{array}{r} {\scriptstyle 1}\ \\ 173 \\ \times\ \ 4 \\ \hline 2 \end{array}$$

The multiplication $4 \times 7 = 28$ is performed, and the 28 is added to the carried 1, yielding 29. The 9 is written below the line in the tens column, and the 2 is carried to the top of the hundreds column.

$$\begin{array}{r} {\scriptstyle 21}\ \\ 173 \\ \times\ \ 4 \\ \hline 92 \end{array}$$

The multiplication $1 \times 4 = 4$ is performed, and the 4 is added to the carried 2, yielding 6. The 6 is written below the line under the hundreds column. The answer is 692.

$$\begin{array}{r} {\scriptstyle 21}\ \\ 173 \\ \times\ \ 4 \\ \hline 692 \end{array}$$

Multiply the following numbers.

**a.** $256 \times 4$  **b.** $543 \times 5$  **c.** $987 \times 8$

## 7

Short multiplications involving decimal numbers require a strategy for placing the decimal point in the answer. The following rule can be used for placing the decimal point:

6. a. 1,024
b. 2,715
c. 7,896

The number of digits to the right of the decimal point in the product (answer) equals the sum of the number of digits to the right of the decimal point in each of the numbers being multiplied.

**Example**

| | | |
|---|---|---|
| 25.8 | ← | one digit |
| × 0.7 | ← | + one digit |
| 18.06 | ← | two digits |

Place the decimal point in each of the following multiplications.

**a.** 45.8 × 7 = 3206  **b.** 4.82 × 0.3 = 1446  **c.** 0.15 × 7 = 105  **d.** 957 × 0.3 = 2871

## 8

Multiply the following numbers.

**a.** 39.8 × 5  **b.** 1.28 × 0.3  **c.** 0.9 × 75

7. **a.** 320.6 **b.** 1.446 **c.** 1.05 **d.** 287.1

## 9

Any multiplication in which both factors have two or more digits is called a long multiplication. For example, these are long multiplications:

54 × 35  15 × 87 =  95 × 30

8. **a.** 199.0 **b.** 0.384 **c.** 67.5

Which of the following are long multiplications? (Circle your selections.)

**a.** 157 × 3  **b.** 28 × 12  **c.** 3 × 17 =  **d.** 35 × 17 =

## 10

The most common method used for performing a long multiplication is called the *long-multiplication algorithm.* It involves the following steps:

9. **b and d**

**Example** 27 × 15

Multiply 7 × 5 = 35. Place the 5 below the ones column and the 3 above the tens column.

```
  3
 27
×15
  5
```

Multiply 2 × 5 = 10 and add the carried 3, yielding 13. The 13 is written below the line.

```
  3
 27
×15
135
```

Multiply the 1 in the tens column times 7 to get the product 7. The 7 is written below the 3 in the tens column.

```
 27
×15
135
 7
```

Multiply the 1 in the tens column times 2 to get the product 2. The 2 is written below the 1 in the hundreds column.

```
  27
 ×15
 ---
 135
 27
```

The numbers 135 and 27 are called *partial products.* The partial products are added to give the final product. The answer is 405.

```
  27
 ×15
 ---
 135
 27
 ---
 405
```

Multiply the following numbers.

**a.** 53 × 18 **b.** 75 × 62 **c.** 741 × 34

## 11

The names of the parts of a long multiplication are shown below:

```
   76  ←— factor
  ×25  ←— factor
  ---
  380  ←— partial product
  152  ←— partial product
 -----
1,900  ←— product
```

**10. a.** 954
**b.** 4,650
**c.** 25,194

## 12

The long-multiplication algorithm is also used for decimal-number multiplications. The strategy for placing the decimal point in the answer is the same as the strategy used in short multiplication:

> The number of digits to the right of the decimal point in the product equals the sum of the number of digits to the right of the decimal point in each of the numbers being multiplied.

**Example**

```
  28.7  ←—  one digit
 ×0.32  ←—  +two digits
 -----
   574
  861
 -----
 9.184  ←—  three digits
```

Multiply the following numbers and place the decimal points in the answers.

**a.** 94.2 × 0.75 **b.** 0.142 × 95 **c.** 92.8 × 0.078

## 13

Any multiplication by 10, 100, 1,000, and so on can be performed by using a shortcut method called the *decimal-point shift.* In this shortcut, the decimal point is shifted to the right the same number of places as there are zeros in the factor. Each multiplication below shows the long multiplication and the decimal-point shift.

**12. a.** 70.65
**b.** 13.49
**c.** 7.2384

**Example**

| *Long Multiplication* | *Decimal-Point Shift* |
|---|---|
| $\begin{array}{r} 0.78 \\ \times\ 10 \\ \hline 00 \\ 78\ \ \\ \hline 7.80 \end{array}$ | $10 \times 0.78 = 7.8$<br>Shift the decimal point one place to the right. |
| $\begin{array}{r} 0.78 \\ \times\ 100 \\ \hline 00 \\ 00\ \ \\ 78\ \ \ \ \\ \hline 78.00 \end{array}$ | $100 \times 0.78 = 78$<br>Shift the decimal point two places to the right. |

Perform the following multiplications using the decimal-point shift.

**a.** $10 \times 32 =$ ______________ **b.** $100 \times 4.38 =$ ______________
**c.** $1{,}000 \times 75.4 =$ ______________ **d.** $10{,}000 \times 4.809 =$ ______________

**14** ________________________________________

The multiplication below can be checked by using a calculator.

$$\begin{array}{r} 907 \\ \times\ 7.6 \\ \hline 5442 \\ 6349\ \ \\ \hline 6{,}893.2 \end{array}$$

| *Enter* | *Press* | *Display* |
|---|---|---|
| 907 | ⊠ | 907 |
| 7.6 | ⊟ | 6893.2 |

$907 \times 7.6 = 6{,}893.2.$

Use a calculator to perform the following multiplications.

**a.** $\begin{array}{r} 4.89 \\ \times 12.2 \\ \hline \end{array}$ **b.** $\begin{array}{r} 48 \\ \times 17 \\ \hline \end{array}$ **c.** $\begin{array}{r} 54.6 \\ \times\ 879 \\ \hline \end{array}$ **d.** $\begin{array}{r} 28.6 \\ \times\ \ 45 \\ \hline \end{array}$

13. **a.** 320
**b.** 438
**c.** 75,400
**d.** 48,090

**15** ________________________________________

Answers to frame 14.

**a.** 59.658 **b.** 816 **c.** 47,993.4 **d.** 1,287

# Exercise Set, Section 1-6

## Multiplication of Whole Numbers and Decimal Numbers

Perform the following multiplications, and check your answers using a calculator.

**1.** $\begin{array}{r} 53 \\ \times\ 3 \\ \hline \end{array}$ **2.** $\begin{array}{r} 72 \\ \times\ 4 \\ \hline \end{array}$ **3.** $\begin{array}{r} 65 \\ \times\ 8 \\ \hline \end{array}$ **4.** $\begin{array}{r} 97 \\ \times\ 7 \\ \hline \end{array}$ **5.** $\begin{array}{r} 6.2 \\ \times 0.3 \\ \hline \end{array}$ **6.** $\begin{array}{r} 0.42 \\ \times\ \ \ 4 \\ \hline \end{array}$ **7.** $\begin{array}{r} 3.6 \\ \times 0.5 \\ \hline \end{array}$ **8.** $\begin{array}{r} 76 \\ \times 0.04 \\ \hline \end{array}$

9. $\begin{array}{r}63\\ \times 13\\ \hline\end{array}$ 10. $\begin{array}{r}75\\ \times 78\\ \hline\end{array}$ 11. $\begin{array}{r}4.7\\ \times 3.5\\ \hline\end{array}$ 12. $\begin{array}{r}0.97\\ \times 0.45\\ \hline\end{array}$ 13. $\begin{array}{r}235\\ \times\ 18\\ \hline\end{array}$ 14. $\begin{array}{r}278\\ \times 146\\ \hline\end{array}$ 15. $\begin{array}{r}3.49\\ \times 0.76\\ \hline\end{array}$ 16. $\begin{array}{r}8.78\\ \times 47.3\\ \hline\end{array}$

17. 92 × 7 = ________
18. 62 × 9 = ________
19. 6.9 × 4 = ________
20. 5.2 × 3 = ________
21. (8.3)(0.4) = ________
22. 9.7 × 0.9 = ________
23. (0.46)(5) = ________
24. 0.13 × 7 = ________
25. (4.6)(0.4) = ________
26. 9.4 × 0.8 = ________
27. (487)(26) = ________
28. 592 × 89 = ________
29. (4.29)(3.71) = ________
30. 6.78 × 0.92 = ________

Use the decimal-point shift for problems 31–40.

31. $\begin{array}{r}67\\ \times 1{,}000\\ \hline\end{array}$ 32. 10 × 47 = ________ 33. $\begin{array}{r}48.30\\ \times\ 100\\ \hline\end{array}$ 34. 10,000 × 73.2 = ________

35. 5.2 × 10 = ________
36. 278 × 10 = ________
37. 0.14 × 100 = ________
38. 41.21 × 100 = ________
39. 0.0183 × 1,000 = ________
40. 1.01 × 1,000 = ________

## Supplementary Exercise Set, Section 1-6

Perform the following multiplications.

1. 47 × 8 =
2. (71)(9) =
3. (7.3)(0.4) =
4. 9.5 × 0.3 =

5. 781 × 428 =
6. (321)(321) =
7. (95.2)(31.3) =
8. 8.71 × 51.2 =

9. 45 × 45 =
10. 4.5 × 45 =
11. (4.5)(4.5) =
12. 0.13 × 4.2 =

13. (3.2)(7.1) =
14. 0.2 × 0.3 =
15. (1.2)(1.02) =
16. (594)(0.43) =

17. 671 × 0.02 =
18. 10 × 48 =
19. (100)(76.2) =
20. (4.85)(1,000) =

# Section 1-7 Division of Whole Numbers and Decimal Numbers

**1**

The statement "4 divided by 2" can be written several ways:

$4 \div 2 \qquad 2\overline{)4} \qquad \frac{4}{2}$

Which of the following means "10 divided by 3"? (Circle your selections.)

**a.** $10 \div 3$ **b.** $10\overline{)3}$ **c.** $\frac{10}{3}$ **d.** $\frac{3}{10}$

**2**

The names of the numbers in a division are shown below.

1. a and c

$$6\overline{)48}^{\,8}$$

The 6 is called the divisor.
The 48 is called the dividend.
The 8 is called the quotient.

**a.** In the problem $45 \div 9 = 5$, the 9 is called the ______________.

**b.** In the problem $7\overline{)56}$ (quotient 8), the 8 is called the ______________.

**c.** In the problem $3\overline{)27}$ (quotient 9), the 27 is called the ______________.

**3**

The most common method of performing divisions is called the *long-division algorithm.*

2. a. divisor
b. quotient
c. dividend

**Example** $6\overline{)3{,}108}$ (quotient 5)

The 6 does not go into 3, so 6 is divided into 31, yielding a quotient of 5. The 5 is written above the 1.

```
    5
6 |3,108
```

The multiplication of $5 \times 6 = 30$ is performed, and the 30 is subtracted from the 31 in the dividend.

```
    5
6 |3,108
   3 0
   ---
     1
```

The 0 in the dividend is brought down and written next to the 1.

```
    5
6 |3,108
   3 0
   ---
     10
```

The 6 is divided into the 10, giving a quotient of 1.

```
    51
6 |3,108
   3 0
   ---
     10
```

The multiplication $1 \times 6 = 6$ is performed, and the 6 is subtracted from the 10.

```
    51
6 |3,108
   3 0
    10
     6
     4
```

The 8 in the dividend is brought down and written next to the 4.

```
    51
6 |3,108
   3 0
    10
     6
    48
```

The 6 is divided into 48, yielding a quotient of 8.

```
   518
6 |3,108
   3 0
    10
     6
    48
```

The multiplication $8 \times 6 = 48$ is performed, and this 48 is subtracted from the earlier 48. The answer is 518.

```
   518
6 |3,108
   3 0
    10
     6
    48
    48
     0
```

## 4

The pattern in the long-division algorithm is:

1. divide;
2. multiply back;
3. subtract;
4. bring down the next number.

**Example** 15 |13,125

Divide:
(131 ÷ 15 = 8)

```
        8
15 |13,125
```

Multiply back:
($15 \times 8 = 120$)

```
        8
15 |13,125
    12 0
```

Subtract:
($131 - 120 = 11$)

```
        8
15 |13,125
    12 0
     1 1
```

Bring down the next number:
(11 becomes 112)

```
        8
15 |13,125
    12 0
     1 12
```

Repeat the entire process.

Divide:
(112 ÷ 15 = 7)

$$\begin{array}{r} 87\phantom{0} \\ 15\overline{)13{,}125} \\ \underline{12\,0\phantom{00}} \\ 1\,12\phantom{0} \end{array}$$

Multiply back:
(15 × 7 = 105)

$$\begin{array}{r} 87\phantom{0} \\ 15\overline{)13{,}125} \\ \underline{12\,0\phantom{00}} \\ 1\,12\phantom{0} \\ \underline{1\,05\phantom{0}} \end{array}$$

Subtract:
(112 − 105 = 7)

$$\begin{array}{r} 87\phantom{0} \\ 15\overline{)13{,}125} \\ \underline{12\,0\phantom{00}} \\ 1\,12\phantom{0} \\ \underline{1\,05\phantom{0}} \\ 7\phantom{0} \end{array}$$

Bring down the next number:
(7 becomes 75)

$$\begin{array}{r} 87\phantom{0} \\ 15\overline{)13{,}125} \\ \underline{12\,0\phantom{00}} \\ 1\,12\phantom{0} \\ \underline{1\,05\phantom{0}} \\ 75 \end{array}$$

Repeat the process again.

Divide:
(75 ÷ 15 = 5)

$$\begin{array}{r} 875 \\ 15\overline{)13{,}125} \\ \underline{12\,0\phantom{00}} \\ 1\,12\phantom{0} \\ \underline{1\,05\phantom{0}} \\ 75 \end{array}$$

Multiply back:
(15 × 5 = 75)

$$\begin{array}{r} 875 \\ 15\overline{)13{,}125} \\ \underline{12\,0\phantom{00}} \\ 1\,12\phantom{0} \\ \underline{1\,05\phantom{0}} \\ 75 \\ \underline{75} \end{array}$$

Subtract:
(75 − 75 = 0)

$$\begin{array}{r} 875 \\ 15\overline{)13{,}125} \\ \underline{12\,0\phantom{00}} \\ 1\,12\phantom{0} \\ \underline{1\,05\phantom{0}} \\ 75 \\ \underline{75} \\ 0 \end{array}$$

There are no more numbers to bring down, so the answer is 875.

Divide the following.

**a.** $28\overline{)7{,}980}$ **b.** $87\overline{)4{,}872}$ **c.** $95\overline{)3{,}420}$

## 5

Divisions of decimal numbers by whole numbers also use the long-division algorithm. The decimal point in the quotient is placed directly above the decimal point in the dividend. This is shown below.

4. a. 285
b. 56
c. 36

```
      18.2
12 |218.4
    12
     98
     96
      24
      24
```

Divide the following.

**a.** 45 |93.15 **b.** 761 |26,787.2

## 6

When a division does not have a whole-number divisor, it is necessary to change the problem to an equivalent division with a whole-number divisor. This is done by shifting the decimal point to the right in both the divisor and the dividend. The decimal point is shifted as many places as needed to make the divisor a whole number. For example:

5. a. 2.07
b. 35.2

3.8 |11.4 is changed to 38 |114

2.6 |18.2 is changed to 26 |182

0.05 |0.0115 is changed to 5 |1.15

Change the following divisions by shifting the decimal points. (You do not need to do the division.)

**a.** 4.3 |21.5 **b.** 7.8 |2.34 **c.** .04 |0.008

## 7

The division below is performed by first converting the problem to one with a whole-number divisor.

6. a. 43 |215
b. 78 |23.4
c. 4 |0.8

**Example** 8.2 |73.8

Convert the problem by shifting the decimal points so that the divisor becomes 82. 82 |738

Divide using the long-division algorithm.

```
     9
82 |738
    738
      0
```

Divide the following by first converting each problem to a division with a whole-number divisor.

**a.** 6.8 |55.08 **b.** 1.05 |3.36

## 8

7. a. 8.1
b. 3.2

In the division below, the last subtraction does not result in a 0 answer. The 6 is called a *remainder*.

$$\begin{array}{r} 8 \\ 8\overline{)70} \\ 64 \\ \hline 6 \end{array}$$

A division of this type can be completed by placing a decimal point after the 0 in the dividend and adding another 0 in the tenths place of the dividend. You can continue adding zeros to the dividend until the subtraction results in a 0 remainder or until you have as many decimal places as desired.

$$\begin{array}{r} 8.75 \\ 8\overline{)70.00} \\ 64\phantom{.00} \\ \hline 6\,0\phantom{0} \\ 5\,6\phantom{0} \\ \hline 40 \\ 40 \\ \hline 0 \end{array}$$

Divide the following.

**a.** $5\overline{)22}$ **b.** $8\overline{)89}$

## 9

8. a. 4.4
b. 11.125

In some divisions, a 0 remainder is never obtained. It is then necessary to round the answer to a given place. The examples below are rounded to the hundredths place.

$$\begin{array}{r} 12.714 \\ 7\overline{)89.000} \\ 7\phantom{9.000} \\ \hline 19\phantom{.000} \\ 14\phantom{.000} \\ \hline 5\,0\phantom{00} \\ 4\,9\phantom{00} \\ \hline 10\phantom{0} \\ 7\phantom{0} \\ \hline 30 \\ 28 \\ \hline 2 \end{array} \longrightarrow 12.71 \qquad \begin{array}{r} 6.888 \\ 9\overline{)62.000} \\ 54\phantom{.000} \\ \hline 8\,0\phantom{00} \\ 7\,2\phantom{00} \\ \hline 80\phantom{0} \\ 72\phantom{0} \\ \hline 80 \\ 72 \\ \hline 8 \end{array} \longrightarrow 6.89$$

In rounding, numbers less than 5 are dropped, and numbers 5 and greater increase the preceding number by 1. Thus, in the first example, the 4 is dropped and the 1 is left as it is because the 4 is less than 5. In the second example, the third 8 is dropped and the second 8 is changed to 9 because the dropped 8 is greater than 5.

## 10

The examples in the last frame were rounded using the following rules:

If the digit immediately to the right of the digit being rounded is 5 or greater:
**a.** 1 is added to the digit being rounded.
**b.** all digits to the right of the digit being rounded are dropped.

If the digit immediately to the right of the digit being rounded is 4 or less:
**a.** the digit being rounded is left unchanged.
**b.** all digits to the right of the digit being rounded are dropped.

More examples of this rounding procedure are shown below.

**Example** Round 8.786 to the hundredths place.

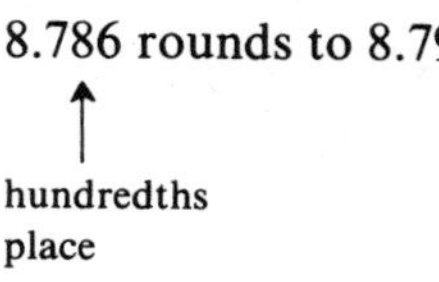


The 6 is the digit immediately to the right of the digit being rounded. Since 6 is greater than 5, 1 is added to the number in the hundredths place.

**Example** Round 27.342 to the tenths place.

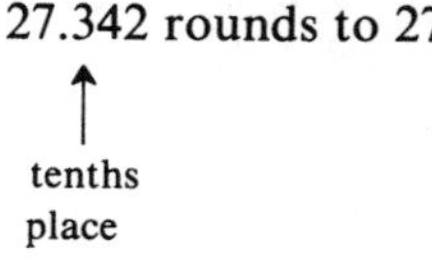


The 4 is the digit immediately to the right of the digit being rounded. Since 4 is less than 5, the digit in the tenths place is left unchanged.

Round the following numbers to the hundredths place.

**a.** 34.781 **b.** 9.095 **c.** 287.461 **d.** 0.0446

## 11

Division can be performed on a calculator. For example, the problem 255 ÷ 15 can be performed as follows:

| *Enter* | *Press* | *Display* |
|---|---|---|
| 255 | ÷ | 255 |
| 15 | = | 17 |

The answer is 17.

Use a calculator to perform these divisions.

**a.** 532 ÷ 28 **b.** 5,985 ÷ 63 **c.** 2,125 ÷ 85

**10. a.** 34.78 **b.** 9.10 **c.** 287.46 **d.** 0.04

## 12

The problem $28\overline{)71.68}$ can be performed on a calculator as follows:

| *Enter* | *Press* | *Display* |
|---|---|---|
| 71.68 | ÷ | 71.68 |
| 28 | = | 2.56 |

The answer is 2.56.

Use a calculator to perform these divisions.

**a.** $1.4\overline{)53.2}$ **b.** $42\overline{)264.6}$ **c.** $24.8\overline{)414.16}$

**11. a.** 19 **b.** 95 **c.** 25

**13**

The following calculator answers are rounded to the hundredths place.

$$\begin{array}{r} 42.47629 \\ 21\overline{)892} \end{array} \text{ rounds to } 42.48$$

$653 \div 58 = 11.25862$ rounds to 11.26

Round the following calculator answers to the hundredths place.

**a.** $793 \div 15 = 52.866666$ **b.** $1091 \div 56 = 19.482142$

**c.** $\begin{array}{r} 47.071428 \\ 14\overline{)659} \end{array}$ **d.** $\begin{array}{r} 5.3483146 \\ 89\overline{)476} \end{array}$

**12. a.** 38
**b.** 6.3
**c.** 16.7

**14**

Perform the following divisions on a calculator, and round the answers to the hundredths place.

**a.** $7.09 \div 3.5 =$ **b.** $48.7 \div 3.28 =$

**c.** $53.4\overline{)947.8}$ **d.** $172\overline{)658}$

**13. a.** 52.87
**b.** 19.48
**c.** 47.07
**d.** 5.35

**15**

Divisions by 10, 100, 1,000, and so on can be done by using the decimal-point-shift method. Using this method, you simply shift the decimal point one place to the *left* in the dividend for every 0 in the divisor. This is illustrated below.

$567 \div 100 = 5.67$
- Two zeros appear in the divisor.
- Shift the decimal two places to the left.

$46.2 \div 10 = 4.62$
- One zero appears in the divisor.
- Shift the decimal point one place to the left.

Use the decimal-point-shift method to do the following divisions.

**a.** $28.7 \div 100$ **b.** $486 \div 1{,}000$ **c.** $14.2 \div 10$

**14. a.** 2.03 (2.0257142)
**b.** 14.85 (14.84756)
**c.** 17.75 (17.749063)
**d.** 3.83 (3.8255813)

**16**

Answers to frame 15.

**a.** 0.287 **b.** 0.486 **c.** 1.42

# Exercise Set, Section 1-7

## Division of Whole Numbers and Decimal Numbers

Perform the following divisions, and check your answers using a calculator. Round all answers to the hundredths place.

**1.** $459 \div 9 =$ ________ **2.** $265 \div 12 =$ ________ **3.** $122.5 \div 4.5 =$ ________

**4.** $62.4 \div 1.2 =$ ________ **5.** $1.6 \div 0.12 =$ ________ **6.** $0.18 \div 0.02 =$ ________

**7.** $14\overline{)98}$ **8.** $20\overline{)582}$ **9.** $25.8\overline{)696.6}$ **10.** $7\overline{)89}$

**11.** $9\overline{)532}$ **12.** $47.8\overline{)430.2}$ **13.** $0.7\overline{)0.38885}$ **14.** $9.5\overline{)49.8085}$

**15.** $2.2\overline{)2.2132}$ **16.** $0.3\overline{)2.97}$ **17.** $1.3\overline{)0.961}$ **18.** $0.12\overline{)0.847}$

**19.** $\frac{42}{6} =$ ______ **20.** $\frac{53}{4} =$ ______ **21.** $\frac{482}{20} =$ ______ **22.** $\frac{698}{2.4} =$ ______

**23.** $\frac{327}{5.2} =$ ______ **24.** $\frac{153}{9.9} =$ ______ **25.** $\frac{28.7}{4} =$ ______ **26.** $\frac{93.9}{8} =$ ______

**27.** $\frac{163.2}{9} =$ ______ **28.** $\frac{147.3}{1.4} =$ ______ **29.** $\frac{963.27}{5.67} =$ ______ **30.** $\frac{826.01}{9.02} =$ ______

**31.** $\frac{1.96}{0.16} =$ ______ **32.** $\frac{0.87}{0.27} =$ ______ **33.** $\frac{0.16}{0.09} =$ ______

Use the decimal-point-shift method for problems 34–39.

**34.** $255 \div 10 =$ ______ **35.** $37.2 \div 1{,}000 =$ ______ **36.** $4.5 \div 100 =$ ______
**37.** $7{,}000{,}000 \div 10{,}000 =$ ______ **38.** $0.956 \div 10 =$ ______ **39.** $1.784 \div 100 =$ ______

# Supplementary Exercise Set, Section 1-7

Perform the following divisions.

**1.** $351 \div 9 =$ ______ **2.** $585 \div 45 =$ ______ **3.** $585 \div 13 =$ ______

**4.** $195 \div 2.5 =$ ______ **5.** $2{,}251.7 \div 25.3 =$ ______ **6.** $0.16 \div 0.04 =$ ______

**7.** 4,500 ÷ 90 = ______

**8.** 360 ÷ 0.012 = ______

**9.** 90,000 ÷ 30,000 = ______

**10.** 90,000 ÷ 300 = ______

**11.** 173.4 ÷ 8 = ______

**12.** 412.2 ÷ 6 = ______

**13.** 75.2 ÷ 3.61 = ______

**14.** 557 ÷ 55.7 = ______

**15.** 398 ÷ 4,200 = ______

**16.** 504.114 ÷ 89.7 = ______

**17.** 0.05 ÷ 10 = ______

**18.** 487 ÷ 100 = ______

**19.** 38.5 ÷ 1,000 = ______

**20.** 28,971 ÷ 10,000 = ______

# Section 1-8 Measurement Concepts

**1**

Because all measurements are approximate or rounded-off numbers, when we use measurements in calculations we must take care to ensure that the answers are reasonable. The calculator does not "know" how to round off an answer from a calculation involving measurements, so the concepts and rules presented in this section will demonstrate how to round off a calculator answer.

**2**

The precision of any measurement is dependent on the smallest subdivision of the instrument used to make the measurement.

**Examples** The measurement 951 is precise to the ones place.
The measurement 10.1 is precise to the tenths place.
The measurement 0.013 is precise to the thousandths place.

What is the precision of these numbers?

**a.** 3.18 ______________________

**b.** 15.3 ______________________

**c.** 0.072 ______________________

**3**

Sometimes a number is precise to a certain place, but the number in that position is 0. The answer must reflect the precision of the measuring device, even if it is necessary to use a 0 in one or more places.

**2. a.** hundredths place
**b.** tenths place
**c.** thousandths place

**Examples** The measurement 29.00 indicates that the measuring device can measure to the hundredths place and that this measurement was a 0 in the hundredths place (and 0 in the tenths place as well).

The measurement 40.100 indicates that the measuring device can measure to the thousandths place.

To what place are these numbers precise?

**a.** 3.10 ______________________
**b.** 0.0700 ______________________
**c.** 42.010 ______________________

## 4

The concept of precision is used only with addition and subtraction and concerns only the number of decimal places in the answer.

3. **a.** hundredths place
**b.** ten-thousandths place
**c.** thousandths place

## 5

In adding or subtracting measurements, an answer cannot be more precise than the least precise measurement used to arrive at the answer.

**Examples**

```
  28.14
+ 10.2     <---- least precise measurement
  38.34    <---- answer must be rounded off to the tenths place
  38.34  (rounded off) = 38.3

  1.00     <---- least precise measurement
+ 0.0093
  1.0093   <---- answer must be rounded off to the hundredths place
  1.0093  (rounded off) = 1.01

  14.62
-  3.0     <---- least precise measurement
  11.62    <---- answer must be rounded off to the tenths place
  11.62  (rounded off) = 11.6
```

Solve these problems.

| **a.** | **b.** | **c.** | **d.** |
|---|---|---|---|
| 104.31 | 58.010 | 34.0 | 27.6512 |
| + 22.111 | + 12.00 | − 12.15 | − 12.34 |

## 6

The concept of precision is used only with the operations of

**a.** ______________________ and **b.** ______________________.

5. **a.** 126.42
**b.** 70.01
**c.** 21.9
**d.** 15.31

## 7

When using the concept of precision, we are interested only in the number of ______________________ in the answer.

6. **a.** addition
**b.** subtraction

## 8

Not all digits in a number are significant; some digits are merely placeholders.

7. decimal places

**Example** In the number 0.0097, all digits to the left of the 9 are placeholders and therefore not significant.

The rules for determining what digits are significant in a number are stated in this and the following frames. The first rule states that all nonzero digits are significant.

**Examples** 897 has three significant digits.
58.62 has four significant digits.

How many significant digits are in these numbers?

**a.** 29 ______ **b.** 3.14 ______ **c.** 123987 ______

## 9

Zeros at the left of a measurement are not significant.

**8. a.** two
**b.** three
**c.** six

**Examples** 0.0097 has only two significant digits: the 9 and the 7.
0.0193 has three significant digits.

How many significant digits are in these numbers?

**a.** 0.001 ______ **b.** 0.0987 ______ **c.** 0.000034 ______

## 10

How many significant digits are in these numbers?

**9. a.** one
**b.** three
**c.** two

**a.** 657 ______ **b.** 0.0657 ______
**c.** 0.009 ______ **d.** 4.26 ______

## 11

Zeros between nonzero digits are significant.

**10. a.** three
**b.** three
**c.** one
**d.** three

**Examples** 903 has three significant digits.
1.04 has three significant digits.
0.0204 has three significant digits.

How many significant digits are in these numbers?

**a.** 406 ______ **b.** 28.07 ______
**c.** 0.00109 ______ **d.** 1.002 ______

## 12

How many significant digits are in these numbers?

**11. a.** three
**b.** four
**c.** three
**d.** four

**a.** 509 ______ **b.** 4.29 ______
**c.** 0.06 ______ **d.** 0.101 ______

## 13

Zeros at the right end of a number, after a decimal point, are significant.

**12. a.** three
**b.** three
**c.** one
**d.** three

**Examples** 0.400 has three significant digits.
0.0090 has two significant digits.
5.000 has four significant digits.

How many significant digits are there in these numbers?

**a.** 0.5000 ______ **b.** 0.0080 ______
**c.** 6.090 ______ **d.** 7.00 ______

## 14

How many significant digits are in the following numbers?

a. 0.0082 ___2___ b. 907 ___3___ c. 0.0104 ___3___
d. 2.01 ___3___ e. 0.30 ___2___ f. 0.010 ___2___
g. 6.090 ___4___ h. 914 ___3___ i. 6.00 ___3___

13. a. four
b. two
c. four
d. three

## 15

Zeros to the right of a nonzero digit but before the decimal point may or may not be significant.

**Examples** 400 (the zeros may be significant).
6,000 (the zeros may be significant).

In this book you may assume that these digits are significant. In other circumstances you will have to be told whether they are significant or not.

14. a. two
b. three
c. three
d. three
e. two
f. two
g. four
h. three
i. three

## 16

In multiplying and dividing measurements, an answer cannot have more significant digits than the measurement with the least number of significant digits used to arrive at the answer.

**Examples** $48.227 \times 1.64 = 79.1$

measurement with the least number of significant digits (→ 1.64)

answer must have the same number of digits as the measurement with the least number of significant digits. (→ 79.1)

The answer has been rounded off to three digits.

$$\frac{52.7}{0.60} = 88$$

answer must have the same number of digits as the measurement with the least number of significant digits. (→ 88)

measurement with the least number of significant digits (→ 0.60)

The answer has been rounded off to two digits.

Here are some problems with answers. Round the answers to the proper number of significant digits.

a. $5.237 \times 6.4 = 33.5168$ ___34___ b. $2.309 \times 35.6 = 82.2004$ ___82.2___

c. $\frac{58.7}{2.4} = 24.45833$ ___24.___ d. $\frac{0.5178}{0.087} = 5.952$ ___6.0___

## 17

a. The concept of precision is used only with the operations of ___addition + subtraction___.

b. Significant digits are used only with the operations of ___multiplication Division___.

16. a. 34
b. 82.2
c. 24
d. 6.0

**17. a.** addition and subtraction
**b.** multiplication and division

## 18

**a.** When using the concept of precision, we are interested only in the number of Decimal places in the answer.

**b.** When using significant digits, we are interested only in the number of Sig. Digits in the answer.

**18. a.** decimal places
**b.** significant digits

## 19

In which of these problems would you use the concept of precision?

**a.** $\begin{array}{r} 2.106 \\ +1.34 \\ \hline \end{array}$ **b.** $\begin{array}{r} 4.98 \\ -2.001 \\ \hline \end{array}$ **c.** $\begin{array}{r} 2.97 \\ \times\ 1.8 \\ \hline \end{array}$

**19. a** and **b**

## 20

In which problems would you use the concept of significant digits?

**a.** $\begin{array}{r} 6.087 \\ \times\ 3.4 \\ \hline \end{array}$ **b.** $\begin{array}{r} 4.22 \\ +8.2 \\ \hline \end{array}$ **c.** $\dfrac{6.97}{2.8}$

**20. a** and **c**

## 21

What happens if you want to use numbers that are not measurements along with numbers that are? Numbers that are not measurements are called *exact numbers.* Exact numbers do not have significant digits and are ignored when determining the number of significant digits in an answer.

**Examples** Divide the measurement 44.88 by 2. (2 is exact.)

four significant digits ⟶ 44.88
no significant digits ⟶ 2

$$\frac{44.88}{2} = 22.44$$ ⟵ The answer has four significant digits.

Multiply the measurement 10.2 by 4. (4 is exact.)

$\begin{array}{r} 10.2 \\ \times\ \ 4 \\ \hline 40.8 \end{array}$

10.2 ⟵ three significant digits
4 ⟵ no significant digits
40.8 ⟵ The answer has three significant digits.

Perform the indicated operations.

**a.** Multiply the measurement 50.6 by 4. 202.4 → 202

**b.** Divide the measurement 8.842 by 2. 4.421.

**21. a.** 202
**b.** 4.421

## 22

Here is a summary of the rules for reporting measurements.

*Adding and subtracting:*

In adding or subtracting measurements, an answer cannot be more precise than the least precise measurement used to arrive at the answer. We are concerned only with the number of decimal places in the answer.

*Multiplying and dividing:*

In multiplying and dividing measurements, an answer cannot have more significant digits than the measurement with the least number of significant digits used to arrive at the answer. We are interested only in the number of significant digits in the answer.

# Section 1-9 Dimensional Analysis

## 1

Both numbers and dimensions must be evaluated to solve technical problems. The process of evaluating dimensions in formulas is called *dimensional analysis.*

The concept most often used in dimensional analysis is:

$$\frac{n}{n} = 1$$

This concept is used in a short-cut technique called *reducing*. Both numbers and dimensions can be reduced.

**Example** $\frac{\not{3}\not{P}V}{\not{3}\not{P}} = V$

The 3s and *P*s cancelled because they appeared in both the numerator and denominator and they take the form:

$$\frac{3}{3} = 1 \text{ and } \frac{P}{P} = 1$$

Here are other examples of reducing.

$$\frac{\overset{2}{\not{10}}L\not{K}}{\not{5}\not{K}} = 2L \qquad \frac{(\overset{2}{\not{6}}\not{K})(4L)}{(\not{3}\not{K})} = 8L$$

Dimensional analysis consists of reducing units of dimensions in a formula to give an answer with the correct dimensional units.

Use dimensional analysis to reduce the following equations. The numbers are exact.

**a.** $\frac{5 \text{ m cm}}{5 \text{ m}} =$ **b.** $\frac{1{,}000 \text{ mg g}}{500 \text{ mg}} =$

**c.** $\frac{500 \text{ cm m}}{100 \text{ cm}} =$ **d.** $\frac{(2L)(4K)}{(2K)} =$

**e.** $\frac{1{,}000 \text{ m}\ell\ \ell}{100 \text{ m}\ell} =$ **f.** $\frac{(10L)(6K)}{5K} =$

## 2

Dimensional analysis can also be used to evaluate formulas written in conversion-factor form.

**Example** Reduce the following.

$$100 \not{\text{cm}} \times \frac{1 \text{ m}}{100 \not{\text{cm}}} = 1 \text{ m}$$

This can be done because

$$100 \text{ cm} \times \frac{1 \text{ m}}{100 \text{ cm}} = \frac{(\not{100} \not{\text{cm}})(1 \text{ m})}{\not{100} \not{\text{cm}}}$$

**1. a.** cm
**b.** 2 g
**c.** 5 m
**d.** $4L$
**e.** $10\ell$
**f.** $12L$

We do not bother to rewrite the conversion-factor form.

**Examples** $2\ \cancel{cm} \times \frac{10\ mm}{1\ \cancel{cm}} = 20\ mm$

$\overset{5}{\cancel{500}}\ \cancel{cm} \times \frac{1\ m}{\cancel{100}\ \cancel{cm}} = 5\ m$

Use dimensional analysis to reduce the following conversion-factor equations.

**a.** $80\ mm \times \frac{1\ cm}{10\ mm} =$

**b.** $2{,}000\ m\ell \times \frac{1\ell}{1{,}000\ m\ell} =$

**c.** $2\ m \times \frac{100\ cm}{1\ m} =$

**d.** $20\ cm \times \frac{10\ mm}{1\ cm} =$

**e.** $500\ g \times \frac{1\ kg}{1{,}000\ g} =$

**f.** $200\ m\ell \times \frac{1\ell}{1{,}000\ m\ell} =$

**3**

---

Here are more-complex conversion-factor equations. They are reduced in the same manner using dimensional analysis.

**Example** $\overset{2}{\cancel{200}}\ \cancel{m\ell} \times \frac{1\ \cancel{kg}}{\cancel{100}\ \cancel{m\ell}} \times \frac{1{,}000\ g}{1\ \cancel{kg}} = 2{,}000\ g$

**2. a.** 8 cm
**b.** $2\ell$
**c.** 200 cm
**d.** 200 mm
**e.** 0.5 kg
**f.** $0.2\ell$

Try these.

**a.** $600\ K \times \frac{1P}{300K} \times \frac{10V}{4P} =$

**b.** $100L \times \frac{10V}{2L} \times \frac{10K}{5V} =$

**4**

---

Answers to frame 3.

**a.** $5V$ **b.** $1{,}000K$

# Section 1-10 Applied Problems

**1**

---

People in technical occupations must think and work in an orderly manner. Having a system for organizing data (pieces of information) is especially helpful in solving word problems. One effective way to organize data is shown below.

*The Five Steps of Problem Solving*

*Step 1.* Determine what is being asked for.
*Step 2.* Determine what information is already known.
*Step 3.* Find a mathematical model that describes the relationship.
*Step 4.* Substitute the data into the model.
*Step 5.* Do the calculations. (A calculator may be used for this step.)

The problem below illustrates the use of the five problem-solving steps.

Voltage meter $A$ reads 10.5 volts, and voltage meter $B$ reads 5.42 volts. What is the total voltage reading of both meters?

*Step 1.* Determine what is being asked for.
Total voltage reading = ?

*Step 2.* Determine what information is already known.
Reading of meter $A$ = 10.5 volts
Reading of meter $B$ = 5.42 volts

*Step 3.* Find a mathematical model that describes the relationship.
Total voltage reading = reading of meter $A$ + reading of meter $B$

*Step 4.* Substitute the data into the model.
Total voltage reading = 10.5 volts + 5.42 volts

*Step 5.* Do the calculations.
Total voltage reading = 15.9 volts (The answer is precise to one decimal place.)

**2**

Calculated answers should be checked by estimation. To estimate an answer, round the numbers used in the calculation and then perform the calculation mentally. The problem solved in frame 1 will be used to illustrate this technique.

Calculated answer: 10.5 volts + 5.42 volts = 15.9 volts
Estimated answer: 11 volts + 5 volts = 16 volts

The estimated and calculated answers are very close, and the calculated answer is probably correct. If there is a big difference between the calculated and estimated answers, do the calculation again.

Obtain a calculated and estimated answer for the following:

| *Calculated* | *Estimated* |
|---|---|
| **a.** $\dfrac{181.5}{61.2} =$ | **b.** $\dfrac{181.5}{61.2} =$ |
| **c.** 48.4 − 12.9 = | **d.** 48.4 − 12.9 = |

**3**

The following problem also can be solved using the five-step procedure. Check your answer by estimation.

A surveyor measures a piece of land and finds it to be 22.58 acres. How many 1.3 acre lots can be made from the 22.58 acres?

*Step 1.* Determine what is being asked for.

**a.** ______________________________

*Step 2.* Determine what information is already known.

**b.** ______________________________

______________________________

*Step 3.* Find a mathematical model that describes the relationship.

**c.** ______________________________

______________________________

**2. a.** $\dfrac{181.5}{61.2} = 2.97$

**b.** $\dfrac{180}{60} = 3$

**c.** 48.4 − 12.9 = 35.5

**d.** 48 − 13 = 35

*Step 4.* Substitute the data into the model.

**d.** ________________________________________

________________________________________

*Step 5.* Do the calculations.

**e.** ________________________________________

________________________________________

## 4

Answers to frame 3.

**a.** number of lots = ?
**b.** total number of acres = 22.58 acres
size of each lot = 1.3 acres/lot (read "acres per lot")

**c.** $\text{number of lots} = \dfrac{\text{total number of acres}}{\text{size of each lot}}$

**d.** $\text{number of lots} = \dfrac{22.58 \text{ acres}}{1.3 \text{ acres/lot}}$

The unit acre cancels because $\dfrac{22.58 \text{ acres}}{1.3 \text{ acres/lot}} = \dfrac{22.58 \text{ acres} \times \text{lot}}{1.3 \text{ acres}}$

**e.** number of lots = 17 lots (Answer contains two significant digits.)

## 5

The following problem can be solved using the five-step procedure. In this problem and all other problems, check your answers by estimation.

A piece of metal weighed 120.7 grams before machining and 98.22 grams afterwards. How much weight did the piece of metal lose?

**a.** ________________________________________

**b.** ________________________________________

________________________________________

**c.** ________________________________________

**d.** ________________________________________

**e.** ________________________________________

________________________________________

## 6

Answers to frame 5.

**a.** amount of weight lost = ?
**b.** original weight = 120.7 grams
final weight = 98.22 grams
**c.** amount of weight lost = original weight − final weight
**d.** amount of weight lost = 120.7 grams − 98.22 grams
**e.** amount of weight lost = 22.5 grams (Does this match your estimated answer?)

**7**

Use the five-step procedure to solve the following problem.

Twelve electric motors are capable of producing 1.33 horsepower each. What would be the total horsepower of all 12 motors? Twelve is an exact number.

**a.** ______________________________

**b.** ______________________________

______________________________

**c.** ______________________________

______________________________

**d.** ______________________________

______________________________

**e.** ______________________________

______________________________

**8**

Answers to frame 7.

**a.** total horsepower = ?
**b.** number of motors = 12
horsepower per motor = 1.33
**c.** total horsepower = number of motors × horsepower/motor
**d.** total horsepower = 12 motors × 1.33 horsepower/motor
**e.** total horsepower = 16.0 horsepower

**9**

Find the length of the missing dimension.

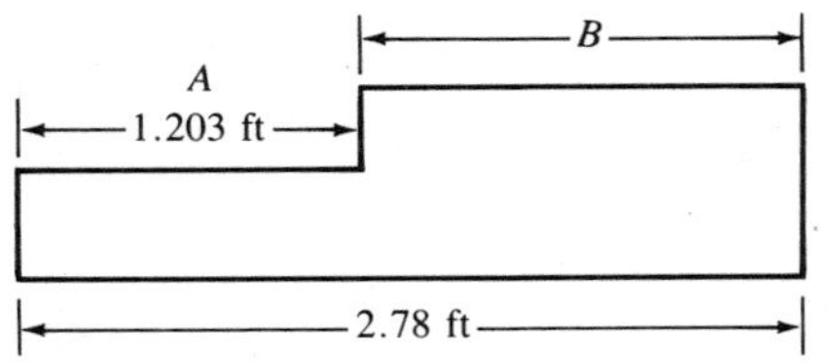

A — 1.203 ft; B; 2.78 ft

**10**

Answers to frame 9.

**a.** length of $B$ = ?
**b.** total length = 2.78 ft
length of $A$ = 1.203 ft
**c.** length of $B$ = total length − length of $A$
**d.** length of $B$ = 2.78 ft − 1.203 ft
**e.** length of $B$ = 1.58 ft

**11**

Many problems in technical fields involve operations with money. Dollars and cents are written in decimal notation; for example, 9 dollars and 27 cents is written as $9.27. Thus, operations with money will involve decimals, and answers to money problems will always have two decimal places.

Find these products.

**a.** $\begin{array}{r} \$5.08 \\ \times \quad 9 \\ \hline \end{array}$

**b.** $\begin{array}{r} \$12.15 \\ \times \quad 7 \\ \hline \end{array}$

## 12

When a calculator is used to perform an operation involving money, it does not automatically report the answer to two decimal places. For example, use a calculator to perform the multiplication $7.35 × 8.

**11. a.** $45.72
**b.** $85.05

| *Enter* | *Press* | *Display* |
|---|---|---|
| 7.35 | [×] | 7.35 |
| 8 | [=] | 58.8 |

The actual answer is $58.80.

Use a calculator to do these operations involving money.

**a.** $25.54 × 5 = **b.** $8.56 × 5 = **c.** 3 )$37.20 =

## 13

A calculator was used to perform the following division involving money: $58.90 ÷ 6

**12. a.** $127.70
**b.** $42.80
**c.** $12.40

| *Enter* | *Press* | *Display* |
|---|---|---|
| 58.90 | [÷] | 58.90 |
| 6 | [=] | 9.8166666 |

The correct answer is $9.82. Notice that the answer must be rounded to the hundredths place (two places to the right of the decimal).

Use a calculator to do these operations involving money.

**a.** $85.21 ÷ 11 = **b.** $120 ÷ 13 =

## 14

Answers to frame 13.

**a.** $7.75 **b.** $9.23

# Exercise Set, Sections 1-8–1-10

## Measurement Concepts

To what place are these numbers precise?

**1.** 14.04 ______ **2.** 309.1 ______ **3.** 0.006 ______ **4.** 29.00 ______

Add or subtract these measurements, and give the answer to the correct decimal place.

**5.** 2.0 + 3.88 = ______ **6.** 6.944 − 3.33 = ______ **7.** 34.000 − 2.15 = ______

How many significant digits are in the following numbers?

**8.** 0.0067 ______ **9.** 409 ______ **10.** 67.02 ______ **11.** 0.0104 ______

**12.** 938 ______ **13.** 34,003 ______ **14.** 0.30 ______ **15.** 0.0120 ______

Round off these answers to the correct number of significant digits.

**16.** $5.386 \times 2.11 = 11.36446$ ______

**17.** $4.090 \times 28.1 = 114.9290$ ______

**18.** $\frac{68.00}{24.0} = 2.83333$ ______

**19.** $\frac{0.6140}{0.093} = 6.60215$ ______

Calculate and round off these answers.

**20.** $21.4 \times 3$ (exact) = ______

**21.** $\frac{6.863}{3 \text{ (exact)}} =$ ______

## Dimensional Analysis

Use dimensional analysis to reduce the following equations.

**22.** $\frac{(1 \text{ m}\ell)(10K)}{(1\text{K})} =$

**23.** $\frac{(30 \text{ atm})(1 \ \ell)}{(10 \text{ atm})} =$

**24.** $400K \times \frac{10\ell}{200K} =$

**25.** $100 \text{ g} \times \frac{1{,}000 \text{ mg}}{50 \text{ g}} =$

**26.** $400 \text{ m}\ell \times \frac{1 \text{ kg}}{100 \text{ m}\ell} \times \frac{1{,}000 \text{ g}}{1 \text{ kg}} =$

**27.** $4\ell \times \frac{5V}{2\ell} \times \frac{10K}{2V} =$

## Applied Problems

Use the five-step problem-solving procedure to solve the following problems. Answers must have the correct number of significant digits and units.

**28.** A metallurgist finds that three samples of metal weigh 82.07 grams, 101.00 grams, and 138.601 grams. What is the total weight of the three samples?
**29.** A piece of metal is 2.00 inches thick. How much metal must be removed to make the piece 1.87 inches thick?
**30.** How far will you travel if you average 53.5 miles/hour for 4.71 hours?
**31.** If 42 sheets of metal form a pile 57.96 inches high, how thick is one sheet?
**32.** In four test runs, a car burned 3.21, 3.07, 3.35, and 3.27 gallons of gasoline. What is the total amount of gasoline burned?
**33.** The total voltage reading from two meters is 8.75 volts. If voltage meter $A$ reads 2.9 volts, what does voltage meter $B$ read?
**34.** If a gallon of gas costs $1.29, what does 37.5 gallons cost?
**35.** If 25 manufactured bolts cost $4.50, what does one bolt cost?
**36.** Find the length of the missing dimension.

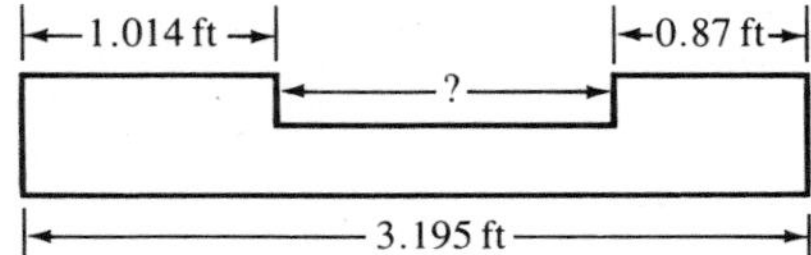

**37.** How many feet of fencing will be needed to enclose this yard?

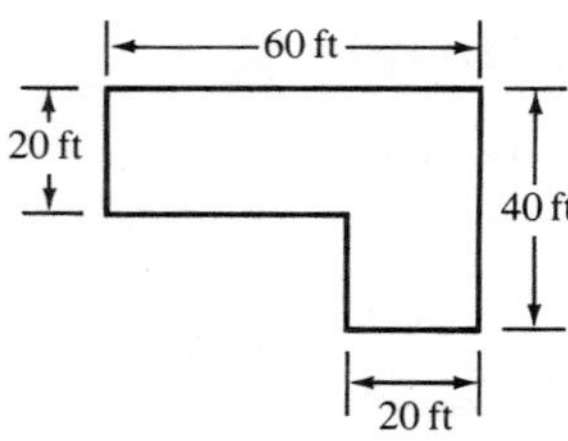


**38.** What is the total number of tiles needed to cover the floor in this room? Each tile is 1 ft by 1 ft.

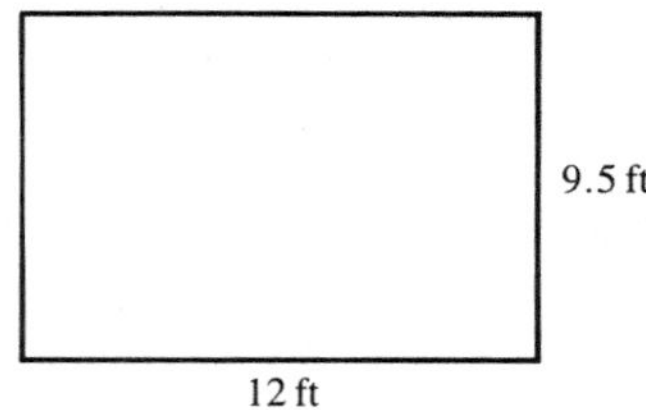


**39.** How long must a roll of carpet be to cover these steps?

**40.** The following table lists the number of parts machined daily by each of four employees over a five-day period.

| *Employee number* | *Day of the week* | | | | | *Weekly totals* |
|---|---|---|---|---|---|---|
| | *Mon.* | *Tues.* | *Wed.* | *Thur.* | *Fri.* | |
| 1 | 127 | 135 | 146 | 128 | 131 | |
| 2 | 118 | 111 | 105 | 120 | 114 | |
| 3 | 142 | 147 | 158 | 149 | 161 | |
| 4 | 121 | 128 | 141 | 146 | 131 | |
| *Daily totals* | | | | | | |

**a.** Calculate the weekly total for each employee.
**b.** Calculate the daily total for each day.
**c.** Add up the weekly totals and the daily totals. If they do not equal each other, you have made a mistake. If you made a mistake, recheck your calculations.

# Supplementary Exercise Set, Sections 1-8–1-10

To what place are these numbers precise?

**1.** 18.69 ______ **2.** 0.900 ______

Add or subtract these measurements, and give the answers to the correct decimal place.

**3.** $4.1 + 5.96 =$ ______ **4.** $8.777 - 2.21 =$ ______ **5.** $0.917 - 0.1 =$ ______

How many significant digits are in the following numbers?

**6.** 0.0042 ______ **7.** 302 ______ **8.** 0.06040 ______

Round these answers to the correct number of significant digits.

**9.** $4.7925 \times 3.12 = 14.9526$ ______ **10.** $\dfrac{0.6030}{1.0745} = 0.561191$ ______

Calculate and round off the answers.

**11.** $38.2 \times 4 \text{ (exact)} =$ ______ **12.** $\dfrac{9.527}{2 \text{ (exact)}} =$ ______

Use dimensional analysis to reduce the following equations.

**13.** $\dfrac{(10\text{ g})(20\,V)}{(5\text{ g})} =$ **14.** $200\,V \times \dfrac{700P}{50\,V} \times \dfrac{10T}{350P} =$

Solve the following problems. Answers must have the correct number of significant digits and units.

**15.** A pipe is 1.375 meters long. What length of the pipe must be cut off to make the pipe 0.87 meters long?
**16.** A chemical technologist finds that four samples weigh 1.967 grams, 1.29 grams, 0.807 grams, and 0.90 grams. What is the total weight of the four samples?
**17.** If one bolt weighs 3.70 ounces, what is the weight of 27 bolts?
**18.** 20.9 acres of land are to be divided into 25 equal lots. How large will each lot be?
**19.** An electric part costs $8.23. How much will 23 parts cost?
**20.** Find the missing dimension.

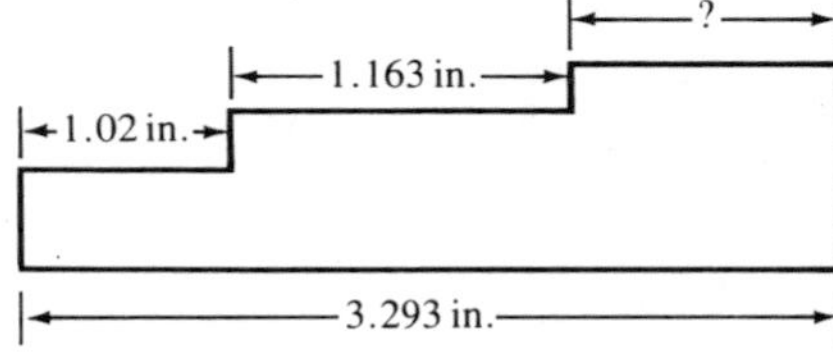

**21.** The following table lists the amount of computer sales for each month, for each salesperson.

| *Salesperson number* | *Month* | | | | *Four-month totals* |
|---|---|---|---|---|---|
| | *Jan.* | *Feb.* | *Mar.* | *Apr.* | |
| 1 | $25,870 | $21,670 | $34,000 | $27,900 | |
| 2 | $18,950 | $14,600 | $27,500 | $31,000 | |
| 3 | $12,500 | $14,300 | $18,750 | $21,000 | |
| 4 | $36,500 | $32,820 | $49,000 | $51,000 | |
| 5 | $32,750 | $30,500 | $42,800 | $49,600 | |
| *One-month totals* | | | | | |

**a.** Calculate the four-month total for each salesperson.
**b.** Calculate the one-month total for each month.
**c.** Add up the four-month totals, and then add the one-month totals. Do the totals equal each other?

**22.** How many feet of fencing are needed to enclose this yard?

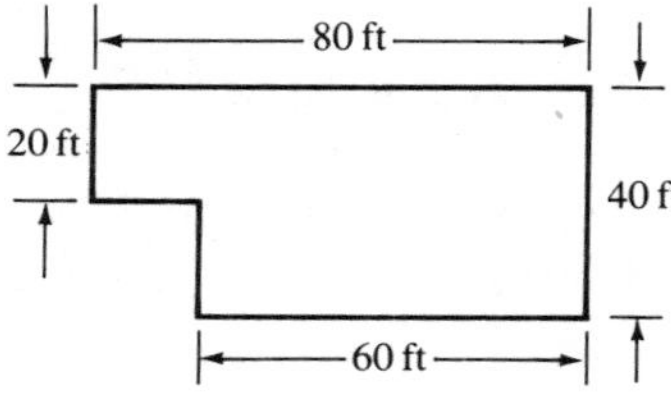


**23.** What is the total number of ceiling tiles needed to cover this ceiling? Each tile is 2 ft by 4 ft.

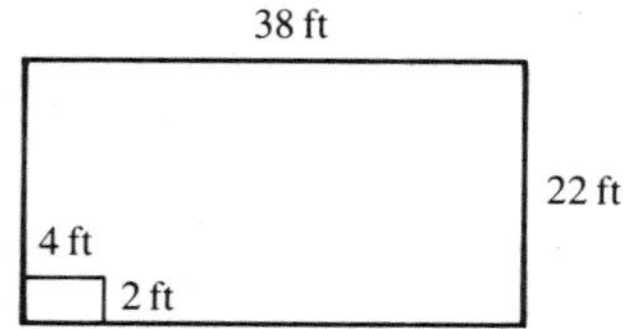


**24.** How long must a roll of carpet to be to cover these steps?

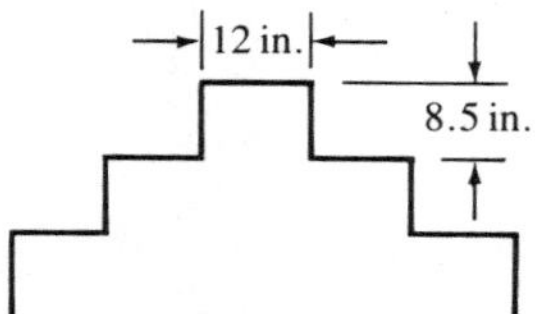

# Summary

1. The place names of whole numbers end in the letter *s* (tens, thousands).
2. A decimal number has a value that falls between two whole numbers.
3. Decimal place names end in *ths* (tenths, thousandths).
4. A decimal number smaller than 1 is read as if it were a whole number with the place name of the last digit added (0.6 is read as "six tenths").
5. A decimal number larger than 1 is read as a whole number, followed by the word "and," followed by the decimal part of the number (4.9 is read as "four and nine tenths").
6. Numbers can be added in any order, but care must be taken to list all digits in their proper columns.
7. Subtraction must be performed in the given order, and digits must be kept in their proper columns. Borrowing may be necessary.
8. Multiplication can be performed in any order.
9. When multiplying decimal numbers, the number of digits to the right of the decimal point in the answer equals the sum of the digits to the right of the decimal point in each of the numbers being multiplied.
10. Division must be performed in the given order.
11. If the divisor is a decimal number, the decimal point must be shifted to the right (the same number of places in both the divisor and dividend) to make the divisor a whole number.
12. An answer to an addition or subtraction problem cannot be more precise than the least precise measurement used to arrive at the answer.
13. An answer to a multiplication or division problem cannot have more significant digits than the measurement with the least number of significant digits used to arrive at the answer.
14. The five steps of problem solving should be used to solve word problems.
    a. Determine what is being asked for.
    b. Determine what information is already known.
    c. Find a mathematical model that describes the relationship.
    d. Substitute the data into the model.
    e. Do the calculations.

# Chapter 1 Self-Test

## Basic Mathematics

What is the place name of the 4 in each of these numbers?

**1.** 2,342,383 ____________ **2.** 23.042 ____________

Write the following numbers.

**3.** Two million, six hundred thousand, fifteen ________________________
**4.** Four hundred four ten-thousandths ________________________

Perform the following additions.

**5.** $\begin{array}{r} 19{,}867 \\ +12{,}482 \\ \hline \end{array}$ **6.** $\begin{array}{r} 82.07 \\ +\ 0.18 \\ \hline \end{array}$ **7.** $\begin{array}{r} 23 \\ +\ 0.7 \\ \hline \end{array}$ **8.** $\begin{array}{l} 38 \\ \ \ 0.042 \\ +\ 2.8 \\ \hline \end{array}$

Perform the following subtractions.

**9.** $\begin{array}{r} 827 \\ -562 \\ \hline \end{array}$ **10.** $\begin{array}{r} 107.3 \\ -\ \ 4.65 \\ \hline \end{array}$ **11.** $\begin{array}{r} 93.8 \\ -\ 0.9 \\ \hline \end{array}$ **12.** $\begin{array}{r} 81.9 \\ -52.89 \\ \hline \end{array}$

Perform the following multiplications.

**13.** $\begin{array}{r} 9.8 \\ \times\ 3 \\ \hline \end{array}$

**14.** $\begin{array}{r} 0.47 \\ \times 0.21 \\ \hline \end{array}$

**15.** $\begin{array}{r} 3.49 \\ \times 2.46 \\ \hline \end{array}$

Perform the following multiplication and division using decimal-point shift.

**16.** $83 \times 10{,}000 =$ __________

**17.** $8{,}650 \div 100 =$ __________

Perform the following divisions. Round your answers to the hundredths place.

**18.** $1.5\overline{)156.5}$

**19.** $2.4\overline{)84}$

**20.** $22\overline{)4.4}$

**21.** To what place is the number 21.00 precise? __________

**22.** How many significant digits are in the number 0.01090? __________

Add or subtract these measurements. Give the answers to the correct decimal places.

**23.** $0.908 + 1.67 =$ __________

**24.** $64.00 - 1.36 =$ __________

Round off these answers to the correct number of significant digits.

**25.** $(4.1659)(0.090) =$ __________

**26.** $\dfrac{4.963}{2\ (\text{exact})} =$ __________

Use dimensional analysis to reduce the following equations.

**27.** $\dfrac{(8\ \text{ft})(48\ \text{in.})}{(4\ \text{ft})} =$ __________

**28.** $10\ \text{g} \times \dfrac{10\ell}{2\ \text{g}} \times \dfrac{10\ \text{mg}}{5\ \ell} =$ __________

Use the five-step problem-solving procedure to obtain answers with the correct units and correct number of significant digits.

**29.** Three operations are required to machine a piece of metal. Operation *A* takes 4.67 minutes, operation *B* requires 3.80 minutes, and operation *C* requires 3.6 minutes. What is the total time required to machine the piece of metal?

**30.** 22.0 quarts of an antifreeze and water solution must be made. If 8.5 quarts of the solution are antifreeze, how much of the solution is water?

**31.** A machine can cut five pieces of metal per minute. How many pieces will be produced in 33.2 minutes?

**32.** The gasoline bill for a trip is \$41.60. If the gasoline cost \$1.28 per gallon, how many gallons were used?

**33.** Find the missing dimension.

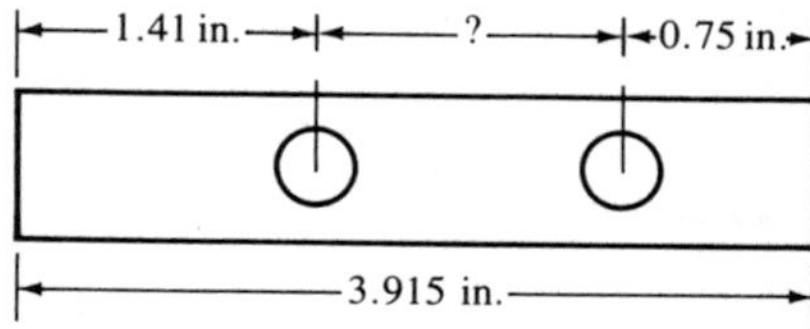

# CHAPTER 2

# Fractions

In many science and technology applications, measurements are often expressed in fractional parts, and a basic understanding of the use of fractions is essential. In this chapter, you will learn to identify fractions and to add, subtract, multiply, and divide using fractions.

## Section 2-1 Introduction to Fractions

**1**

A fraction is a number that is used to compare a part with a whole.

The circle in Figure 2-1, for example, is divided into four parts. One part is shaded. The fraction that represents this circle is 1/4.

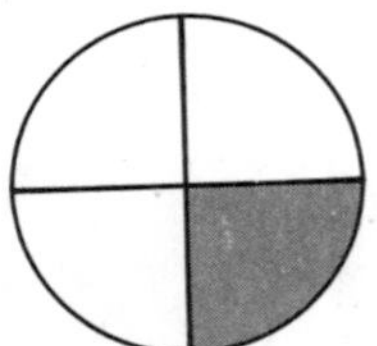

**Figure 2-1**

$\frac{1}{4}$ ← 1: number of shaded parts; 4: total number of parts

A fraction may be written vertically, as $\frac{1}{4}$, or horizontally, as 1/4.

Write the fraction that represents the circle in Figure 2-2. ______

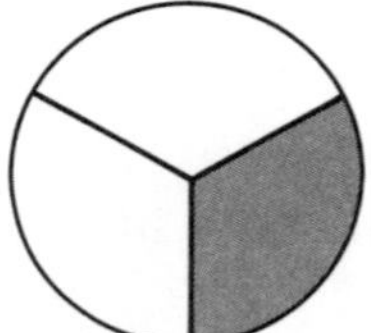

**Figure 2-2**

**2**

The parts of a fraction have these technical names:

$\frac{4}{5}$ ← 4: numerator; 5: denominator

**1.** $\frac{1}{3}$

In the fraction $\frac{7}{8}$,

**a.** the numerator is ______.

**b.** the denominator is ______.

**3**

The name of a fraction begins with the name of the numerator and ends with the name of the denominator.

**2. a.** 7
**b.** 8

**Examples** The fraction $\frac{3}{4}$ is read as "three-fourths."

numerator → three; denominator → fourths

The fraction seven eighths is written as $\frac{7}{8}$.

**a.** $\frac{5}{7}$ is read as ____________.

**b.** eleven-fourteenths is written as ________.

**c.** one-fifth is written as ________.

**4**

A fraction also can be used to indicate division. For example:

**3. a.** five-sevenths
**b.** $\frac{11}{14}$
**c.** $\frac{1}{5}$

$\frac{10}{2}$ $\quad 10 \div 2 \quad 2\overline{)10} \quad 10/2$

all mean "10 divided by 2."

Write the fractions that represent these divisions.

**a.** 8 divided by 3 ________

**b.** $15 \div 7$ ________

**c.** $4\overline{)51}$ ________

**5**

A fraction can be converted to a decimal by performing the indicated division.

**4. a.** $\frac{8}{3}$
**b.** $\frac{15}{7}$
**c.** $\frac{51}{4}$

The fraction $\frac{3}{4}$ means "3 divided by 4," or $4\overset{0.75}{\overline{)3.00}}$.

This same conversion can be done using a calculator.

| *Enter* | *Press* | *Display* |
|---|---|---|
| 3 | ÷ | 3 |
| 4 | = | 0.75 |

Therefore, $\frac{3}{4} = 0.75$.

Use a calculator to convert the following fractions to decimals. (Round to thousandths.)

a. $\frac{5}{8}$ b. $\frac{1}{4}$ c. $\frac{7}{4}$ d. $\frac{2}{3}$

## 6

Two fractions that are equal in value are called *equivalent fractions.* A fraction can be changed to an equivalent fraction by multiplying the numerator and denominator by the same number. For example:

$$\frac{3}{4} = \frac{3 \times 2}{4 \times 2} = \frac{6}{8}$$

Therefore,

$$\frac{3}{4} = \frac{6}{8}$$

Change each of these fractions to equivalent fractions with denominators of 8.

a. $\frac{1}{4}$ b. $\frac{1}{2}$ c. $\frac{3}{4}$

5. a. 0.625
   b. 0.250
   c. 1.750
   d. 0.667

## 7

In the example below, 3/4 is to be changed to an equivalent fraction with a denominator of 16.

$$\frac{3}{4} = \frac{?}{16}$$

To find the number that both the numerator and the denominator should be multiplied by, divide 16 by 4: 16 ÷ 4 = 4. Perform the multiplication.

$$\frac{3 \times 4}{4 \times 4} = \frac{12}{16}$$

Complete each of the following problems.

a. $\frac{1}{2} = \frac{}{8}$ b. $\frac{1}{3} = \frac{}{9}$ c. $\frac{3}{4} = \frac{}{32}$

6. a. $\frac{2}{8}\left(\frac{1 \times 2}{4 \times 2}\right)$
   b. $\frac{4}{8}\left(\frac{1 \times 4}{2 \times 4}\right)$
   c. $\frac{6}{8}\left(\frac{3 \times 2}{4 \times 2}\right)$

## 8

A fraction can be reduced when the same number can be divided out of both the numerator and the denominator. For example, 6/8 can be reduced because both 6 and 8 can be divided by 2.

$$\frac{6 \div 2}{8 \div 2} = \frac{3}{4}$$

Reduce the following fractions.

a. $\frac{14}{16}$ b. $\frac{6}{12}$ c. $\frac{10}{16}$ d. $\frac{2}{8}$

7. a. 4
   b. 3
   c. 24

**9**

A fraction is reduced to *lowest terms* when no number (other than 1) can be divided out of both the numerator and the denominator. For example:

$\frac{2}{3}$ is in lowest terms because no number can be divided out of both 2 and 3.

$\frac{6}{9}$ is not in lowest terms because 3 can be divided out of both 6 and 9.

Which of the following fractions are in lowest terms? (Circle your selections.)

a. $\frac{3}{4}$ b. $\frac{3}{6}$ c. $\frac{5}{8}$ d. $\frac{9}{12}$

8. a. $\frac{7}{8}$ b. $\frac{1}{2}$ c. $\frac{5}{8}$ d. $\frac{1}{4}$

**10**

A fraction with a numerator smaller than the denominator is called a *proper fraction.* The fractions 1/4, 3/8, and 7/15 are all proper fractions.

Which of the following are proper fractions? (Circle your selections.)

a. $\frac{3}{8}$ b. $\frac{5}{4}$ c. $\frac{2}{3}$ d. $\frac{9}{7}$

9. a and c

**11**

A fraction with a numerator as large as or larger than the denominator is called an *improper fraction.* The fractions 3/2, 5/5, and 8/5 are all improper fractions.

Which of the following are improper fractions? (Circle your selections.)

a. $\frac{3}{4}$ b. $\frac{4}{3}$ c. $\frac{5}{2}$ d. $\frac{1}{2}$

10. a and c

**12**

An improper fraction can be converted to a *mixed number* (a whole number plus a fraction) by dividing the denominator into the numerator and writing the remainder as a fraction. For example:

$$\frac{7}{5} \qquad 5\overline{)7}^{\,1\frac{2}{5}} \quad \begin{matrix} 5 \\ \overline{2} \end{matrix}$$

2 ← remainder
5 ← divisor

Therefore,

$$\frac{7}{5} = 1\frac{2}{5}$$

Convert the following improper fractions to mixed numbers.

a. $\frac{8}{5}$ b. $\frac{5}{2}$ c. $\frac{7}{4}$ d. $\frac{10}{3}$

11. b and c

### 13

A mixed number can be changed to an improper fraction by multiplying the denominator by the whole number and adding that product to the numerator of the fraction. This new number is then written over the original denominator to make the improper fraction.

**Example** $2\frac{7}{8} = \frac{(8 \times 2) + 7}{8} = \frac{16 + 7}{8} = \frac{23}{8}$

Convert the following mixed numbers to improper fractions.

**a.** $3\frac{1}{2}$ **b.** $1\frac{9}{16}$ **c.** $4\frac{7}{8}$ **d.** $25\frac{3}{4}$.

12. a. $1\frac{3}{5}$
b. $2\frac{1}{2}$
c. $1\frac{3}{4}$
d. $3\frac{1}{3}$

### 14

Answers to frame 13.

**a.** $\frac{7}{2}$ **b.** $\frac{25}{16}$ **c.** $\frac{39}{8}$ **d.** $\frac{103}{4}$

# Exercise Set, Section 2-1

## Introduction to Fractions

Use a calculator to convert the following fractions to equivalent decimals. (Round to the hundredths place.)

**1.** $\frac{7}{16} =$ **2.** $\frac{21}{29} =$ **3.** $\frac{14}{31} =$ **4.** $\frac{2}{3} =$

**5.** $\frac{2}{5} =$ **6.** $\frac{1}{3} =$ **7.** $\frac{3}{4} =$ **8.** $\frac{17}{18} =$

Change each of these fractions to an equivalent fraction with a denominator of 32.

**9.** $\frac{1}{2} =$ **10.** $\frac{3}{4} =$ **11.** $\frac{7}{8} =$ **12.** $\frac{3}{16} =$

Change each of these fractions to an equivalent fraction with a denominator of 48.

**13.** $\frac{3}{12} =$ **14.** $\frac{3}{4} =$ **15.** $\frac{1}{16} =$ **16.** $\frac{5}{24} =$

Reduce the following fractions.

**17.** $\frac{14}{18} =$ **18.** $\frac{5}{10} =$ **19.** $\frac{11}{33} =$ **20.** $\frac{2}{16} =$

**21.** $\frac{12}{20} =$ **22.** $\frac{35}{40} =$ **23.** $\frac{3}{27} =$ **24.** $\frac{22}{66} =$

Convert the following improper fractions to mixed numbers.

**25.** $\frac{9}{5} =$ **26.** $\frac{7}{3} =$ **27.** $\frac{25}{4} =$ **28.** $\frac{10}{2} =$

**29.** $\frac{19}{6} =$ **30.** $\frac{38}{7} =$ **31.** $\frac{33}{5} =$ **32.** $\frac{25}{3} =$

Convert the following mixed numbers to improper fractions.

**33.** $2\frac{1}{2} =$ **34.** $1\frac{1}{8} =$ **35.** $21\frac{2}{3} =$ **36.** $14\frac{7}{8} =$

**37.** $1\frac{5}{6} =$ **38.** $2\frac{2}{7} =$ **39.** $12\frac{2}{3} =$ **40.** $22\frac{7}{9} =$

## Supplementary Exercise Set, Section 2-1

Change each fraction below to an equivalent decimal. (Round to the hundredths place.)

**1.** $\frac{5}{16}$ **2.** $\frac{15}{19}$ **3.** $\frac{3}{4}$ **4.** $\frac{7}{8}$ **5.** $\frac{1}{3}$

Reduce each of the following fractions.

**6.** $\frac{14}{16}$ **7.** $\frac{6}{8}$ **8.** $\frac{5}{10}$ **9.** $\frac{35}{45}$ **10.** $\frac{75}{125}$

Convert the following improper fractions to mixed numbers.

**11.** $\frac{3}{2}$ **12.** $\frac{15}{9}$ **13.** $\frac{25}{5}$ **14.** $\frac{41}{20}$ **15.** $\frac{50}{4}$

Convert the following mixed numbers to improper fractions.

**16.** $3\frac{1}{2}$ **17.** $4\frac{1}{8}$ **18.** $21\frac{3}{4}$ **19.** $7\frac{1}{3}$ **20.** $6\frac{2}{3}$

## Section 2-2 Multiplication and Division of Fractions

**1**

Multiplication of two fractions involves the following steps:

**a.** Multiply the numerators.
**b.** Multiply the denominators.
**c.** Reduce the answer to lowest terms.

This procedure is shown in the following examples.

$$\frac{3}{4} \times \frac{1}{2} = \frac{3 \times 1}{4 \times 2} = \frac{3}{8}$$

$$\frac{3}{4} \times \frac{2}{3} = \frac{3 \times 2}{4 \times 3} = \frac{6}{12} = \frac{1}{2}$$

Multiply the following fractions.

**a.** $\frac{7}{8} \times \frac{1}{3} =$ **b.** $\frac{1}{16} \times \frac{4}{5} =$

**c.** $\frac{3}{4} \times \frac{1}{4} =$ **d.** $\frac{7}{16} \times \frac{1}{7} =$

## 2

In some cases it is possible to reduce the fractions before multiplying. This can be done by using a shortcut method called *cancelling*.

**1. a.** $\frac{7}{24}$ **b.** $\frac{1}{20}$ **c.** $\frac{3}{16}$ **d.** $\frac{1}{16}$

$$\frac{1}{\underset{4}{\cancel{16}}} \times \frac{\overset{1}{\cancel{4}}}{5} = \frac{1}{20}$$

Both 4 and 16 can be divided by 4. Therefore, the multiplication can be reduced by dividing 4 out of the numerator and the denominator *before* multiplying.

Complete the following multiplications.

**a.** $\frac{\overset{1}{\cancel{3}}}{4} \times \frac{1}{\underset{1}{\cancel{3}}} =$ **b.** $\frac{1}{\underset{1}{\cancel{2}}} \times \frac{\overset{2}{\cancel{4}}}{5} =$ **c.** $\frac{7}{\underset{4}{\cancel{8}}} \times \frac{\overset{1}{\cancel{2}}}{3} =$

## 3

Complete the following multiplications.

**2. a.** $\frac{1}{4}$ **b.** $\frac{2}{5}$ **c.** $\frac{7}{12}$

**a.** $\frac{1}{12} \times \frac{2}{3} =$ **b.** $\frac{7}{8} \times \frac{4}{5} =$ **c.** $\frac{1}{16} \times \frac{8}{9} =$

## 4

In some cases, all four numbers can be cancelled.

**3. a.** $\frac{1}{18}$ **b.** $\frac{7}{10}$ **c.** $\frac{1}{18}$

$$\frac{\overset{1}{\cancel{3}}}{\underset{2}{\cancel{4}}} \times \frac{\overset{1}{\cancel{2}}}{\underset{1}{\cancel{3}}} = \frac{1}{2}$$

Complete the following multiplications.

**a.** $\frac{\overset{1}{\cancel{5}}}{\underset{4}{\cancel{8}}} \times \frac{\overset{1}{\cancel{2}}}{\underset{3}{\cancel{15}}} =$ **b.** $\frac{\overset{1}{\cancel{2}}}{\underset{1}{\cancel{3}}} \times \frac{\overset{3}{\cancel{9}}}{\underset{5}{\cancel{10}}} =$ **c.** $\frac{\overset{1}{\cancel{3}}}{\underset{1}{\cancel{4}}} \times \frac{\overset{1}{\cancel{4}}}{\underset{3}{\cancel{9}}} =$

**5**

To multiply a whole number times a fraction, it is best to think of the whole number as a fraction with a denominator of 1. For instance:

$$3 \times \frac{1}{4} \text{ can be written as } \frac{3}{1} \times \frac{1}{4}$$

Write the following whole numbers as fractions with denominators of 1, and perform the multiplications.

**a.** $4 \times \frac{2}{5} =$ **b.** $7 \times \frac{1}{8} =$ **c.** $6 \times \frac{3}{16} =$

**6**

Complete the following multiplications.

**a.** $5 \times \frac{9}{10} =$ **b.** $\frac{1}{2} \times 7 =$

**c.** $\frac{3}{4} \times 32 =$ **d.** $16 \times \frac{6}{8} =$

**7**

It is important to note that, when multiplying a whole number with a proper fraction, the answer will always be a number *smaller* than the whole number being multiplied.

**Example** $\frac{1}{2} \times 5 = \frac{5}{2} = 2\frac{1}{2}$

The answer, 2 1/2, is less than the whole number, 5. (When multiplying two whole numbers, the answer is always a number *larger* than either of the whole numbers being multiplied: 3 × 5 = 15.)

**8**

Division by a fraction is performed by inverting the second fraction and multiplying.

$$\frac{3}{4} \div \frac{1}{2} = \frac{3}{4} \times \frac{2}{1} = \frac{6}{4} = \frac{3}{2}$$

Complete the following divisions.

**a.** $\frac{7}{8} \div \frac{1}{4} = \frac{7}{8} \times \frac{4}{1} =$ **b.** $\frac{5}{6} \div \frac{1}{3} = \frac{5}{6} \times \frac{3}{1} =$

**c.** $\frac{1}{12} \div \frac{2}{3} = \frac{1}{12} \times \frac{3}{2} =$ **d.** $\frac{6}{8} \div \frac{1}{2} = \frac{6}{8} \times \frac{2}{1} =$

**4. a.** $\frac{1}{12}$
**b.** $\frac{3}{5}$
**c.** $\frac{1}{3}$

**5. a.** $\frac{4}{1} \times \frac{2}{5} = \frac{8}{5}$ or $1\frac{3}{5}$
**b.** $\frac{7}{1} \times \frac{1}{8} = \frac{7}{8}$
**c.** $\frac{6}{1} \times \frac{3}{16} = \frac{9}{8}$

**6. a.** $\frac{9}{2} = 4\frac{1}{2}$
**b.** $\frac{7}{2} = 3\frac{1}{2}$
**c.** $\frac{24}{1} = 24$
**d.** $\frac{12}{1} = 12$

**9**

Complete the following divisions.

a. $\frac{2}{3} \div \frac{4}{5} =$

b. $\frac{3}{16} \div \frac{2}{3} =$

c. $\frac{3}{4} \div \frac{3}{8} =$

8. a. $\frac{7}{2} = 3\frac{1}{2}$

b. $\frac{5}{2} = 2\frac{1}{2}$

c. $\frac{1}{8}$

d. $\frac{6}{4} = 1\frac{1}{2}$

**10**

Division by a whole number can also be performed by inverting and multiplying. The whole number is made into a fraction with a denominator of 1.

$$\frac{2}{3} \div 4 = \frac{2}{3} \div \frac{4}{1} = \frac{2}{3} \times \frac{1}{4} = \frac{2}{12} = \frac{1}{6}$$

Complete the following divisions.

a. $\frac{1}{5} \div 5 =$

b. $\frac{2}{5} \div 4 =$

c. $\frac{6}{11} \div 3 =$

9. a. $\frac{10}{12} = \frac{5}{6}$

b. $\frac{9}{32}$

c. $\frac{8}{4} = 2$

**11**

Answers to frame 10.

a. $\frac{1}{25}$

b. $\frac{1}{10}$

c. $\frac{2}{11}$

# Exercise Set, Section 2-2

## Multiplication and Division of Fractions

Multiply the following fractions.

1. $\frac{1}{8} \times \frac{3}{4} =$
2. $\frac{7}{8} \times \frac{1}{5} =$
3. $\frac{3}{4} \times \frac{2}{3} =$
4. $\frac{7}{11} \times \frac{1}{7} =$
5. $\frac{1}{16} \times \frac{8}{9} =$
6. $\frac{1}{2} \times \frac{2}{3} =$
7. $\frac{3}{24} \times \frac{8}{12} =$
8. $\frac{8}{48} \times \frac{6}{16} =$
9. $\frac{11}{33} \times \frac{3}{9} =$
10. $\frac{5}{16} \times \frac{3}{5} =$
11. $\frac{5}{24} \times \frac{8}{15} =$
12. $\frac{9}{48} \times \frac{12}{27} =$
13. $\frac{11}{30} \times \frac{6}{33} =$
14. $\frac{7}{27} \times \frac{9}{14} =$
15. $\frac{5}{18} \times \frac{3}{15} =$
16. $\frac{3}{8} \times 7 =$
17. $9 \times \frac{5}{8} =$
18. $5 \times \frac{2}{3} =$
19. $\frac{1}{7} \times 7 =$
20. $9 \times \frac{2}{3} =$
21. $15 \times \frac{4}{5} =$

Divide the following fractions.

**22.** $\frac{1}{2} \div \frac{2}{3} =$ **23.** $\frac{3}{4} \div \frac{7}{16} =$ **24.** $\frac{9}{10} \div \frac{1}{5} =$ **25.** $\frac{5}{8} \div \frac{8}{5} =$

**26.** $\frac{2}{3} \div \frac{4}{6} =$ **27.** $\frac{5}{8} \div \frac{10}{2} =$ **28.** $\frac{5}{6} \div \frac{1}{3} =$ **29.** $\frac{7}{30} \div \frac{14}{15} =$

**30.** $\frac{7}{24} \div \frac{3}{12} =$ **31.** $\frac{2}{3} \div 4 =$ **32.** $\frac{5}{8} \div 5 =$ **33.** $\frac{5}{6} \div 4 =$

**34.** $\frac{5}{16} \div 10 =$ **35.** $10 \div \frac{1}{2} =$ **36.** $15 \div \frac{1}{3} =$ **37.** $20 \div \frac{2}{5} =$

**38.** $30 \div \frac{5}{6} =$ **39.** $48 \div \frac{3}{24} =$

## Supplementary Exercise Set, Section 2-2

Perform the following multiplications and divisions.

**1.** $\frac{1}{4} \times \frac{3}{8} =$ **2.** $\frac{5}{8} \times \frac{1}{2} =$ **3.** $\frac{1}{4} \times \frac{2}{3} =$ **4.** $\frac{3}{4} \times \frac{2}{3} =$

**5.** $\frac{1}{15} \times \frac{5}{8} =$ **6.** $\frac{3}{4} \times \frac{1}{6} =$ **7.** $5 \times \frac{1}{10} =$ **8.** $5 \times \frac{2}{3} =$

**9.** $\frac{1}{2} \div \frac{1}{4} =$ **10.** $\frac{3}{4} \div \frac{1}{4} =$ **11.** $\frac{9}{10} \div \frac{3}{4} =$ **12.** $\frac{2}{3} \div \frac{1}{2} =$

**13.** $\frac{3}{4} \div 6 =$ **14.** $5 \div \frac{5}{8} =$ **15.** $\frac{7}{8} \div 2 =$ **16.** $2 \div \frac{7}{8} =$

**17.** $\frac{7}{8} \times 4 =$ **18.** $4 \times \frac{7}{8} =$ **19.** $\frac{7}{8} \div 4 =$ **20.** $4 \div \frac{7}{8} =$

## Section 2-3 Addition and Subtraction of Fractions

**1**

Fractions can be added or subtracted if they have like denominators. Fractions with like denominators are added or subtracted by adding or subtracting the numerators and leaving the denominators unchanged.

**Examples** $\frac{2}{7}+\frac{3}{7}=\frac{5}{7}$ $\quad \frac{9}{11}-\frac{5}{11}=\frac{4}{11}$

Perform the following additions and subtractions.

a. $\frac{1}{5}+\frac{3}{5}=$ b. $\frac{2}{3}-\frac{1}{3}=$

c. $\frac{3}{9}+\frac{5}{9}=$ d. $\frac{6}{7}-\frac{3}{7}=$

## 2

When adding or subtracting fractions, it is necessary to reduce the answer to lowest terms. In the addition

$$\frac{1}{8}+\frac{3}{8}=\frac{4}{8}$$

4 can be divided into both the numerator and the denominator. Thus, the answer can be reduced as follows:

$$\frac{\overset{1}{\cancel{4}}}{\underset{2}{\cancel{8}}}=\frac{1}{2}$$

Therefore, $\frac{1}{8}+\frac{3}{8}=\frac{1}{2}$.

In the subtraction

$$\frac{3}{4}-\frac{1}{4}=\frac{2}{4}$$

2 can be divided into both the numerator and the denominator. Thus, the answer can be reduced as follows:

$$\frac{\overset{1}{\cancel{2}}}{\underset{2}{\cancel{4}}}=\frac{1}{2}$$

Perform the following additions and subtractions, and reduce the answers.

a. $\frac{1}{4}+\frac{3}{4}=$ b. $\frac{7}{8}-\frac{5}{8}=$ c. $\frac{9}{16}+\frac{1}{16}=$ d. $\frac{5}{6}-\frac{1}{6}=$

1. a. $\frac{4}{5}$ b. $\frac{1}{3}$ c. $\frac{8}{9}$ d. $\frac{3}{7}$

## 3

Fractions with unlike denominators must be converted to fractions with like denominators before they can be added or subtracted. This is done by rewriting each fraction as an equivalent fraction, using the same number as a denominator. For example, to add

$$\frac{1}{2}+\frac{1}{4}$$

the $\frac{1}{2}$ must be rewritten as an equivalent fraction with a denominator of 4.

2. a. 1 b. $\frac{1}{4}$ c. $\frac{5}{8}$ d. $\frac{2}{3}$

$$\frac{1}{2} = \frac{1}{2} \times \frac{2}{2} = \frac{2}{4}$$

Therefore, $\frac{2}{4}$ is equivalent to $\frac{1}{2}$ and can be used in the addition. The original addition is then rewritten.

$$\frac{1}{2} + \frac{1}{4} \text{ is equivalent to } \frac{2}{4} + \frac{1}{4}$$

The addition can now be performed.

$$\frac{2}{4} + \frac{1}{4} = \frac{3}{4}$$

Perform the following additions and subtractions by first rewriting one of the fractions with the common denominator.

**a.** $\frac{3}{8} + \frac{1}{16} =$ **b.** $\frac{1}{4} + \frac{1}{8} =$ **c.** $\frac{9}{16} - \frac{1}{2} =$

## 4

The equivalent fractions listed below can be helpful in finding common denominators.

**Example** 1/2 + 3/16 can be written as 8/16 + 3/16, because 1/2 = 8/16.

| | | |
|---|---|---|
| 2/16 = 1/8 | 10/16 = 5/8 | 4/8 = 1/2 |
| 4/16 = 1/4 | 12/16 = 3/4 | 6/8 = 3/4 |
| 6/16 = 3/8 | 14/16 = 7/8 | 2/4 = 1/2 |
| 8/16 = 1/2 | 2/8 = 1/4 | |

Rewrite each addition, and add.

**a.** 13/16 + 5/8 = **b.** 3/4 + 1/8 =

**c.** 1/2 + 1/4 = **d.** 1/2 + 7/16 =

**3. a.** $\frac{7}{16}$ **b.** $\frac{3}{8}$ **c.** $\frac{1}{16}$

## 5

The fractions most often used in science and industry are listed below, along with their decimal equivalents.

| | | |
|---|---|---|
| 1/32 = 0.0312 | 3/8 = 0.3750 | 11/16 = 0.6875 |
| 1/16 = 0.0625 | 13/32 = 0.4062 | 23/32 = 0.7188 |
| 3/32 = 0.0938 | 7/16 = 0.4375 | 3/4 = 0.7500 |
| 1/8 = 0.1250 | 15/32 = 0.4688 | 25/32 = 0.7812 |
| 5/32 = 0.1562 | 1/2 = 0.5000 | 13/16 = 0.8125 |
| 3/16 = 0.1875 | 17/32 = 0.5312 | 27/32 = 0.8438 |
| 7/32 = 0.2188 | 9/16 = 0.5625 | 7/8 = 0.8750 |

**4. a.** $\frac{23}{16}$ or $1\frac{7}{16}$ **b.** $\frac{7}{8}$ **c.** $\frac{3}{4}$ **d.** $\frac{15}{16}$

| | | |
|---|---|---|
| 1/4 = 0.2500 | 19/32 = 0.5938 | 29/32 = 0.9062 |
| 9/32 = 0.2812 | 5/8 = 0.6250 | 15/16 = 0.9375 |
| 5/16 = 0.3125 | 21/32 = 0.6562 | 31/32 = 0.9688 |
| 11/32 = 0.3438 | | |

## 6

When one denominator is not a multiple of the other it is more difficult to find a common denominator.

*Addition when one denominator is a multiple of the other:*

$\frac{1}{4} + \frac{3}{8} =$

$\frac{2}{8} + \frac{3}{8} = \frac{5}{8}$

Only one fraction had to be changed to an equivalent fraction.

*Addition when one denominator is not a multiple of the other:*

$\frac{1}{4} + \frac{2}{3} =$

$\frac{3}{12} + \frac{8}{12} = \frac{11}{12}$

Both fractions had to be changed to equivalent fractions with a new denominator. This new denominator is the smallest number that is divisible by both 4 and 3.

Change each of these fractions to equivalent fractions with common denominators.

**a.** $\frac{1}{3}, \frac{3}{4}$ **b.** $\frac{1}{5}, \frac{1}{4}$ **c.** $\frac{1}{3}, \frac{1}{8}$

## 7

Perform the following additions.

**a.** $\frac{1}{3} + \frac{3}{4} =$ **b.** $\frac{1}{5} + \frac{1}{4} =$

**c.** $\frac{1}{3} + \frac{1}{8} =$ **d.** $\frac{2}{3} + \frac{1}{5} =$

**6. a.** $\frac{4}{12}, \frac{9}{12}$
**b.** $\frac{4}{20}, \frac{5}{20}$
**c.** $\frac{8}{24}, \frac{3}{24}$

## 8

Fractions with unlike denominators can be added by first converting to equivalent decimals.

**Example** First convert the fractions to equivalent decimals using the calculator.

$\frac{3}{4} + \frac{7}{8}$ can be converted to 0.75 + 0.875.

Then perform the addition on the calculator.

0.75 + 0.875 = 1.625

**7. a.** $\frac{13}{12}$ or $1\frac{1}{12}$
**b.** $\frac{9}{20}$
**c.** $\frac{11}{24}$
**d.** $\frac{13}{15}$

Perform the following additions and subtractions by first converting the fractions to equivalent decimals.

**a.** $\frac{1}{5}+\frac{3}{4}=$ **b.** $\frac{1}{8}+\frac{3}{16}=$ **c.** $\frac{7}{8}-\frac{7}{16}=$

**9**

Each of the following decimals was rounded to the thousandths place in order to perform the subtraction.

$$\frac{2}{3}-\frac{1}{7}=0.667-0.143=0.524$$

Perform the following additions and subtractions by first converting the fractions to equivalent decimals. (Round each decimal to the thousandths place.)

**a.** $\frac{1}{3}+\frac{1}{8}=$ **b.** $\frac{5}{9}-\frac{1}{2}=$ **c.** $\frac{2}{3}+\frac{1}{16}=$

**8. a.** $0.20+0.75=0.95$
**b.** $0.125+0.1875=0.3125$
**c.** $0.875-0.4375=0.4375$

**10**

Answers to frame 9.

**a.** $0.333+0.125=0.458$ **b.** $0.556-0.500=0.056$ **c.** $0.667+0.063=0.730$

# Exercise Set, Section 2-3

## Addition and Subtraction of Fractions

Perform the following additions and subtractions.

**1.** $\frac{1}{7}+\frac{3}{7}=$ **2.** $\frac{9}{16}+\frac{2}{16}=$ **3.** $\frac{3}{15}+\frac{4}{15}=$ **4.** $\frac{7}{12}+\frac{4}{12}=$

**5.** $\frac{1}{9}+\frac{3}{9}=$ **6.** $\frac{3}{14}+\frac{8}{14}=$ **7.** $\frac{1}{2}+\frac{1}{4}=$ **8.** $\frac{1}{3}+\frac{1}{6}=$

**9.** $\frac{1}{10}+\frac{2}{5}=$ **10.** $\frac{3}{4}+\frac{1}{8}=$ **11.** $\frac{1}{3}+\frac{4}{9}=$ **12.** $\frac{3}{12}+\frac{1}{4}=$

**13.** $\frac{2}{21}+\frac{2}{7}=$ **14.** $\frac{5}{24}+\frac{3}{8}=$ **15.** $\frac{9}{7}-\frac{5}{7}=$ **16.** $\frac{7}{16}-\frac{5}{16}=$

**17.** $\frac{11}{15}-\frac{6}{15}=$ **18.** $\frac{8}{24}-\frac{5}{24}=$ **19.** $\frac{1}{2}-\frac{1}{4}=$ **20.** $\frac{2}{3}-\frac{1}{6}=$

**21.** $\frac{4}{5}-\frac{3}{10}=$ **22.** $\frac{1}{4}-\frac{1}{8}=$ **23.** $\frac{2}{3}-\frac{3}{8}=$ **24.** $\frac{7}{8}-\frac{1}{2}=$

**25.** $\frac{15}{16}-\frac{3}{32}=$ **26.** $\frac{5}{6}-\frac{5}{24}=$

Perform the following additions and subtractions by first converting the fractions to equivalent decimals. (Round the answers to two decimal places.)

**27.** $\frac{2}{3} - \frac{3}{8} =$ **28.** $\frac{7}{16} + \frac{1}{8} =$ **29.** $\frac{7}{16} + \frac{1}{3} =$ **30.** $\frac{7}{8} - \frac{1}{2} =$

## Supplementary Exercise Set, Section 2-3

Perform the following additions and subtractions.

**1.** $\frac{1}{5} + \frac{2}{5} =$ **2.** $\frac{1}{4} + \frac{2}{4} =$ **3.** $\frac{1}{8} + \frac{3}{8} =$ **4.** $\frac{1}{3} + \frac{2}{3} =$

**5.** $\frac{1}{2} + \frac{1}{4} =$ **6.** $\frac{3}{8} + \frac{1}{4} =$ **7.** $\frac{1}{4} + \frac{2}{3} =$ **8.** $\frac{1}{3} + \frac{3}{4} =$

**9.** $\frac{3}{5} - \frac{1}{5} =$ **10.** $\frac{3}{4} - \frac{1}{4} =$ **11.** $\frac{7}{8} - \frac{3}{8} =$ **12.** $\frac{2}{3} - \frac{1}{3} =$

**13.** $\frac{7}{8} - \frac{3}{4} =$ **14.** $\frac{7}{8} - \frac{1}{3} =$ **15.** $\frac{3}{4} - \frac{1}{8} =$ **16.** $\frac{2}{3} - \frac{3}{8} =$

**17.** $\frac{2}{3} + \frac{1}{8} =$ **18.** $\frac{5}{16} + \frac{1}{8} =$ **19.** $\frac{7}{8} - \frac{1}{5} =$ **20.** $\frac{7}{10} + \frac{1}{2} =$

## Section 2-4 Addition and Subtraction of Fractions Using the Least Common Denominator

**1**

In Section 2-3, additions and subtractions of fractions were performed when the denominators were equal or when it was easy to find equivalent fractions with common denominators. More difficult fractions were added or subtracted by using a calculator to convert the fractions to equivalent decimals.

This section will introduce the method of finding the least common denominator to add or subtract fractions. To use this method, we must first discuss the concept of *prime factoring*.

**2**

Recall from Chapter 1 that multiplication consists of factors being multiplied together to get a product. In the multiplication below, the 4 and 9 are factors and the 36 is the product.

$$(4)(9) = 36$$

(4) and (9): factors; 36: product

**a.** In (15)(3) = 45, 45 is called a ________________________.

**b.** In (9)(10) = 90, 9 is called a ________________________.

**c.** In (5)(7)(6) = 210, 5, 7, and 6 are called ________________.

## 3

A factor of a given number is a number that will divide into that given number and have a remainder of zero.

2. **a.** product
**b.** factor
**c.** factors

**Example** 3 is a factor of 12 because

$$\begin{array}{r|l} & \ \ 4 \\ 3 & 12 \\ & \underline{12} \\ & \ \ 0 \end{array} \qquad \text{remainder of 0}$$

5 is *not* a factor of 12 because

$$\begin{array}{r|l} & \ \ 2 \\ 5 & 12 \\ & \underline{10} \\ & \ \ 2 \end{array} \qquad \text{remainder of 2}$$

**a.** List all of the factors of 12:

______, ______, ______, ______, ______, ______

**b.** List all the factors of 30:

______, ______, ______, ______, ______, ______, ______, ______

## 4

Answers to frame 3.

**a.** 1, 2, 3, 4, 6, 12
**b.** 1, 2, 3, 5, 6, 10, 15, 30

*Notice:* Both 1 and 12 are factors of 12.
Both 1 and 30 are factors of 30.

## 5

The factoring rule for the number 2 states that if the last digit of a number is even (a multiple of 2), then 2 is a factor of that number.

**Example** 2 is a factor of 598, because 8 (the last digit) is a multiple of 2.

2 is not a factor of 599, because 9 (the last digit) is *not* a multiple of 2.

**a.** Is 2 a factor of 387? ____________

**b.** Is 2 a factor of 752? ____________

**c.** Is 2 a factor of 1,298? ____________

**d.** Is 2 a factor of 45,200? ____________

## 6

Answers to frame 5.

**a.** no **b.** yes **c.** yes **d.** yes; when a number ends in zero, it is a multiple of 2, because $2 \times 5 = 10$.

**7**

The factoring rule for the number 3 states that if the sum of the digits of a number is a multiple of 3, then 3 is a factor of that number.

> **Example** 3 is a factor of 87, because $8 + 7 = 15$ and 15 is a multiple of 3 ($3 \times 5 = 15$).
>
> 3 is a factor of 5,121, because $5 + 1 + 2 + 1 = 9$ and 9 is a multiple of 3 ($3 \times 3 = 9$).
>
> 3 is *not* a factor of 7,255, because $7 + 2 + 5 + 5 = 19$ and 19 is *not* a multiple of 3.

**a.** Is 3 a factor of 78? ______

**b.** Is 3 a factor of 451? ______

**c.** Is 3 a factor of 2,951? ______

**d.** Is 3 a factor of 13,333? ______

**8**

Answers to frame 7.

**a.** yes ($7 + 8 = 15$) **b.** no ($4 + 5 + 1 = 10$)
**c.** no ($2 + 9 + 5 + 1 = 17$) **d.** no ($1 + 3 + 3 + 3 + 3 = 13$)

**9**

The factoring rule for the number 5 states that if the last digit of a number is 0 or 5, then 5 is a factor of that number.

> **Example** 5 is a factor of 500, because the last digit is 0.
>
> 5 is *not* a factor of 146, because the last digit is 6 (not 5 or 0).
>
> 5 is a factor of 145, because the last digit is 5.

**a.** Is 5 a factor of 6,790? ______

**b.** Is 5 a factor of 50,501? ______

**c.** Is 5 a factor of 295? ______

**d.** Is 5 a factor of 576? ______

**10**

Answers to frame 9.

**a.** yes **b.** no **c.** yes **d.** no

**11**

A prime number is a number greater than 1 that has only itself and 1 as factors.

> **Examples** 5 is a prime number, because its only factors are 5 and 1.
>
> 2 is a prime number, because its only factors are 2 and 1.
>
> 4 is *not* a prime number, because it can be factored into $(2)(2) = 4$. (It can also be factored into 4 and 1, but these are not its only factors; therefore it is *not* prime.)

**a.** Is 9 a prime number? ____________

**b.** Is 10 a prime number? ____________

**c.** Is 11 a prime number? ____________

**d.** Is 12 a prime number? ____________

## 12

Answers to frame 11.

**a.** no ($3 \times 3 = 9$) **b.** no ($2 \times 5 = 10$) **c.** yes **d.** no ($3 \times 4 = 12$)

## 13

Answer "yes" or "no" to the following:

**a.** Is 13 a prime number? ____________

**b.** Is 15 a prime number? ____________

**c.** Is 17 a prime number? ____________

**d.** Is 19 a prime number? ____________

**e.** Is 21 a prime number? ____________

**f.** Is 23 a prime number? ____________

## 14

Answers to frame 13.

**a.** yes **b.** no ($3 \times 5 = 15$) **c.** yes
**d.** yes **e.** no ($3 \times 7 = 21$) **f.** yes

## 15

List the first nine prime numbers.

______, ______, ______, ______, ______, ______, ______, ______, ______

## 16

Answers to frame 15.

2, 3, 5, 7, 11, 13, 17, 19, 23

## 17

To find least common denominators, it is necessary to use prime factors. A prime factor of a given number is any factor that is also prime.

**Example** 36 can be factored as follows:

$36 = (6)(6)$ These factors are *not* prime, because each 6 can be factored further into $3 \times 2$.

$36 = (3 \times 2)(3 \times 2)$

$36 = (3)(3)(2)(2)$ All prime factors.

**a.** Factor 210 into prime factors:

$210 = (21)(10)$
$210 = (\quad)(\quad)$
$210 =$

**b.** Factor 356 into prime factors:

$356 = (2)(178)$
$356 = (2)(\quad)(\quad)$

## 18

Answers to frame 17.

**a.** $210 = (3)(7)(2)(5)$ **b.** $356 = (2)(2)(89)$

## 19

The least common denominator, or LCD, of two or more denominators is the smallest number that is a multiple of each of the denominators. This concept is used to add and subtract fractions with different denominators when a common denominator is not easily seen.

To add $\frac{1}{6} + \frac{1}{15}$, it is necessary to find the smallest denominator that is a multiple of 6 and 15. This is done by prime factoring 6 and 15 and placing the factors in order, as follows:

$$6 = (2)(3)$$
$$15 = (3)(5)$$

Next, the similar factors are placed above and below each other.

$$6 = (2)(3)$$
$$15 = \quad\ \ (3)(5)$$

The least common denominator is found by using each factor the maximum number of times it appears in each denominator.

$$\text{LCD} = (2)(3)(5)$$
$$\text{LCD} = 30$$

Therefore, the LCD of 6 and 15 is 30. That is, 30 is the smallest number that has both 6 and 15 as factors.

**a.** Find the LCD of 14 and 12.

*Step 1.* 14 = (  )(  )
12 = (  )(  )(  )

*Step 2.* 14 = (  )      (  )
12 = (  )(  )(  )

*Step 3.* LCD = (  )(  )(  )(  )
LCD =

**b.** Find the LCD of 9 and 15.

## 20

Answers to frame 19.

**a.**
$14 = (2) \qquad (7)$
$12 = (2)(2)(3)$
$\text{LCD} = (2)(2)(3)(7)$
$\text{LCD} = 84$

**b.**
$9 = (3)(3)$
$15 = (3) \qquad (5)$
$\text{LCD} = (3)(3)(5)$
$\text{LCD} = 45$

**21**

We can now add $\frac{1}{6} + \frac{1}{15}$ by using the least common denominator 30. Each fraction can be changed to an equivalent fraction with a denominator of 30.

$$\frac{1}{6} = \frac{?}{30} \qquad \frac{1}{15} = \frac{?}{30}$$

$$\frac{1 \times 5}{6 \times 5} = \frac{5}{30} \qquad \frac{1 \times 2}{15 \times 2} = \frac{2}{30}$$

$$\frac{5}{30} + \frac{2}{30} = \frac{7}{30}$$

**a.** Using 84 as the LCD, add $\frac{5}{12} + \frac{1}{14}$.

$$\frac{5}{12} = \frac{?}{84} \qquad \frac{1}{14} = \frac{?}{84}$$

$$\frac{5 \times 7}{12 \times 7} = \frac{\phantom{0}}{84} \qquad \frac{1 \times 6}{14 \times 6} = \frac{\phantom{0}}{84}$$

$$\frac{\phantom{0}}{84} + \frac{\phantom{0}}{84} =$$

**b.** Using 45 as the LCD, subtract $\frac{1}{9} - \frac{1}{15}$.

$$\frac{1}{9} = \frac{?}{45} \qquad \frac{1}{15} = \frac{?}{45}$$

$$\frac{1 \times 5}{9 \times 5} = \frac{\phantom{0}}{45} \qquad \frac{1 \times 3}{15 \times 3} = \frac{\phantom{0}}{45}$$

$$\frac{\phantom{0}}{45} - \frac{\phantom{0}}{45} =$$

**22**

Answers to frame 21.

**a.** $\frac{35}{84} + \frac{6}{84} = \frac{41}{84}$ **b.** $\frac{5}{45} - \frac{3}{45} = \frac{2}{45}$

# Exercise Set, Section 2-4

## Addition and Subtraction of Fractions Using the Least Common Denominator

Find the least common denominator for each of the following pairs of fractions.

**1.** $\frac{1}{8}, \frac{2}{9}$ **2.** $\frac{7}{8}, \frac{2}{3}$ **3.** $\frac{3}{4}, \frac{1}{8}$

**4.** $\frac{2}{3}, \frac{1}{2}$ **5.** $\frac{2}{7}, \frac{1}{5}$ **6.** $\frac{1}{2}, \frac{5}{9}$

Perform the following additions and subtractions.

7. $\frac{1}{10} + \frac{5}{6}$

8. $\frac{3}{4} - \frac{1}{8}$

9. $\frac{1}{6} + \frac{4}{9}$

10. $\frac{4}{5} - \frac{1}{9}$

11. $\frac{9}{16} - \frac{1}{2}$

12. $\frac{11}{16} + \frac{3}{4}$

Find the prime factors of the following numbers.

**13.** 75 **14.** 28 **15.** 93 **16.** 29

**17.** 125 **18.** 3,000 **19.** 45 **20.** 81

# Supplementary Exercise Set, Section 2-4

Find the least common denominator for each of the following pairs of fractions.

1. $\frac{1}{7}, \frac{3}{14}$

2. $\frac{5}{8}, \frac{2}{3}$

3. $\frac{3}{4}, \frac{1}{12}$

4. $\frac{1}{6}, \frac{1}{2}$

5. $\frac{2}{9}, \frac{1}{5}$

6. $\frac{1}{2}, \frac{2}{11}$

Perform the following additions and subtractions.

7. $\frac{1}{5} + \frac{3}{6}$

8. $\frac{3}{4} - \frac{1}{12}$

9. $\frac{1}{3} + \frac{2}{9}$

10. $\frac{13}{16} - \frac{1}{9}$

11. $\frac{3}{16} + \frac{3}{4}$

12. $\frac{15}{16} - \frac{1}{3}$

Find the prime factors of the following numbers.

**13.** 90 **14.** 36 **15.** 63 **16.** 39

**17.** 225 **18.** 4,000 **19.** 35 **20.** 162

# Section 2-5 Applied Problems

**1**

The five steps used in Chapter 1 to solve whole number and decimal number problems can also be used to solve problems with fractions. The problem below illustrates the use of these five steps with fractions.

Board *A*, which is 5/8 inch thick, is nailed to board *B*, which is 1/2 inch thick. What is the total thickness of the two boards?

*Step 1.* Determine what is asked for.
total thickness of board = ?

*Step 2.* Determine what information is already known.
board $A$ = 5/8 inch thick
board $B$ = 1/2 inch thick

*Step 3.* Write the mathematical model that describes the relationship.
total thickness = thickness $A$ + thickness $B$

*Step 4.* Substitute the data into the model.
total thickness = 5/8 inch + 1/2 inch

*Step 5.* Do the calculations.
total thickness = 9/8 inches or 1 1/8 inches

**2**

Solve this problem using the five steps just demonstrated.

A pipe that is 4 1/2 inches long is to be cut into pieces that are each 3/16 inch long. How many pieces will there be?

*Step 1.* ____________________

*Step 2.* ____________________

____________________

*Step 3.* ____________________

*Step 4.* ____________________

*Step 5.* ____________________

**3**

Answers to frame 2.

*Step 1.* number of pieces
*Step 2.* total length = 4 1/2 inches
length of piece = 3/16 inch/piece
*Step 3.* number of pieces = total length ÷ length/piece
*Step 4.* number of pieces $= \dfrac{4\ 1/2 \text{ inches}}{3/16 \text{ inch/piece}}$ (inches and inch cancelled)
*Step 5.* number of pieces $= \dfrac{9}{2} \times \dfrac{16}{3}$ or 24 pieces

# Exercise Set, Section 2-5

## Applied Problems

Use the five-step procedure to solve the following problems.

1. Tube *A* is 9 1/8 inches long, and tube *B* is 6 3/4 inches long. What is their total length?
2. A board is 7 3/16 ft long. If 2 1/2 ft are removed, how much of the board is left?
3. Three samples of metal weigh 1/3 oz, 1/4 oz, and 1/2 oz. What is the total weight of the metal samples? Give your answer as a decimal number.
4. A section of pipe 2 1/4 ft long is welded to a section 3 1/8 ft long. A length 1 3/16 ft long is then cut from the welded pipe. How long is the remaining section of pipe?

5. A chemical technician adds 3/4 ounce of solution *A* to 1/3 ounce of solution *B*. What is the total volume of the solution? Give your answer in decimals.
6. A metallurgical technician must divide a 3/4-gram sample into 1/16-gram units. How many 1/16-gram units will there be?
7. How many grams of a chemical are needed to make 20 capsules of 3/4 gram each?
8. A machine can cut 1 7/8 inches of metal per minute. How many inches of metal can it cut in 15 1/2 minutes?
9. What is the width of the pipe wall?

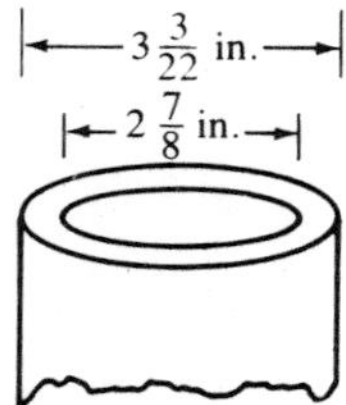


10. Find the missing dimension.

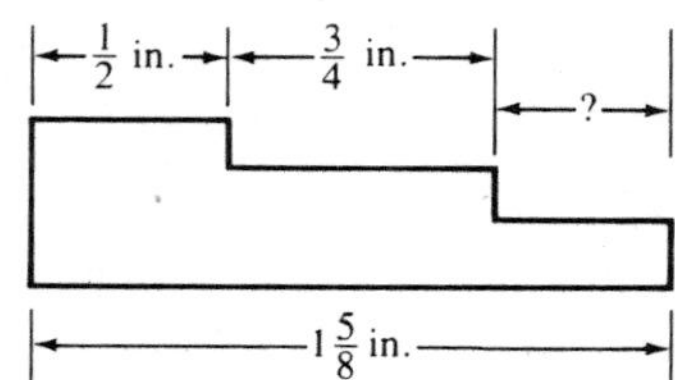


11. The table shown here is to have a Formica top. How many square feet of Formica are needed to cover the table? The formula for finding the area of a table is given.

$$\text{area of table} = \text{length} \times \text{width}$$

Length $= 3\frac{1}{2}$ ft

Width $= 1\frac{3}{4}$ ft

12. A fence is to be built around the property shown. How many feet of fencing are needed?

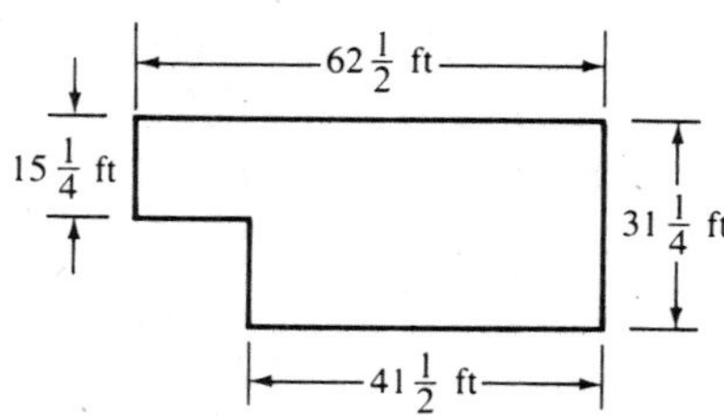


13. The rungs on the ladder below are 1 1/2 feet apart. How high is the ladder?

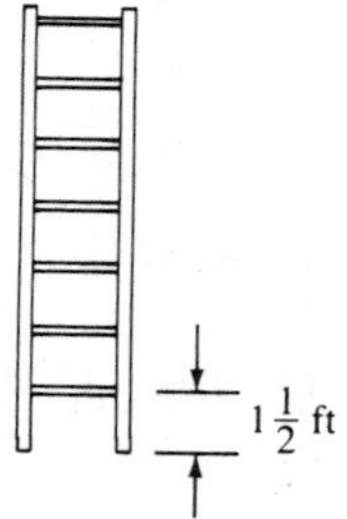

14. A concrete-block wall is made with 12-in.-high blocks. If the mortar joints are 3/4 in. thick, how high is the wall?

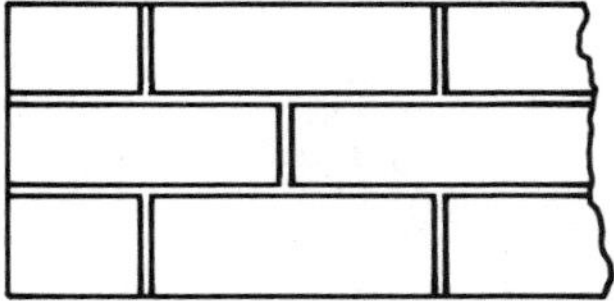

15. Find the total thickness of this wall.

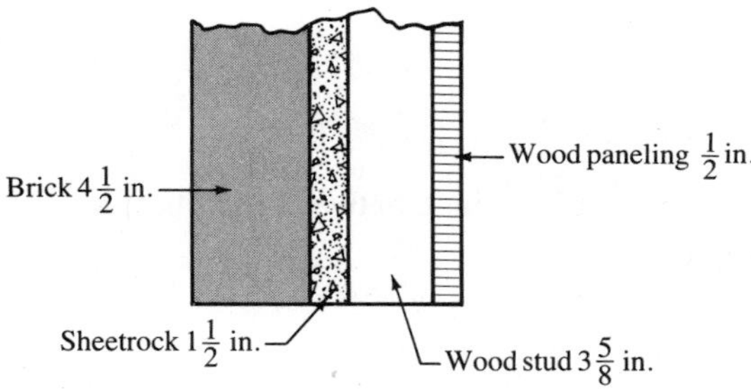


## Supplementary Exercise Set, Section 2-5

Use the five-step procedure to solve the following problems.

1. A container holds 7/8 gallon of solvent. If 1/2 gallon is used, how much is left in the container?
2. A metal alloy is made by adding 1/3 oz of metal *A* to 3/4 oz of metal *B*. What is the total weight of the alloy? Give your answer as a decimal.
3. A machinist must trim 1/5 of the weight off a 15/16-oz piece of metal. How much weight has to be trimmed off?
4. How many 4 1/8-inch-long bolts can be cut from a 37 1/8-inch-long piece of metal?
5. What is the total volume of a mixture of 1/8 gallon of solvent *X* mixed with 2 1/4 gallons of solvent *Y*? Give your answer as a decimal number.
6. 8 2/3 acres of land are divided into 2/3-acre lots. How many lots will there be?
7. A car gets 15 1/2 miles per gallon. How far can it go on 8/14 gallon of gas?
8. A screw has 16 threads per inch. How many threads are there in 2 1/2 inches?
9. Find the missing dimension.

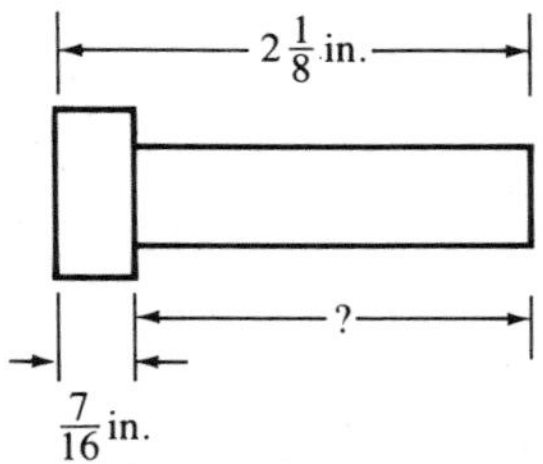


10. Find the missing dimension.

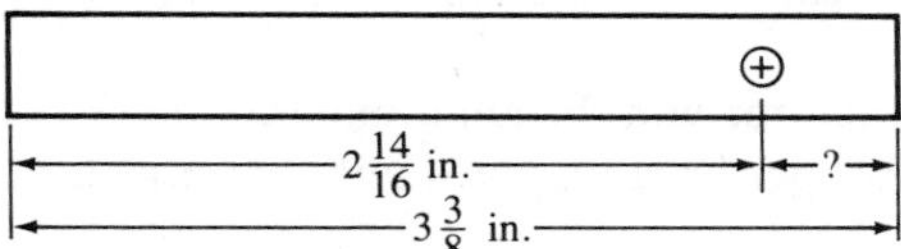

**11.** In the drawing below, each thread is 1/16 inch apart. Find the missing dimension.

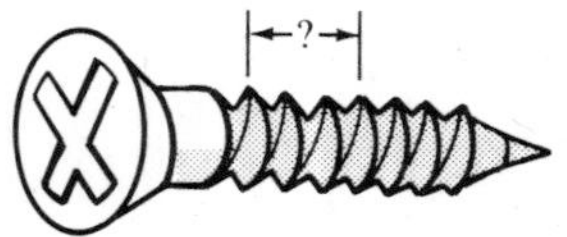


**12.** Find the missing dimension.

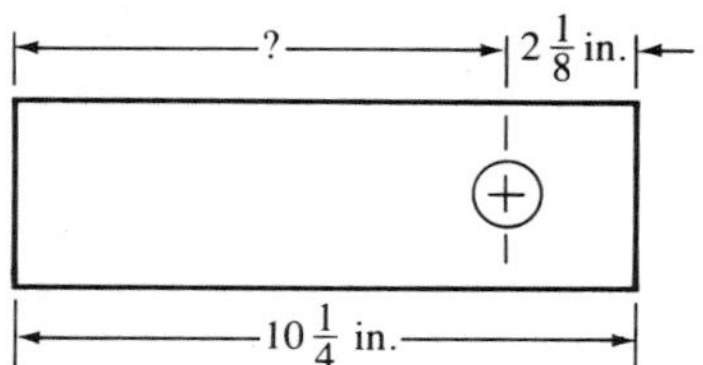


**13.** A bar of metal is to have four holes, each 9/8 inches wide. There are to be 3/4 inch between each hole and 1/2 inch from the ends to the first and last holes. How long will the metal bar be?

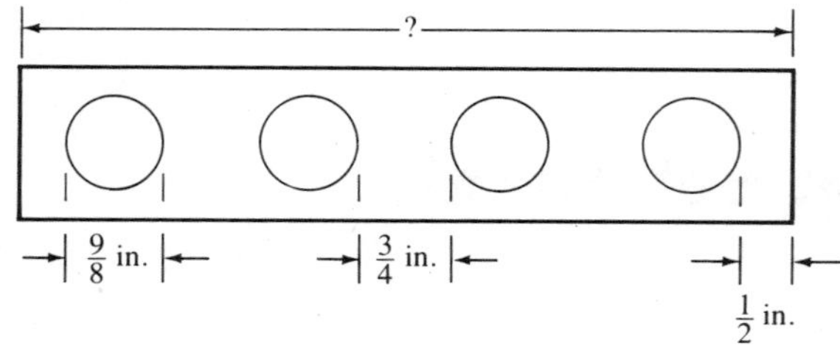


**14.** How many square yards of carpeting are needed to cover this floor? Area = length × width.

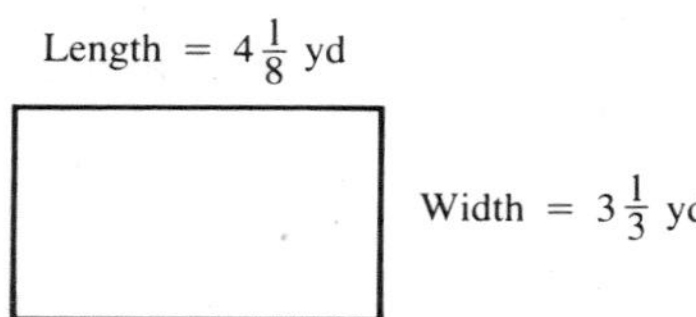


**15.** The following pieces were cut from an 18-inch metal bar. How much of the bar is left?

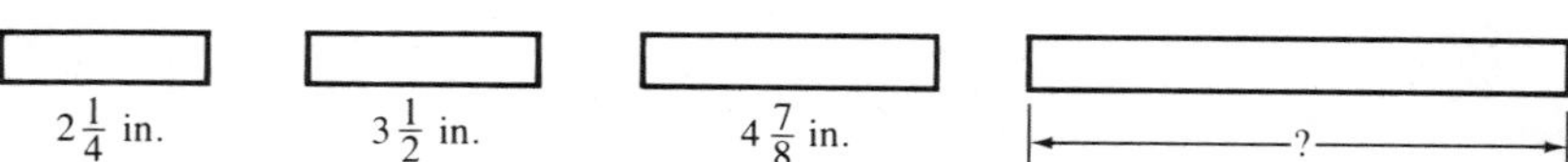


# Summary

1. A fraction is a number that compares a part with a whole.
2. A fraction is named by first stating the name of the numerator (top number) and then the name of the denominator (bottom number).
3. A fraction is converted to a decimal by dividing the numerator by the denominator.
4. Two fractions that are equal are called equivalent fractions. A fraction can be changed to its equivalent by multiplying the numerator and the denominator by the same number.
5. A fraction is reduced when the same number is divided out of the numerator and the denominator.
6. A proper fraction has a smaller numerator than denominator. An improper fraction has a numerator as large or larger than the denominator.
7. An improper fraction can be converted to a mixed number by dividing the numerator by the denominator.

8. Fractions must have like denominators to be added or subtracted. Add or subtract the numerators, leaving the denominators unchanged. All answers are reduced to lowest terms.
9. Fractions with unlike denominators must be converted to fractions with like denominators before adding or subtracting.
10. To multiply fractions, cancel first; then multiply the numerators together and then multiply the denominators together. Reduce the answer to lowest terms.
11. To divide fractions, first invert the divisor and then follow the rules for multiplication.

# Chapter 2 Self-Test

## Fractions

Convert these fractions to equivalent decimals.

**1.** $\frac{5}{18}$ **2.** $\frac{11}{30}$ **3.** $\frac{13}{28}$ **4.** $\frac{3}{8}$

Change each of these fractions to an equivalent fraction with a denominator of 30.

**5.** $\frac{1}{5}$ **6.** $\frac{5}{6}$ **7.** $\frac{3}{10}$ **8.** $\frac{2}{15}$

Reduce the following fractions.

**9.** $\frac{14}{16}$ **10.** $\frac{4}{10}$ **11.** $\frac{11}{44}$ **12.** $\frac{4}{16}$

Convert the following fractions to mixed numbers.

**13.** $\frac{9}{4}$ **14.** $\frac{8}{3}$ **15.** $\frac{23}{4}$

Convert the following mixed numbers to improper fractions.

**16.** $2\frac{1}{3}$ **17.** $1\frac{3}{8}$ **18.** $17\frac{2}{3}$

Perform the following additions and subtractions.

**19.** $\frac{1}{5}+\frac{3}{5}=$ **20.** $\frac{2}{3}+\frac{1}{6}=$ **21.** $\frac{11}{16}-\frac{7}{16}=$ **22.** $\frac{2}{3}-\frac{1}{8}=$

Perform the following additions and subtractions by first changing the fractions to equivalent decimals.

**23.** $\frac{5}{8}-\frac{1}{2}=$ **24.** $\frac{5}{16}+\frac{1}{3}=$

Multiply the following fractions, reducing your answers to lowest terms.

**25.** $\frac{1}{8} \times \frac{1}{4} =$

**26.** $\frac{2}{16} \times \frac{8}{9} =$

**27.** $\frac{4}{3} \times \frac{3}{8} =$

**28.** $\frac{1}{8} \times 8 =$

**29.** $12 \times \frac{2}{3} =$

**30.** $15 \times \frac{2}{3} =$

Divide the following fractions.

**31.** $\frac{1}{6} \div \frac{2}{3} =$

**32.** $\frac{9}{10} \div \frac{3}{5} =$

**33.** $\frac{3}{4} \div \frac{9}{16} =$

**34.** $10 \div \frac{2}{5} =$

**35.** $\frac{5}{16} \div 5 =$

**36.** $\frac{3}{8} \div 5 =$

Use the five-step method to solve these problems.

**37.** A pipe, 6 1/2 ft long, has two pieces of 3/4 ft and 1 7/8 ft cut from it. How long is the remaining piece of pipe?

**38.** Three electrical appliances have power ratings of 1 7/8 watts, 2 1/2 watts, and 4 3/4 watts. What is the total power used by these appliances?

**39.** A 1 3/4-gram sample is divided into 1/16-gram samples. How many 1/16-gram samples will there be?

**40.** A 1 1/8-in. bolt is cut from a metal rod. How long must the rod be to cut 16 such bolts from it?

# CHAPTER 3

# Basic Concepts of Algebra

Many relationships in technology are expressed in formula form. Since principles of algebra are needed to evaluate formulas, this chapter will present operations with signed numbers and an introduction to equation solving. Techniques of formula evaluation will also be included.

## Section 3-1 Signed Numbers

**1**

Only positive numbers have been used in this book so far. Such numbers can be illustrated on a number line as follows:

**Example** The numbers 3 and 7 have been located on the number line.

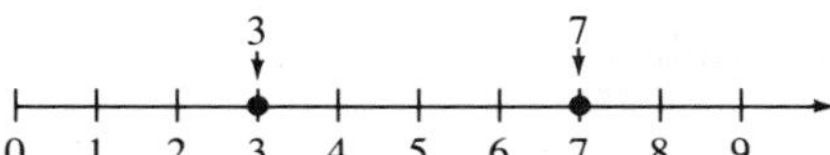

Locate 2 and 5 on the number line.

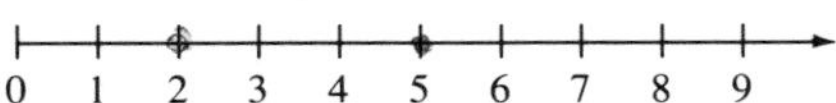

**2**

Answers to frame 1.

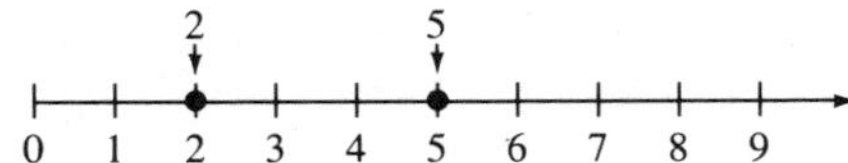

**3**

The number lines in the previous frames are incomplete. For each positive number on the number line, there is a corresponding negative number. A number line should, therefore, include a negative side.

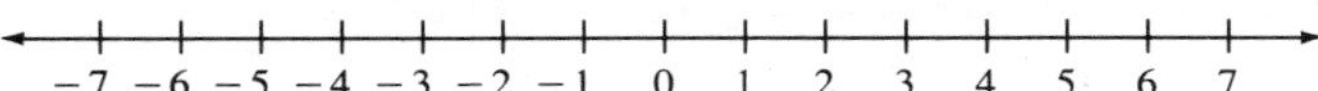

Notice that the positive numbers are to the right of 0 and the negative numbers to the left of 0.

Locate $-6$ and $-1$ on the number line below.

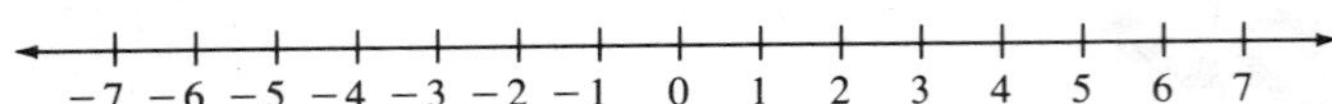


**4**

Answers to frame 3.

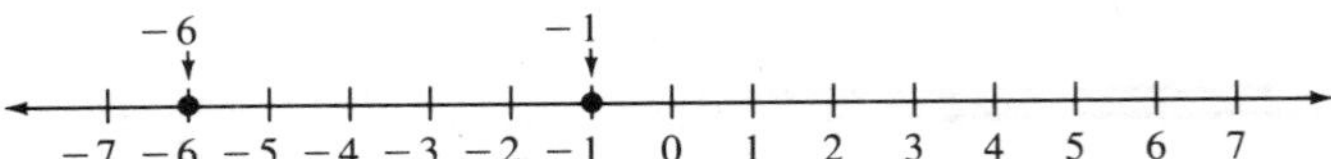


**5**

When comparing numbers, the number farther to the right on the number line is always the larger number. For example, the number 5 is larger than $-2$, because 5 is to the right of $-2$ on the number line.

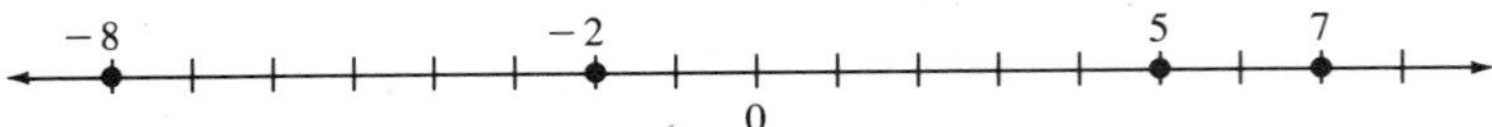


**a.** Which is larger, $-8$ or $-2$?
**b.** Which is larger, $-8$ or 7?

**6**

The absolute value of a signed number (a positive or negative number) is the distance that number is from 0 on the number line. Only the distance is important. It is not important whether the number is to the left or to the right of 0.

5. **a.** $-2$
**b.** 7

**Example** The number $-3$ is 3 units to the left of 0. The absolute value of $-3$ is 3.

The number $+3$ is 3 units to the right of 0. The absolute value of $+3$ is 3.

The symbol for absolute value is $| \ |$. Therefore, the statements above can be written as:

$$|-3| = 3 \qquad \text{and} \qquad |+3| = 3$$

Complete the following.

**a.** $|+17| =$ **b.** $|-95| =$ **c.** $|-4| =$

**7**

Answers to frame 6.

**a.** 17 **b.** 95 **c.** 4

# Section 3-2 Addition and Subtraction of Signed Numbers

**1**

Addition and subtraction of positive numbers are simple. Negative numbers, however, require different methods of addition and subtraction. Some of these operations are easily done in arithmetic:

$(+3) + (+4) = +7$ is the same as $3 + 4 = 7$

$(+5) - (+2) = +3$ is the same as $5 - 2 = 3$

Perform the following operations.

**a.** $(+7) + (+5) =$ **b.** $(+17) - (+3) =$ **c.** $(+4) + (+4) =$

**2**

---

The following operations cannot be done so easily:

1. **a.** 12 **b.** 14 **c.** 8

$(-4) + (+2) = -2$

$(-8) + (-4) = -12$

It is therefore necessary to apply a set of rules for adding and subtracting signed numbers. The set of rules for adding two signed numbers with the same sign follows.

*When adding two numbers with the same sign:*
1. The sign of the sum is the same as the sign of the numbers being added.
2. The absolute value of the sum is the sum of the absolute values of the addends.

**Example** $(-3) + (-4) = -7$

The sign is negative because the signs of both addends are negative.

The sum is 7 because the sum of the absolute values of 3 and 4 is 7.

Add these:

**a.** $(+3) + (+4) =$ **b.** $(-5) + (-9) =$ **c.** $(-7) + (-14) =$

**3**

---

The rules for adding two numbers with different signs follow.

2. **a.** +7 **b.** −14 **c.** −21

*When adding two numbers with different signs:*
1. The sign of the sum is the same as the sign of the addend with the larger absolute value.
2. The absolute value of the sum is the difference of the absolute values of the addends.

**Example** $(-9) + (+5) = -4$

The sign is negative because −9 has a larger absolute value than +5.

The sum is 4 because the difference of the absolute values of 9 and 5 is 4.

Add these:

**a.** $(-3) + (+7) =$ **b.** $(+4) + (-9) =$ **c.** $(+18) + (-7) =$

**4**

---

Two numbers are called *additive inverses* if their sum is 0. The numbers −5 and +5 are additive inverses because $(-5) + (+5) = 0$. In general, any number with a negative sign is the additive inverse of that same number with a positive sign.

3. **a** +4 **b.** −5 **c.** +11

Complete the following.

**a.** $(+3) + (\ \ ) = 0$ **b.** $(-7) + (\ \ ) = 0$ **c.** $(+9) + (-9) =$ ____

## 5

Write the additive inverses of the following numbers.

**a.** $+5$ ____ **b.** $-7$ ____ **c.** $+15$ ____

**d.** $-75$ ____ **e.** $+281$ ____ **f.** $-38$ ____

**4. a.** $-3$ **b.** $+7$ **c.** $0$

## 6

The concept of additive inverses is used in the subtraction of signed numbers. Any subtraction can be converted to an equivalent addition by adding the additive inverse of the number being subtracted. The following examples illustrate this principle:

| *Subtraction* | *Equivalent Addition* | *Answer* |
|---|---|---|
| $(+5) - (+2)$ | $(+5) + (-2)$ | $+3$ |
| additive inverses | | |
| $(-3) - (-2)$ | $(-3) + (+2)$ | $-1$ |
| additive inverses | | |
| $(-7) - (+6)$ | $(-7) + (-6)$ | $-13$ |
| additive inverses | | |
| $(+9) - (-7)$ | $(+9) + (+7)$ | 16 |
| additive inverses | | |

**5. a.** $-5$ **b.** $+7$ **c.** $-15$ **d.** $+75$ **e.** $-281$ **f.** $+38$

## 7

The rules for subtracting two signed numbers follow.

*To subtract two signed numbers:*

1. Convert the subtraction to an equivalent addition.
2. Add, using the rules for signed numbers.

**Examples** **a.** $(-2) - (-4) = (-2) + (+4)$

change signs

**b.** $(-6) - (+2) = (-6) + (-2)$

change signs

Change the following subtraction problems to addition problems.

**a.** $(+3) - (+9)$ ____ **b.** $(+7) - (-7)$ ____

**c.** $(-3) - (+4)$ ____ **d.** $(-8) - (-7)$ ____

## 8

We can now go through the steps of changing subtraction to addition and using addition rules to solve the problem.

**Examples** **a.** $(-9) - (-3)$

$(-9) + (+3)$ ← change to addition

$(-9) + (+3) = -6$ ← use addition rules

**7. a.** $(+3) + (-9)$ **b.** $(+7) + (+7)$ **c.** $(-3) + (-4)$ **d.** $(-8) + (+7)$

**b.** $(+6) - (+3)$
$(+6) + (-3)$ ← change to addition
$(+6) + (-3) = +3$ ← use addition rules

Solve these subtraction problems using the steps described in frames 2 and 3.

**a.** $(+4) - (+8) =$ **b.** $(+6) - (-2) =$

**c.** $(-3) - (+8) =$ **d.** $(-8) - (-9) =$

**9**

Answers to frame 8.

**a.** $-4$ **b.** $+8$ **c.** $-11$ **d.** $+1$

# Exercise Set, Sections 3-1–3-2

## Signed Numbers

Circle the larger number.

**1.** 0 or 8 **2.** $-3$ or 2 **3.** $-2$ or 7 **4.** $-8$ or $-6$

Find the following absolute values.

**5.** $|15| =$ **6.** $|-8| =$ **7.** $|2| =$ **8.** $|-74| =$

## Addition and Subtraction of Signed Numbers

Perform the following additions and subtractions.

**9.** $(+5) + (+2) =$
**10.** $(+7) + (+11) =$
**11.** $(+5) + (+9) =$
**12.** $(+12) + (+18) =$
**13.** $(-6) + (-11) =$
**14.** $(-2) + (-15) =$
**15.** $(-11) + (-24) =$
**16.** $(-17) + (-35) =$
**17.** $(+7) + (-3) =$
**18.** $(-15) + (+15) =$
**19.** $(-28) + (+18) =$
**20.** $(+6) + (-7) =$
**21.** $(+7) - (+3) =$
**22.** $(+11) - (+7) =$
**23.** $(+4) - (+15) =$
**24.** $(+8) - (+11) =$
**25.** $(-6) - (+3) =$
**26.** $(-12) - (+4) =$
**27.** $(-3) - (+9) =$
**28.** $(-2) - (+7) =$
**29.** $(-11) - (-5) =$
**30.** $(-6) - (-8) =$
**31.** $(-27) - (-12) =$
**32.** $(-18) - (-22) =$
**33.** $(+98) + (-98) =$
**34.** $(+98) - (+98) =$
**35.** $(+32) - (+41) =$
**36.** $(+62) + (+4) =$
**37.** $(-17) - (+45) =$
**38.** $(-85) - (-71) =$
**39.** $(-53) + (+54) =$
**40.** $8 - 10 =$

# Supplementary Exercise Set, Sections 3-1–3-2

Perform the following additions and subtractions.

**1.** $(+3) + (+5) =$
**2.** $(+7) + (-2) =$
**3.** $(+11) + (-15) =$
**4.** $(-15) + (+3) =$
**5.** $(-7) + (-15) =$
**6.** $(-8) - (-25) =$
**7.** $(+8) + (-95) =$
**8.** $(+11) - (-75) =$
**9.** $(+41) + (-9) =$
**10.** $(+75) - (-9) =$
**11.** $(-18) + (-18) =$
**12.** $(-41) - (-40) =$
**13.** $(-16) - (+16) =$
**14.** $(-16) + (+16) =$
**15.** $(+16) + (-16) =$
**16.** $(500) + (-500) =$
**17.** $(-75) + (+76) =$
**18.** $(-75) + (+74) =$
**19.** $8 - 11 =$
**20.** $15 - 4 =$

# Section 3-3 Multiplication and Division of Signed Numbers

## 1

In the problem $7 \times 5 = 35$, the 7 and the 5 are both factors and 35 is the product.

**a.** In $9 \times 3 = 27$, the factors are ______ and ______.

**b.** In $6 \times 8 = 48$, the product is ______.

**1. a.** 9 and 3
**b.** 48

## 2

The multiplication $9 \times 8$ can be written using various notation. For example:

$9(8)$
$(9)(8)$
$9 \cdot 8$
$9 \times 8$

Do the following operations.

**a.** $3(9) =$ **b.** $5 \times 7 =$ **c.** $4 \cdot 2 =$ **d.** $(6)(7) =$

**2. a.** 27
**b.** 35
**c.** 8
**d.** 42

## 3

The rules for multiplying two signed numbers are:

1. If both factors have the same sign, the product is positive.
2. If the factors have different signs, the product is negative.

Write each product, using the rules above.

**a.** $(-6)(7) =$ **b.** $(-3)(-4) =$ **c.** $7 \times 5 =$

**3. a.** $-42$
**b.** $+12$
**c.** $+35$

## 4

Perform the following multiplications.

**a.** $(-9)(-7) =$ **b.** $3(-4) =$
**c.** $(-6)(6) =$ **d.** $(-5)(-5) =$

**4. a.** $+63$
**b.** $-12$
**c.** $-36$
**d.** $+25$

## 5

To multiply three or more factors, multiply two factors at a time. For example, to multiply $(2)(-3)(-5)$, first multiply $(2)(-3) = -6$. Then the multiplication becomes $(-6)(-5) = +30$.

Therefore, $(2)(-3)(-5) = (-6)(-5) = +30$

Multiply the following.

**a.** $(3)(-7)(4) = (\quad)(4) =$ ______

**b.** $(-5)(-9)(-3) = (\quad)(-3) =$ ______

**c.** $(-3)(-4)(9) =$ ______

**d.** $(-8)(5)(-2) =$ ______

**6**

A multiplication of four factors is performed below.

$$(-7)(-3)(4)(-2) = (+21)(4)(-2) = (84)(-2) = -168$$

Multiply the following.

**a.** $(-5)(+2)(-4)(9) = (\quad)(-4)(9) = (\quad)(9) =$ ______

**b.** $(7)(-4)(-3)(-5) =$ ______

**c.** $(-8)(-2)(-5)(-6) =$ ______

5. **a.** $(-21)(4) = -84$
**b.** $(45)(-3) = -135$
**c.** $+108$
**d.** $+80$

**7**

If one of the factors is 0, the product is always 0.

**Example** $(7)(0) = 0$
$(3)(-5)(0)(-8) = 0$

Perform the following multiplications.

**a.** $(-2)(0)(7) =$ ______ **b.** $(-8)(-9)(0)(-57) =$ ______

6. **a.** $(-10)(-4)(9) = (40)(9) = 360$
**b.** $-420$
**c.** $+480$

**8**

The rules for dividing two signed numbers are:

1. If the divisor and dividend both have the same sign, the quotient is positive.
2. If the divisor and dividend have different signs, the quotient is negative.

Write each quotient, using the rules above.

**a.** $\frac{35}{7} =$ **b.** $\frac{-48}{-6} =$ **c.** $\frac{+45}{-9} =$ **d.** $\frac{-56}{7} =$

7. **a.** 0
**b.** 0

**9**

Perform the following divisions.

**a.** $\frac{-72}{-8} =$ **b.** $\frac{81}{-9} =$ **c.** $\frac{90}{-3} =$ **d.** $\frac{-45}{-15} =$

8. **a.** 5
**b.** 8
**c.** $-5$
**d.** $-8$

**10**

Answers to frame 9.

**a.** $+9$ **b.** $-9$ **c.** $-30$ **d.** $+3$

# Section 3-4 Using the Calculator to Perform Operations with Signed Numbers

**1**

The calculator can be used to perform operations with signed numbers. The key used for entering negative numbers is:

[+/-]

The addition below can be done using a calculator.

$(-5) + 4 =$

| *Enter* | *Press* | *Display* |
|---|---|---|
| 5 | [+/-] [+] | $-5$ |
| 4 | [=] | $-1$ |

Therefore, $(-5) + 4 = -1$.

Use a calculator to perform the following additions.

**a.** $(-9) + 7 =$ **b.** $(-58) + 19 =$

**2**

The addition below can be performed by using a calculator.

**1. a.** $-2$ **b.** $-39$

$(+9) + (-7) =$

| *Enter* | *Press* | *Display* |
|---|---|---|
| 9 | [+] | 9 |
| 7 | [+/-] [=] | 2 |

Therefore, $(+9) + (-7) = 2$.

Use a calculator to perform the following additions.

**a.** $7 + (-8) =$ **b.** $95 + (-40) =$

**3**

The addition below is performed by using a calculator.

**2. a.** $-1$ **b.** $+55$

$(-8) + (-7) =$

| *Enter* | *Press* | *Display* |
|---|---|---|
| 8 | [+/-] [+] | $-8$ |
| 7 | [+/-] [=] | $-15$ |

Therefore, $(-8) + (-7) = -15$.

Use a calculator to perform the following additions.

**a.** $(-9) + (-4) =$ **b.** $(-87) + (-12) =$

**4**

The subtraction below is performed by using a calculator. The [+/-] key is used to enter a negative number in subtractions, as in additions.

**3. a.** $-13$ **b.** $-99$

$(5) - (-7) =$

| *Enter* | *Press* | *Display* |
|---|---|---|
| 5 | [−] | 5 |
| 7 | [+/-] [=] | 12 |

Therefore, $5 - (-7) = 12$.

Use a calculator to perform the following subtractions.

**a.** $(7) - (-2) =$ **b.** $(-8) - (-5) =$ **c.** $(-9) - (4) =$

**5**

The multiplication below is performed by using a calculator.

4. **a.** 9
**b.** $-3$
**c.** $-13$

$(-5)(8) =$

| *Enter* | *Press* | *Display* |
|---|---|---|
| 5 | [+/-] [×] | $-5$ |
| 8 | [=] | $-40$ |

Therefore, $(-5)(8) = -40$.

Use a calculator to perform the following multiplications.

**a.** $(-9)(8) =$ **b.** $(-97)(18) =$

**6**

The multiplication below is performed by using a calculator.

5. **a.** $-72$
**b.** $-1{,}746$

$(9)(-7) =$

| *Enter* | *Press* | *Display* |
|---|---|---|
| 9 | [×] | 9 |
| 7 | [+/-] [=] | $-63$ |

Therefore, $(9)(-7) = -63$.

Use a calculator to perform the following multiplications.

**a.** $(7)(-8) =$ **b.** $(95)(-42) =$

**7**

The multiplication below is performed by using a calculator.

6. **a.** $-56$
**b.** $-3{,}990$

$(-7)(-6) =$

| *Enter* | *Press* | *Display* |
|---|---|---|
| 7 | [+/-] [×] | $-7$ |
| 6 | [+/-] [=] | 42 |

Therefore, $(-7)(-6) = +42$.

Use a calculator to perform the following multiplications.

**a.** $(-8)(-4) =$ **b.** $(-31)(-47) =$

**8**

The same procedures are used to perform a division with a calculator. The key used to enter a negative number is again:

7. **a.** $+32$
**b.** $+1{,}457$

[+/-]

The division below is performed by using a calculator.

$$\frac{-48}{-6} =$$

| *Enter* | *Press* | *Display* |
|---|---|---|
| 48 | [+/-] [÷] | −48 |
| 6 | [+/-] [=] | 8 |

Therefore, $\frac{-48}{-6} = 8$.

Use a calculator to perform the following divisions.

**a.** $\frac{-64}{8} =$ **b.** $\frac{-400}{-20} =$ **c.** $\frac{180}{-45} =$ **d.** $\frac{-3{,}200}{-40} =$

**9**

Answers to frame 8.

**a.** −8 **b.** 20 **c.** −4 **d.** 80

# Exercise Set, Sections 3-3–3-4

## Multiplication and Division of Signed Numbers

Perform the following multiplications and divisions.

**1.** $(+3)(-9) =$ **2.** $(-10)(+8) =$ **3.** $(+6)(+7) =$ **4.** $(-9)(-9) =$

**5.** $(+8)(+9) =$ **6.** $(-20)(+5) =$ **7.** $(-4)(-7) =$ **8.** $(+9)(-4) =$

**9.** $(-12)(-10) =$ **10.** $\frac{36}{6} =$ **11.** $\frac{48}{-4} =$ **12.** $\frac{-144}{6} =$

**13.** $\frac{-93}{-3} =$ **14.** $\frac{49}{7} =$ **15.** $\frac{39}{-13} =$ **16.** $\frac{-66}{11} =$

**17.** $\frac{-84}{-4} =$ **18.** $\frac{-42}{7} =$

## Using the Calculator to Perform Operations with Signed Numbers

Perform the following operations with a calculator.

**19.** $(-5) + (-7) =$ **20.** $(-8) - (-12) =$ **21.** $(15) + (-35) =$

**22.** $(98) - (-22) =$ **23.** $(-41) - (41) =$ **24.** $(-8) + (58) =$

**25.** $(-7)(12) =$ **26.** $(-90)(-45) =$ **27.** $(92)(815) =$

**28.** $\frac{3{,}692}{-52} =$ **29.** $\frac{-480}{60} =$ **30.** $\frac{-35}{-5} =$

**31.** $(8)(-13) =$ **32.** $(7)(19) =$ **33.** $(-12)(-3) =$

**34.** $(-28)(5) =$ **35.** $\frac{-135}{27} =$ **36.** $\frac{231}{-7} =$

**37.** $\frac{-462}{-11} =$ **38.** $\frac{630}{-84} =$ **39.** $\frac{-77}{14} =$ **40.** $\frac{1{,}411}{83} =$

# Supplementary Exercise Set, Sections 3-3–3-4

Perform the following signed number operations.

**1.** $(-2)(-4) =$ **2.** $(+3)(-7) =$ **3.** $(+8)(+11) =$

**4.** $(+15)(-7) =$ **5.** $(-85)(+2) =$ **6.** $(-9)(-10) =$

**7.** $(+36) - (-4) =$ **8.** $(+14) - (+7) =$ **9.** $(-100) - (-20) =$

**10.** $\frac{-75}{-25} =$ **11.** $\frac{18}{-2} =$ **12.** $\frac{-414}{-23} =$

**13.** $(328) - (-41) =$ **14.** $(-396) - (-36) =$ **15.** $(+1{,}170) - (+25) =$

**16.** $(-245)(+23) =$ **17.** $(-57)(-57) =$ **18.** $(-10)(-10) =$

**19.** $(-10)(-10)(-10) =$ **20.** $(-10)(-10)(-10)(-10) =$

# Section 3-5 Introduction to Equations

**1**

An equation is a mathematical sentence that contains an equals sign. An equation may be either true or false. For example, $3 + 5 = 8$ is a true statement, but $3 + 5 = 7$ is a false one.

**a.** Is $9 + 5 = 14$ true or false? ___________

**b.** Is $6 - 8 = -4$ true or false? ___________

**2**

When a number is unknown, a letter is used to represent that number in an equation. For example, the equation $5 + a = 9$ means "five plus some unknown number equals nine." The equation $5 + a = 9$ can be either true or false, depending on what value $a$ represents:

If $a = 3$, the equation $5 + a = 9$ becomes $5 + 3 = 9$, which is false.

If $a = 4$, the equation $5 + a = 9$ becomes $5 + 4 = 9$, which is true.

**1. a.** true
**b.** false

**3**

The *root* of an equation is the number that, when substituted for the letter, will make the equation true.

Find the root of each of the following equations.

**a.** $(3)(n) = 15$ **b.** $4 + y = 7$ **c.** $t - 32 = 8$

**d.** $2 + a = 5$ **e.** $9 - b = 7$ **f.** $x/2 = 4$

3. a. 5
b. 3
c. 40
d. 3
e. 2
f. 8

**4**

Two equations are said to be equivalent equations when they have the same root. For example,

$$2x + 5 = 9 \qquad 2x = 4 \qquad x = 2$$

are all equivalent equations because they have the same root (2):

$2x + 5 = 9$ is true when 2 is substituted for $x$.
$2x = 4$ is true when 2 is substituted for $x$.
$x = 2$ is true when 2 is substituted for $x$.

Which of the equations below are equivalent equations? (Circle your selections.)

**a.** $3x = 9$ **b.** $3x = 6$ **c.** $5 + x = 8$ **d.** $4 + x = 11$

4. a and c (both have a root of 3)

**5**

The concept of equivalent equations is useful when trying to find the root of a given equation, which is known as solving an equation. Solving equations involves finding equivalent equations that are simpler than the original one. The next topic in this chapter will deal with the laws that can be used in finding equivalent equations.

# Section 3-6 The Associative, Commutative, and Distributive Laws

**1**

The associative, commutative, and distributive laws are important when simplifying equations.

**Associative Law of Addition** $(a + b) + c = a + (b + c)$
This law is demonstrated by the following addition:

$$(3 + 4) + 8 = 3 + (4 + 8)$$
$$7 + 8 = 3 + 12$$
$$15 = 15$$

Notice that 15 is the sum when $3 + 4$ is added and then 8 is added; 15 is also the sum when $4 + 8$ is added and then 3 is added.

Use the associative law of addition to complete the following.

**a.** $(5 + 6) + 3 = 5 + (\_\_\_\_ + \_\_\_\_)$

**b.** $9 + (x + 7) = (9 + \_\_\_\_) + \_\_\_\_$

**2**

**Associative Law of Multiplication** $(ab)c = a(bc)$
This law is demonstrated by the following multiplication:

$$(5 \times 8) \times 3 = 5 \times (8 \times 3)$$
$$40 \times 3 = 5 \times 24$$
$$120 = 120$$

Notice that (40)(3) and (5)(24) both equal 120.

Use the associative law of multiplication to complete the following.

**a.** $(3 \times 5) \times 8 = 3(\_\_\_\_ \times \_\_\_\_)$ **b.** $4(3h) = (\_\_\_\_ \times \_\_\_\_)h$

**1. a.** $5 + (6 + 3)$
**b.** $(9 + x) + 7$

**3**

**Commutative Law of Addition** $a + b = b + a$
This law is demonstrated below.

$$5 + 6 = 6 + 5$$
$$11 = 11$$

Thus, this law states that two numbers can be added in any order.

Use the commutative law of addition to complete the following.

**a.** $18 + 3 = 3 + \_\_\_\_$ **b.** $h + 7 = 7 + \_\_\_\_$

**c.** $5 + x = \_\_\_\_ + 5$ **d.** $17 + k = \_\_\_\_ + \_\_\_\_$

**2. a.** $3(5 \times 8)$
**b.** $(3 \times 4)h$

**4**

**Commutative Law of Multiplication** $ab = ba$
This law is demonstrated below.

$$3 \times 9 = 9 \times 3$$
$$27 = 27$$

Thus, this law states that two numbers can be multiplied in any order.

Use the commutative law of multiplication to complete the following.

**a.** $18 \times 3 = 3 \times \_\_\_\_$ **b.** $xy = \_\_\_\_$

**c.** $hk = k\_\_\_\_$ **d.** $12 \times 7 = \_\_\_\_ \times \_\_\_\_$

**3. a.** $3 + 18$
**b.** $7 + h$
**c.** $x + 5$
**d.** $k + 17$

**5**

**Distributive Law of Multiplication over Addition** $a(b + c) = ab + ac$
This law is illustrated below.

$$5(3 + 4) = 5(3) + 5(4)$$
$$5(7) = 15 + 20$$
$$35 = 35$$

**4. a.** $3 \times 18$
**b.** $yx$
**c.** $kh$
**d.** $7 \times 12$

Use the distributive law of multiplication over addition to complete the following.

**a.** $5(x + 8) =$ **b.** $h(3 + 4) =$

**c.** $10(3 + 2x) =$ **d.** $3(4 + 8) =$

**6**

---

These laws can be used to simplify algebraic expressions as follows:

| | |
|---|---|
| Simplify: | $3(x + 4) + x$ |
| *Step 1.* Distributive law | $3x + 12 + x$ |
| *Step 2.* Commutative law of addition | $3x + x + 12$ |
| *Step 3.* Associative law | $(3x + x) + 12$ |
| *Step 4.* Add like terms. | $4x + 12$ |

Simplify the following expressions.

**a.** $5(x + 4) + 3$ **b.** $8(3 + x) + 2x$

**5. a.** $5x + 40$
**b.** $3h + 4h$
**c.** $30 + 20x$
**d.** $12 + 24$

**7**

---

Simplify the following expressions.

**a.** $3(x + 5) + 2x$ **b.** $9(1 + x) + 8$

**c.** $4(x + 2) + 18$ **d.** $h(7 + 2) + h$

**e.** $3(x + 1) + x(4 + 2)$ **f.** $5(k - 1) + 5$

**6. a.** $5x + 23$
**b.** $24 + 10x$

**8**

---

Answers to frame 7.

**a.** $15 + 5x$ **b.** $17 + 9x$ **c.** $4x + 26$
**d.** $10h$ **e.** $9x + 3$ **f.** $5k$

# Exercise Set, Sections 3-5–3-6

## Introduction to Equations

Find the root of the following equations.

**1.** $6 + a = 8$ **2.** $7 - x = 2$ **3.** $5x = 15$ **4.** $x + 5 = 15$

**5.** $2x = 10$ **6.** $2x + 1 = 11$ **7.** $11 + y = 16$ **8.** $24 - x = 7$

**9.** $3x = 21$ **10.** $y + 20 = 32$ **11.** $4x = 28$ **12.** $13 - y = 5$

**13.** $3x + 12 = 17$ **14.** $3x - 2 = 16$ **15.** $2x - 5 = 15$

## The Associative, Commutative, and Distributive Laws

Simplify the following algebraic expressions.

**16.** $5(x + 2) + 9$ **17.** $7(x + 5) + 4x$ **18.** $3(x + 9) - 9$

**19.** $5(x + 3) + 4x$ **20.** $2(4x + 8) - 2x$ **21.** $4(2x + 2) - 6x$

**22.** $2(x + 4) + 3(x + 5)$ **23.** $3(2h + 2) + h(3 + 7)$ **24.** $6(4x + 5) - 4x$

**25.** $7(3x + 2) - 5$ **26.** $3(4x - 12) - 2x$ **27.** $2(x + 4) - 6x$

**28.** $(x + 3) - 2(x + 1)$ **29.** $2(2x - 3) + x(1 + 4)$ **30.** $4(x - 6) + 3(2x + 1)$

**31.** $4(2x + 6) - 3(2x + 1)$

# Supplementary Exercise Set, Sections 3-5–3-6

Find the root of each of the following equations.

**1.** $5 + H = 7$ **2.** $15 - K = 10$ **3.** $5w = 45$ **4.** $3 + y = 0$

**5.** $x - 5 = -2$ **6.** $3x = 66$ **7.** $2x - 1 = 9$ **8.** $3x + 5 = 20$

**9.** $H + 5 = 5$ **10.** $2h + 7 = 7$

Simplify the following algebraic expressions.

**11.** $3(x + 2) + 7$ **12.** $9(x - 5) + 45$ **13.** $7(H + 1) - 5H$

**14.** $5(k + 5) + k$ **15.** $10(x + 10) - 100$ **16.** $5 + 3(x + 5)$

**17.** $2(x + 1) + 3(x - 1)$ **18.** $5(2x + 1) - 2(5x + 1)$ **19.** $3(x + 1) - 3(x + 1)$

**20.** $5(H + 2) - 3(H - 3) + H$

# Section 3-7 Multiplication Axiom for Solving Equations

## 1

Two numbers are called *reciprocals* if their product is 1. Some examples of reciprocals are:

$$3 \times \frac{1}{3} = \frac{3}{3} = 1 \qquad \frac{5}{8} \times \frac{8}{5} = 1$$

$$\left(\frac{3}{4}\right)\left(\frac{4}{3}\right) = 1 \qquad \left(\frac{1}{17}\right)\left(17\right) = 1$$

Complete the following reciprocals.

**a.** $\left(\frac{2}{3}\right)\left(\quad\right) = 1$ **b.** $\left(5\right)\left(\quad\right) = 1$

**c.** $\left(\frac{1}{3}\right)\left(\quad\right) = 1$ **d.** $\left(\frac{7}{8}\right)\left(\frac{8}{7}\right) =$

## 2

In general, the reciprocal of a fraction can be expressed as that fraction with the numerator and denominator interchanged. For example, the reciprocal of 9/10 is 10/9.

Complete the following.

**a.** The reciprocal of $\frac{7}{8}$ is ______.

**b.** The reciprocal of $\frac{11}{16}$ is ______.

**c.** The reciprocal of $\frac{12}{5}$ is ______.

**1. a.** $\left(\frac{2}{3}\right)\left(\frac{3}{2}\right) = 1$

**b.** $\left(5\right)\left(\frac{1}{5}\right) = 1$

**c.** $\left(\frac{1}{3}\right)\left(3\right) = 1$

**d.** $\left(\frac{7}{8}\right)\left(\frac{8}{7}\right) = 1$

## 3

The reciprocal of a whole number can be expressed as that number written as a fraction with 1 as the numerator. For example, the reciprocal of 5 is 1/5; the reciprocal of 89 is 1/89.

Complete the following.

**a.** The reciprocal of 9 is ______.

**b.** The reciprocal of 37 is ______.

**c.** The reciprocal of 159 is ______.

**2. a.** $\frac{8}{7}$

**b.** $\frac{16}{11}$

**c.** $\frac{5}{12}$

## 4

In any product of a number and a letter, the number is referred to as the *coefficient* of the letter. For example, in the product $7h$, 7 is the coefficient of $h$.

Complete the following.

**a.** In the product $19x$, ______ is the coefficient of $x$.

**3. a.** $\frac{1}{9}$

**b.** $\frac{1}{37}$

**c.** $\frac{1}{159}$

**b.** In the product $-79x$, $-79$ is the coefficient of ______.

**c.** In the product $-34h$, ______ is the coefficient of $h$.

**5**

The following axiom is used in solving equations.

**4. a.** 19
**b.** $x$
**c.** $-34$

**Multiplication Axiom for Solving Equations.** If both sides of an equation are multiplied by the same nonzero quantity, the new equation is equivalent to the original equation.

This axiom is used with reciprocals to solve equations as follows. To solve

$$6x = 30$$

both sides are multiplied by the reciprocal of the coefficient of the $x$.

The coefficient of $x$ is 6.

The reciprocal of 6 is $\frac{1}{6}$.

Therefore, both sides are multiplied by 1/6 as follows:

$$\left(\frac{1}{6}\right)\left(6x\right) = \left(\frac{1}{6}\right)\left(30\right)$$

$$\left(\frac{6}{6}\right)\left(x\right) = \frac{30}{6}$$

$$x = 5$$

The equation $x = 5$ is equivalent to the equation $6x = 30$. Therefore, the root of $6x = 30$ is 5.

**6**

The process of solving an equation involves finding an equivalent equation of the form $x =$ some number. An equation of the type $ax = b$ is solved by using the multiplication axiom. Both sides of the equation are multiplied by the reciprocal of the coefficient of the $x$. For example, to solve

$$3x = 9$$

both sides of the equation are multiplied by the reciprocal of the coefficient of $x$.

$$\left(\frac{1}{3}\right)\left(3x\right) = \left(9\right)\left(\frac{1}{3}\right)$$

$$\left(\frac{3}{3}\right)\left(x\right) = \left(\frac{9}{3}\right)$$

$$x = 3$$

The root of $3x = 9$ is 3.

**7**

The multiplication axiom was used to solve the following equation:

$$7x = 56$$

The coefficient of $x$ is 7.
The reciprocal of 7 is 1/7.

$$\left(\frac{1}{7}\right)\left(7x\right) = \left(56\right)\left(\frac{1}{7}\right)$$

Therefore, both sides are multiplied by 1/7.

$$\left(\frac{7}{7}\right)\left(x\right) = \left(\frac{56}{7}\right)$$

$$x = 8$$

The root of $7x = 56$ is 8.

Find the root of the following equations using the multiplication axiom.

**a.** $12x = 96$ **b.** $9x = 225$ **c.** $5m = 45$

**d.** $10y = 60$ **e.** $11c = 121$ **f.** $9f = 72$

**8**

Answers to frame 7.

**a.** 8 **b.** 25 **c.** 9 **d.** 6 **e.** 11 **f.** 8

# Section 3-8 Addition Axiom for Solving Equations

**1**

As mentioned earlier, two numbers are additive inverses if their sum is 0. Some examples of additive inverses are:

$$(-5) + 5 = 0$$

$$418 + (-418) = 0$$

$$(+19) + (-19) = 0$$

$$\frac{3}{4} + \left(-\frac{3}{4}\right) = 0$$

$$(-2.5) + (+2.5) = 0$$

Complete the following additive inverses.

**a.** $(-8) + (\quad) = 0$ **b.** $(15) + (\quad) = 0$

**c.** $\frac{1}{3} + \frac{-1}{3} =$ ______ **d.** $(\quad) + (7.6) = 0$

**2**

In general, the additive inverse of a number is that number with the opposite sign. Thus, additive inverses are sometimes referred to as opposites. For example, the additive inverse of $+38$ is $-38$.

**1. a.** 8
**b.** $-15$
**c.** 0
**d.** $-7.6$

Complete the following.

**a.** The additive inverse of $-78$ is ______.

**b.** The additive inverse of $\frac{3}{8}$ is ______.

**3**

---

The following axiom is used in solving equations.

**2. a.** $+78$

**b.** $\frac{-3}{8}$

**Addition Axiom.** When the same quantity is added to both sides of an equation, the new equation is equivalent to the original equation.

This axiom is used with additive inverses to solve equations as follows:

$x + 5 = 8$ — The additive inverse of 5 is $(-5)$.

$x + 5 + (-5) = 8 + (-5)$ — Therefore, $(-5)$ is added to both sides.

$x + 0 = 3$

$x = 3$

The equation $x = 3$ is equivalent to $x + 5 = 8$. Therefore, the root of $x + 5 = 8$ is 3.

**4**

---

As mentioned earlier, the process of solving an equation involves finding an equivalent equation of the form $x =$ some number.

An equation of the type $x + a = b$ is solved by using the addition axiom. The additive inverse of the number on the same side of the equation as $x$ is added to both sides. For example, to solve

$x + 9 = 15$ — $-9$ is added to both sides of the equation.

$x + 9 + (-9) = 15 + (-9)$

$x + 0 = 6$

$x = 6$ — The root of $x + 9 = 15$ is 6.

**5**

---

The addition axiom was used to solve the following equation:

$$a + 30 = 10$$
$$a + (30) + (-30) = 10 + (-30)$$
$$a + 0 = -20$$
$$a = -20$$

The root of $a + 30 = 10$ is $-20$.

Find the root of the following equations using the addition axiom.

**a.** $x + 32 = 17$

**b.** $h + 4 = 52$

**c.** $f + 10 = 45$

**d.** $c + 17 = 41$

**6**

The addition axiom can be used for the following equation:

$$k - 15 = 7$$

**5. a.** $-15$
**b.** 48
**c.** 35
**d.** 24

The additive inverse of $(-15)$ is $(+15)$. Thus,

$$k - 15 + 15 = 7 + 15$$
$$k + 0 = 22$$
$$k = 22$$

Use the addition axiom to find the root of the following equations.

**a.** $x - 25 = 50$ **b.** $m - 4 = -2$

**c.** $f - 32 = 40$ **d.** $k - 7 = 17$

**7**

Answers to frame 6.

**a.** 75 **b.** 2 **c.** 72 **d.** 24

# Section 3-9 Strategies for Solving Equations

**1**

An efficient strategy for solving equations involves these steps:

1. Simplify each side of the equation.
2. Use the addition axiom to get the letter terms together on one side.
3. Use the addition axiom to get the number terms together on the other side.
4. Simplify the equation.
5. Use the multiplication axiom to get a coefficient of 1 for the letter term.

Each of these steps can be used to get equivalent equations that are simpler than the original equation until, finally, an equivalent equation of the type $x =$ "the root" is obtained.

**2**

The strategy in the previous frame is used to solve the following equation.

| | |
|---|---|
| $2x + 6 = 10$ | The equation is already simplified, and the letter terms are together on one side. |
| $2x + 6 + (-6) = 10 + (-6)$ | Therefore, the first step is step 3: use the addition axiom to get the number terms together. |
| $2x + 0 = 4$ | The number terms are together. |
| $2x = 4$ | Simplify the equation. |

$$\left(\frac{1}{2}\right)\left(2x\right) = \left(\frac{1}{2}\right)\left(4\right)$$ Use the multiplication axiom to get a coefficient of 1 for $x$.

$$1x = \frac{4}{2}$$ Simplify.

$$x = 2$$ The root is 2.

Solve the following.

**a.** $4x - 2 = 10$ **b.** $7h + 5 = 40$

**c.** $5c - 6 = 9$ **d.** $9x - 45 = 0$

**2. a.** $x = 3$
**b.** $h = 5$
**c.** $c = 3$
**d.** $x = 5$

## 3

Solve equation **b** using the strategies from equation **a.**

**a.**

$$3x + 6 = x + 10$$
$$3x + 6 + (-x) = x + 10 + (-x)$$
$$2x + 6 = 10$$
$$2x + 6 + (-6) = 10 + (-6)$$
$$2x = 4$$
$$\left(\frac{1}{2}\right)\left(2x\right) = \left(\frac{1}{2}\right)\left(4\right)$$
$$x = 2$$

**b.** $4x - 2 = x + 7$

## 4

Answer to frame 3.

**b.**

$$4x - 2 + (-x) = x + 7 + (-x)$$
$$3x - 2 = 7$$
$$3x - 2 + 2 = 7 + 2$$
$$3x = 9$$
$$\left(\frac{1}{3}\right)\left(3x\right) = \left(\frac{1}{3}\right)\left(9\right)$$
$$x = 3$$

## 5

Solve these equations.

**a.** $2x + 2 = x + 7$ **b.** $5h + 1 = h + 5$

**c.** $3f - 6 = f - 2$ **d.** $4x + 8 = x + 11$

## 6

Now consider a more complex equation.

5. a. $x = 5$
b. $h = 1$
c. $f = 2$
d. $x = 1$

| | |
|---|---|
| $3(x + 2) = x + 10$ | Original equation. |
| $3x + 6 = x + 10$ | Simplify (distributive law). |
| $3x + 6 + (-x) = x + 10 + (-x)$ | Addition axiom (gather $x$ terms). |
| $2x + 6 = 10$ | Simplify (combine like terms). |
| $2x + 6 + (-6) = 10 + (-6)$ | Addition axiom. |
| $2x = 4$ | Simplify (combine like terms). |
| $\left(\frac{1}{2}\right)\left(2x\right) = \left(\frac{1}{2}\right)\left(4\right)$ | Multiplication axiom. |
| $1x = \frac{4}{2}$ | Simplify. |
| $x = 2$ | The root is 2. |

Solve the following equations.

**a.** $5(x - 1) = x + 7$ **b.** $6(h + 1) = h - 14$

**c.** $3(x - 2) = x + 8$ **d.** $9(f + 1) = 10f$

## 7

Answers to frame 6.

**a.** $x = 3$ **b.** $h = -4$ **c.** $x = 7$ **d.** $f = 9$

# Exercise Set, Sections 3-7–3-9

## Multiplication Axiom for Solving Equations

Find the root of the following equations using the multiplication axiom.

**1.** $2x = 26$ **2.** $15h = -30$ **3.** $3k = 54$ **4.** $-2m = 10$

**5.** $6x = 108$ **6.** $38x = 38$ **7.** $-3y = 33$ **8.** $12m = -48$

**9.** $1.2x = 3.6$ **10.** $13y = 65$ **11.** $4.8x = 33.6$ **12.** $11.2y = 67.2$

## Addition Axiom for Solving Equations

Find the root of the following equations using the addition axiom.

**13.** $x + 5 = 11$ **14.** $k + 8 = -2$ **15.** $h + 2.5 = 2.5$

**16.** $3 + x = 10$ **17.** $m - 7 = 11$ **18.** $x - 8 = -5$

**19.** $x + 22 = -5$

**20.** $y - 9 = 18$

**21.** $m - 7 = -6$

**22.** $h + 5 = -5$

**23.** $x - 7.6 = -3.4$

**24.** $y + 3.7 = 8.9$

## Strategies for Solving Equations

Solve these equations.

**25.** $3x + 2 = 17$

**26.** $4h - 2 = -10$

**27.** $5x - 7 = 18$

**28.** $3y + 5 = -16$

**29.** $2x + 4 = x + 9$

**30.** $4x - 3 = x + 3$

**31.** $7x + 8 = 2x - 7$

**32.** $8y - 12 = 2y + 12$

# Supplementary Exercise Set, Sections 3-7–3-9

Solve the following equations.

**1.** $3H = 30$

**2.** $16x = -32$

**3.** $5K = 6$

**4.** $-12h = 60$

**5.** $1.5M = 12$

**6.** $13.1y = 107.42$

**7.** $H + 18 = 20$

**8.** $K + 1.2 = 4$

**9.** $W - 8 = -6$

**10.** $3 + x = 0$

**11.** $H + 2.8 = -7.2$

**12.** $x - 13.5 = 7.2$

**13.** $2x + 8 = 18$

**14.** $3x + 20 = -1$

**15.** $4H - 2 = 30$

**16.** $9x + 1 = 5x + 5$

**17.** $3h + 5 = 19h + 5$

**18.** $15W = 2W + 130$

**19.** $17x + 2 = x + 18$

**20.** $4.2x + 7.8 = 2.5x - 9.2$

# Section 3-10 Applied Problems

**1**

Solving problems with the use of a calculator or computer requires that the information given in a word problem be put into the form of an algebraic equation. The steps of problem solving presented previously can be used.

**Example** The height of a building is found by multiplying 9 times its width. Write an algebraic equation for the height of the building.

*Step 1.* Determine what is being asked for.
Height of building = ?
In an algebraic expression, a letter would be assigned to this quantity. The letter $H$ will be used for the quantity *height*.

*Step 2.* What information is needed to solve for the height?
The width of the building will have to be known in order to solve for the height of the building. The letter $W$ will be assigned to the *width.*

*Step 3.* Find the algebraic equation that describes the relationship. In this case, the height $(H) = 9$ times the width $(W)$ or $H = 9W$.

Write an algebraic equation for the following word problem.

The total resistance in a series circuit is the resistance in the first resistor plus twice the resistance in the second resistor.

*Step 1.* Determine what is being asked for.
*Step 2.* What information is needed to solve for the quantity in Step 1?
*Step 3.* Write the algebraic equation describing the relationship.

**2**

Answers to frame 1.

*Step 1.* Total resistance = ?
The symbol $R_T$ is used for total resistance.

$R_T = ?$

*Step 2.* The information needed to solve for the total resistance $(R_T)$ is the resistance in the first resistor and the resistance in the second resistor.

Resistance in the first resistor = $R_1$
Resistance in the second resistor = $R_2$

*Step 3.* Write the equation.
Total resistance = resistance in the first resistor plus twice the resistance in the second resistor.

$R_T = R_1 + 2R_2$

**3**

Write an equation for the following.

The net pay of an employee is the gross pay, minus the withholding tax and minus the social security tax.

**4**

Answers to frame 3.

*Step 1.* Net pay = ?
NP = ?
*Step 2.* Gross pay $(GP)$
Withholding tax $(WT)$
Social Security tax $(SST)$
*Step 3.* NP $= GP - WT - SST$

*Note:* Other symbols than the ones suggested here can be used. The symbols should relate to the quantity being described.

# Exercise Set, Section 3-10

## Applied Problems

Write algebraic equations for the word problems. Use symbols that relate to the quantity being described.

1. An electrician finds that the buildings she is going to wire need four times as much type-2 wire as type-1 wire. Write an equation for the total wire she will need.

2. The volume of a tank is its length times width times height. Show this in an equation.

3. A technician mixes a water-and-alcohol solution in which 3/4 of the total is water. Write an equation for the amount of alcohol in the solution.

4. One employee receives a certain amount of pay. A second employee receives 1.2 times as much as the first employee. Write an equation for the total pay of both employees.

5. The total weight of two metal castings is the weight of casting 1 plus the weight of casting 2. Casting 2 is 1/3 of the total weight. Write an equation to find the weight of casting 1.

6. A student wishes to average three test grades. Write an equation to find the average.

7. There are 1.6 kilometers in 1 mile. The total number of kilometers equals 1.6 times the number of miles. Write an equation to show this.

8. Write an equation to find the total cost of four tires if the cost of one tire is the price plus sales tax.

9. The applied voltage ($V_a$) in a system is equal to the sum of the voltage drops across three resistors ($V_1$, $V_2$, and $V_3$). Write an equation to show this.

10. The area of a floor is the length times the width. The length of the floor is three times the width. Write an equation for the area of the floor using the width only.

## Supplementary Exercise Set, Section 3-10

Write algebraic equations for the word problems.

1. The current in an electrical circuit is the voltage divided by the resistance.

2. Fahrenheit temperature is found by multiplying the Celsius temperature by 1.8 and adding 32.

3. To convert kilograms (kg) to pounds (lb), multiply the number of kilograms by 2.2. Write an equation to find the total number of pounds.

4. The area of room $B$ is twice that of room $A$. Write an equation for the total area of room $A$ and $B$.

5. A concrete-block wall is 5 blocks high. Each block is $L$ inches high. There are also four mortar joints, each 1-inch wide, between the blocks. What is the total height of this wall?

# Summary

1. The absolute value of a number is the distance the number is from 0 on the number line.
2. The rules for adding signed numbers are:
   a. When adding numbers with like signs (all positive or all negative), add the absolute value of the numbers to find the sum and use the common sign of the numbers being added for the sign of the sum.
   b. When adding numbers with unlike signs, subtract the absolute value of the smaller number from the absolute value of the larger number and use the sign of the larger number.
3. To subtract signed numbers, convert the subtraction to an equivalent addition and proceed as in addition.
4. The rules for multiplying two signed numbers are:
   a. If the factors have the same sign, the product is positive.
   b. If the factors have unlike signs, the product is negative.
5. When dividing two numbers having the same sign, the answer is positive. If the numbers to be divided have different signs, the answer is negative.
6. The root of an equation is the number that, when substituted for an unknown letter, makes the equation true. Equivalent equations have the same root.
7. The associative, commutative, and distributive laws are used to simplify equations.
8. The reciprocal of a fraction is that fraction with the numerator and denominator interchanged.
9. When both sides of an equation are multiplied by the same nonzero quantity, the new equation is equivalent to the original equation.
10. When the same quantity is added to both sides of an equation, the new equation is equivalent to the original equation.
11. The strategy for solving equations is:
    a. Simplify each side.
    b. Get the letter terms together on one side.
    c. Get the number terms together on the other side.
    d. Simplify again.
    e. Use the multiplication axiom to get a coefficient of 1 for the letter term.
12. To put a word problem in equation form:
    a. Determine what is asked for.
    b. What information is needed?
    c. Find the algebraic equation that describes the relationship.

# Chapter 3 Self-Test

## Basic Concepts of Algebra

Circle the larger number.

**1.** $0$ or $-4$ **2.** $-5$ or $7$

What is the absolute value of these numbers?

**3.** $|-36|$ **4.** $|18|$

Perform the following additions and subtractions.

**5.** $(+3) + (+7) =$ **6.** $(-8) + (-4) =$ **7.** $(+5) + (-9) =$
**8.** $(-21) + (+6) =$ **9.** $(+21) - (-11) =$ **10.** $(+18) - (+9) =$
**11.** $(-30) - (-12) =$ **12.** $(-42) - (+14) =$

Perform the following multiplications and divisions.

**13.** $(+4)(-8) =$ **14.** $(-11)(-3) =$ **15.** $(+4)(+9) =$

16. $\dfrac{36}{-3} =$

17. $\dfrac{-144}{12} =$

18. $\dfrac{78}{8} =$

Perform the following with a calculator.

19. $(-41) + (+23) =$

20. $(-82) - (-24) =$

21. $\dfrac{-480}{80} =$

22. $(-90)(-35) =$

Find the root of each of the following equations.

23. $4 + x = 12$

24. $15 - a = 9$

25. $5x = 25$

26. $3x + 2 = 20$

Simplify the following algebraic expressions.

27. $4(x + 3) + 6$

28. $2(x + 5) + 3x$

29. $3(x + 2) + 2(x + 4)$

30. $3(4h + 2) + h(2 + 5)$

Find the root of each of the following equations.

31. $2x = 28$

32. $11y = -22$

Find the root of each of these equations.

33. $x + 6 = 13$

34. $y + 20 = -2$

35. $m - 5 = 17$

36. $x - 11 = -4$

Solve these equations.

37. $4x + 5 = 21$

38. $4x + 8 = 2x + 18$

39. The voltage in an electrical system is the resistance ($R$) times the current ($I$). Write an equation to show this.

40. Write an equation for the weight of machine part $A$, if the total weight of three machine parts equals the weight of part $A$ plus part $B$ plus part $C$.

# CHAPTER 4

# Fractional Equations and Formulas

Because workers in science, industry, and technology frequently need to use formulas, this chapter teaches you how to evaluate a formula and solve for a specific unknown. The topics covered in this chapter include fractional equations and formula rearrangement.

## Section 4-1 Fractional Equations and Proportions

**1**

A fractional equation is any equation that contains an algebraic fraction. All of the following are fractional equations:

$$\frac{x}{8} = 5 \qquad \frac{2x}{7} = 6 \qquad \frac{5}{7} = \frac{x}{14} \qquad \frac{x}{54} = \frac{8}{10}$$

Which of the following are fractional equations? (Circle your selections.)

**a.** $\frac{x}{7} = 3$ **b.** $3x + 5 = 7$ **c.** $y = 7x + 3$ **d.** $\frac{3}{4} = \frac{x}{20}$

**2**

Before solving fractional equations, it is necessary to discuss the numerical coefficient of an algebraic expression such as $x/5$. The expression $x/5$ can be written as

**1. a and d**

$$\left(\frac{1}{5}\right)\left(x\right)$$

Therefore, the numerical coefficient of $x/5$ is $1/5$.

**a.** The numerical coefficient of $\frac{x}{9}$ is ________.

**b.** The numerical coefficient of $\frac{h}{15}$ is ________.

**c.** The numerical coefficient of $\frac{2x}{3}$ is ________.

**3**

A fractional equation can be solved using the multiplication axiom.

Solve: $\frac{x}{3} = 15$

The coefficient of the $x$ term is $1/3$. Therefore, the reciprocal of the coefficient of the $x$ term is 3, so both sides of the equation are multiplied by 3.

$$\left(3\right)\left(\frac{x}{3}\right) = 3\left(15\right)$$

$$\left(\frac{3}{3}\right)x = 45$$

$$x = 45$$

Solve the following fractional equations.

**a.** $\frac{k}{5} = 15$ **b.** $\frac{h}{40} = 3.5$

**2. a.** $\frac{1}{9}$
**b.** $\frac{1}{15}$
**c.** $\frac{2}{3}$

**4**

The equation $3h/7 = 18$ is solved below.

$$\frac{3h}{7} = 18$$

$$\left(\frac{3}{7}\right)\left(h\right) = 18$$

$$\left(\frac{7}{3}\right)\left(\frac{3}{7}\right)\left(h\right) = \left(\frac{7}{3}\right)\left(18\right)$$

$$1\,(h) = \left(\frac{7}{\cancel{3}_{1}}\right)\left(\frac{\cancel{18}^{6}}{1}\right)$$

$$h = 42$$

Use the same steps to solve these equations.

**a.** $\frac{5x}{2} = 20$ **b.** $\frac{7k}{8} = 4.2$

**3. a.** 75
**b.** 140

**5**

Solve the following equations.

**a.** $\frac{x}{9} = 2$ **b.** $\frac{3x}{4} = 15$ **c.** $\frac{9m}{2} = 9$

**4. a.** 8
**b.** 4.8

**6**

The fractional equation shown below is a special type called a *proportion.* (Proportions will be discussed in detail in Chapter 5.)

$$\frac{x}{9} = \frac{2}{7}$$

This type of equation is so common that a special shortcut called *cross-multiplication* is used to solve it.

**5. a.** $x = 18$
**b.** $x = 20$
**c.** $m = 2$

To illustrate the advantage of cross-multiplication, the equation below is solved by two methods. The first method involves using the multiplication axiom; the second involves the cross-multiplication shortcut.

*Method 1:*
*Multiplication Axiom*

$$\frac{x}{9}=\frac{2}{7}$$

$$\left(7\right)\left(9\right)\left(\frac{x}{9}\right)=\left(7\right)\left(9\right)\left(\frac{2}{7}\right)$$

$$\left(7\right)\left(\frac{9}{9}\right)\left(x\right)=\left(\frac{7}{7}\right)\left(9\right)\left(2\right)$$

$$\left(7\right)\left(1\right)\left(x\right)=\left(1\right)\left(9\right)\left(2\right)$$

$$7x=18$$

$$\left(\frac{1}{7}\right)\left(7x\right)=\left(\frac{1}{7}\right)\left(18\right)$$

$$x=\frac{18}{7}$$

*Method 2:*
*Cross-Multiplication*

$$\frac{x}{9}=\frac{2}{7}$$

$$7x=\left(9\right)\left(2\right)$$

$$7x=18$$

$$\left(\frac{1}{7}\right)\left(7x\right)=\left(\frac{1}{7}\right)\left(18\right)$$

$$x=\frac{18}{7}$$

**7**

---

The cross-multiplication shortcut involves multiplying the numerator of one fraction by the denominator of the other fraction and setting that product equal to the product of the denominator of the first fraction times the numerator of the second. This is illustrated below.

**Examples**

$$\frac{2}{y}=\frac{7}{8} \qquad\qquad \frac{3}{2t}=\frac{4}{7}$$

$$\frac{2}{y} \times \frac{7}{8} \qquad\qquad \frac{3}{2t} \times \frac{4}{7}$$

$$(7)(y)=(2)(8) \qquad\qquad (4)(2t)=(3)(7)$$

$$7y=16 \qquad\qquad 8t=21$$

Remember to multiply first in the direction that gives you the variable (letter term) on the left side of the equation.

This method of cross-multiplication is called *clearing the fractions.* Use the same process to clear the fractions in the following equations.

**a.** $\frac{3}{h}=\frac{9}{5}$ **b.** $\frac{7}{2}=\frac{3x}{5}$

**8**

---

Clearing the fractions is the first step in solving a proportion. After the fractions are cleared, the equation is solved using the multiplication axiom. The proportion below is solved using the cross-multiplication shortcut. Use the same method to solve the proportion **b**.

**7. a.** $9h = 15$
**b.** $6x = 35$

**a.** $\frac{3}{k} = \frac{8}{7}$ **b.** $\frac{5}{x} = \frac{2}{3}$

clearing the fractions → $\frac{3}{k} \times \frac{8}{7}$

$8k = (3)(7)$

$8k = 21$

$k = \frac{21}{8}$

## 9

The proportion $6/5 = 2/3x$ is solved below. Use the same steps to solve the other proportions.

8. **b.** $x = \frac{15}{2}$

**a.** $\frac{6}{5} = \frac{2}{3x}$ **b.** $\frac{10}{7} = \frac{4}{5x}$ **c.** $\frac{9}{10} = \frac{3}{2h}$

$(6)(3x) = (5)(2)$

$18x = 10$

$x = \frac{10}{18}$

$x = \frac{5}{9}$

*Note:* The answer $\frac{10}{18}$ is reduced to $\frac{5}{9}$.

## 10

Solve the following proportions.

9. **b.** $x = \frac{14}{25}$ $\left(\text{reduced from } \frac{28}{50}\right)$

**c.** $h = \frac{5}{3}$ $\left(\text{reduced from } \frac{30}{18}\right)$

**a.** $\frac{9}{4} = \frac{x}{6}$ **b.** $\frac{3h}{10} = \frac{1}{15}$ **c.** $\frac{1}{2k} = \frac{5}{9}$

## 11

The calculator can be used to solve proportions and is especially useful if the proportion contains decimal numbers. The calculator is used to solve a proportion as follows:

10. **a.** $x = \frac{27}{2}$

**b.** $h = \frac{2}{9}$

**c.** $k = \frac{9}{10}$

**a.** First solve the proportion for $x$.

$$\frac{3.80}{x} = \frac{5.28}{17.15}$$

$$(5.28)\,x = (3.80)(17.15)$$

$$x = \frac{(3.80)(17.15)}{(5.28)}$$

**b.** Then calculate the answer using the calculator.

| *Enter* | *Press* | *Display* |
|---|---|---|
| 3.80 | ⊠ | 3.80 |
| 17.15 | ÷ | 65.17 |
| 5.28 | = | 12.342803 |

The answer is then rounded to 12.3.

**12**

Use a calculator to solve the following proportions. (Assume that these are measurements, and round to the correct number of significant digits.)

**a.** $\frac{7.5}{h} = \frac{5.7}{28.2}$ **b.** $\frac{15.28}{4.15} = \frac{k}{42.67}$

**13**

Answers to frame 12.

**a.** 37 **b.** 157

# Exercise Set, Section 4-1

## Fractional Equations and Proportions

Solve the following fractional equations and proportions.

**1.** $\frac{x}{3} = 5$ **2.** $\frac{k}{2} = 11$ **3.** $\frac{y}{4} = -8$ **4.** $\frac{x}{-3} = -4$

**5.** $\frac{m}{3.9} = 1.2$ **6.** $\frac{k}{-1.6} = 2.6$ **7.** $\frac{x}{-4.2} = -3.6$ **8.** $\frac{2h}{5} = 10$

**9.** $\frac{3y}{5} = 9$ **10.** $\frac{4m}{7} = 8$ **11.** $\frac{2.2x}{4} = 4.4$ **12.** $\frac{1.1y}{3} = 3.3$

**13.** $\frac{x}{2} = \frac{5}{4}$ **14.** $\frac{1}{h} = \frac{7}{10}$ **15.** $\frac{3}{4} = \frac{k}{8}$ **16.** $\frac{11}{5} = \frac{7}{y}$

**17.** $\frac{4}{y} = \frac{7}{8}$ **18.** $\frac{5}{2t} = \frac{6}{7}$ **19.** $\frac{5}{x} = \frac{9}{2}$ **20.** $\frac{2}{7} = \frac{4}{5k}$

**21.** $\frac{4}{10x} = \frac{4}{5}$ **22.** $\frac{2}{3} = \frac{8}{6x}$ **23.** $\frac{3}{8} = \frac{2}{6m}$ **24.** $\frac{3}{7} = \frac{3}{7x}$

**25.** $\frac{4.8}{h} = \frac{2.0}{7.5}$ **26.** $\frac{9.2}{3.4} = \frac{0.1}{t}$ **27.** $\frac{v}{20} = \frac{2.5}{1.9}$ **28.** $\frac{200}{350} = \frac{v}{400}$

**29.** $\frac{k}{0.02} = \frac{50}{20}$ **30.** $\frac{151}{h} = \frac{405}{218}$ **31.** $\frac{75}{28} = \frac{t}{19}$ **32.** $\frac{3.85}{2.81} = \frac{7.25}{3t}$

# Supplementary Exercise Set, Section 4-1

Solve the following fractional equations and proportions.

**1.** $\frac{x}{5} = 1.5$ **2.** $\frac{k}{7} = 1$ **3.** $\frac{m}{2.8} = -7.1$ **4.** $\frac{2h}{7.1} = -5$

**5.** $\frac{x}{2} = \frac{5}{4}$ **6.** $\frac{3}{m} = \frac{6}{17}$ **7.** $\frac{3}{5} = \frac{h}{6}$ **8.** $\frac{9}{7} = \frac{3}{w}$

**9.** $\frac{2w}{10} = \frac{5}{3}$ **10.** $\frac{x}{2.5} = \frac{3.5}{5}$ **11.** $\frac{1.7}{9} = \frac{m}{3.6}$ **12.** $\frac{20.2}{h} = \frac{5.2}{7.0}$

**13.** $\frac{85}{100} = \frac{n}{28}$ **14.** $\frac{14}{100} = \frac{12}{n}$ **15.** $\frac{7}{100} = \frac{n}{50.3}$ **16.** $\frac{n}{100} = \frac{45}{62}$

**17.** $\frac{k}{0.03} = \frac{4.2}{1.5}$ **18.** $\frac{3.8}{h} = \frac{380}{51}$ **19.** $\frac{8}{15} = \frac{n}{4.2}$ **20.** $\frac{7}{19} = \frac{147}{x}$

# Section 4-2 Introduction to Formulas

**1**

A formula is a mathematical shorthand for describing a relationship between numbers. A typical formula is the Celsius-to-Fahrenheit conversion formula shown below.

$$°F = 1.8°C + 32$$

This is a shorthand way of stating the relationship between degrees Celsius and degrees Fahrenheit.

We can convert 20 degrees Celsius to degrees Fahrenheit by substituting into the formula and evaluating.

Find °F when °C = 20

$$°F = 1.8(20) + 32$$

$$°F = 36 + 32$$

$$°F = 68$$

This process is called *formula evaluation.*

## 2

Some common formulas from science, industry, and the technologies are shown below.

Dilution Formula
$C_1V_1 = C_2V_2$

Boyle's Gas Law
$P_1V_1 = P_2V_2$

Charles' Gas Law
$$\frac{V_1}{T_1} = \frac{V_2}{T_2}$$

General Gas Law
$$\frac{P_1V_1}{T_1} = \frac{P_2V_2}{T_2}$$

Hydraulic Pressure
$$P = \frac{F}{A}$$

S.A.E. Horsepower
$$HP = \frac{hd^2}{2.5}$$

Weight of a Plastic Charge
$$W = 0.036sv$$

Specific Gravity
$$S = \frac{W_a}{W_a - W_w}$$

Flexural Strength
$$S = \frac{3pl}{2bd^2}$$

Power
$P = I^2R$

Force
$$F = ma$$

Energy
$$E = \frac{mv^2}{2} + mgh$$

Cardiac Output
$$Q = \frac{V}{Ca - Cv}$$

Resistance
$$\frac{1}{R_t} = \frac{1}{R_1} + \frac{1}{R_2}$$

Ideal-Gas Equation
$PV = nRT$

Celsius-to-Kelvin Conversion
$K = {}^\circ C + 273$

Fahrenheit-to-Celsius Conversion
$${}^\circ C = \frac{({}^\circ F - 32)}{1.8}$$

Celsius-to-Fahrenheit Conversion
$${}^\circ F = 1.8{}^\circ C + 32$$

Molding Pressure
$$P = \frac{2{,}500 \text{ psi} \times A}{2{,}000 \text{ lb}}$$

Brake Horsepower
$$BHP = \frac{DPN}{792{,}000}$$

Volatile Content in Plastic
$$V = \frac{W_o - W_d}{W_0}$$

Tensile Strength
$$T = \frac{\ell}{A}$$

Melting Rate in Welding
$$MR = aI + bLI^2$$

Potential Energy
$PE = mg(h_2 - h_1)$

Kinetic Energy
$$KE = \frac{mv^2}{2}$$

Absorption Spectroscopy
$$a = \frac{A}{bc}$$

Magnetic Induction
$$B = \frac{mi}{6.28d}$$

Slope-Intercept Form of a Straight Line
$$y = mx + b$$

**3**

---

Some of the formulas in the preceding frame involved subscripts. A subscript is a small number or letter written slightly below and to the right of a letter in a formula. For example, in the term $X_k$, $k$ is a subscript. In the term $V_1$, 1 is a subscript. A subscript is used to distinguish one variable (letter) from another in a formula. Therefore, $V_1$ and $V_2$ represent different numbers, just as $x$ and $y$ represent different numbers.

**4**

---

A subscript should not be confused with an exponent. An exponent is a number written to the right and slightly *above* a number or letter. The exponent indicates how many times that number or letter should be multiplied times itself.

**Example** $5^2$ means $5 \cdot 5 = 25$

$M^2 = M \cdot M$

$2^3 = 2 \cdot 2 \cdot 2 = 8$

$k^3 = k \cdot k \cdot k$

Indicate which of the following are examples of exponents and which are examples of subscripts.

**a.** $X^3$ ______ **b.** $V_2$ ______ **c.** $P^2$ ______

**d.** $P_2$ ______ **e.** $b_H$ ______ **f.** $A^k$ ______

**5**

---

When an exponent occurs in a formula, it is evaluated as follows:

**Example** Evaluate $X^3$ when $X = 5$.

$X^3 = 5^3 = 5 \cdot 5 \cdot 5 = 125$

Complete the following.

**a.** Evaluate $Y^2$ when $Y = 7$.
$Y^2 =$

**b.** Evaluate $V^3$ when $V = 4$.
$V^3 =$

**4. a.** exponent
**b.** subscript
**c.** exponent
**d.** subscript
**e.** subscript
**f.** exponent

**6**

---

Answers to frame 5.

**a.** $Y^2 = 49$ **b.** $V^3 = 64$

# Section 4-3 Formula Evaluation

**1**

---

Formulas of the type $A = BX + C$ are quite common in technical areas. A formula of this type can be evaluated as follows:

**Example** Using the formula $A = BX + C$, find $A$ when $B = 5$, $C = 3$, and $X = 2$.

$A = 5(2) + 3$
$A = 10 + 3$
$A = 13$

**a.** Using $y = mx + b$, find $y$ when $m = 2$, $x = 3$, and $b = 5$.

**b.** Using $°F = 1.8°C + 32$, find $°F$ when $°C = 30$.

## 2

Formulas of the type $A = BX + C$ can be evaluated using a calculator. This practice is especially helpful when the formula contains decimal numbers.

1. a. 11
   b. 86

**Example** In the formula $y = mx + b$, the calculator is used to find $y$ when $m = 0.76$, $x = 2.05$, and $b = 4.215$.

$y = (0.76)(2.05) + 4.215$

| *Enter* | *Press* | *Display* |
|---|---|---|
| 0.76 | [×] | 0.76 |
| 2.05 | [+] | 1.558 |
| 4.215 | [=] | 5.773 |

Therefore, $y = 5.8$ (correct significant digits).

Use a calculator to evaluate the following formulas:

a. Using $°F = 1.8°C + 32$, find °F when $°C = 22.5$.
b. Using $K = °C + 273$, find $K$ when $°C = 28.42$.
c. Using $y = mx + b$, find $y$ when $m = 2.50$, $x = 15.0$, and $b = 4.28$.
d. Using $°F = 1.8°C + 32$, find °F when $°C = -10$.

## 3

Formulas involving divisions are also used in science and industry. This type of formula can be evaluated as follows:

2. a. 72.5
   b. 301.42
   c. 41.78
   d. 14

In the formula $a = \frac{A}{bc}$, find $a$ when $A = 20$, $b = 2$, and $c = 5$.

$$a = \frac{20}{(2)(5)}$$

$$a = \frac{20}{10}$$

$$a = 2$$

a. Using $X = \frac{A}{CD}$, find $X$ when $A = 6$, $C = 3$, and $D = 8$.

b. Using $Y_1 = \frac{X}{2Y_2}$, find $Y_1$ when $X = 10$ and $Y_2 = 5$.

## 4

Formulas involving divisions can be evaluated using a calculator.

3. a. 0.25
   b. 1

Given the formula $A = \frac{a}{bc}$, the calculator can be used to find $A$ when $a = 0.165$, $b = 2.0$, and $c = 0.75$.

$$A = \frac{0.165}{(2)(0.75)}$$

| *Enter* | *Press* | *Display* |
|---|---|---|
| 0.165 | [÷] | 0.165 |
| 2.0 | [÷] | 0.0825 |
| 0.75 | [=] | 0.11 |

Therefore, $A = 0.11$.

**a.** Using $C = \frac{A}{ab}$, find $C$ when $A = 0.141$, $a = 0.342$, and $b = 1.00$.

**b.** Using $T = \frac{P}{P_o}$, find $T$ when $P = 3.075$, and $P_o = 2.05$.

## 5

The formula below requires both a subtraction and a division.

**4. a.** 0.412
**b.** 1.50

Given $°C = \frac{(°F - 32)}{1.8}$, find °C when °F = 50.

(*Note:* Always do the operations in parentheses first.)

$$°C = \frac{(50 - 32)}{1.8}$$

$$°C = \frac{18}{1.8}$$

$$°C = 10$$

This formula can be evaluated on the calculator as follows:

$$°C = \frac{(50 - 32)}{1.8}$$

| *Enter* | *Press* | *Display* |
|---|---|---|
| 50 | [−] | 50 |
| 32 | [=] [÷] | 18 |
| 1.8 | [=] | 10 |

$°C = 10$

**a.** Using $°C = \frac{(°F - 32)}{1.8}$, find °C when °F = 78.

**b.** Using $Z = \frac{(x - m)}{s}$, find $Z$ when $x = 5.2$, $m = 4.8$, and $s = 2$.

## 6

The following formula must be evaluated by performing the operations in parentheses first.

**5. a.** 26
**b.** 0.2

**Example** Given $PE = mg(h_2 - h_1)$, find $PE$ when $m = 5.00$, $g = 32.0$, $h_2 = 6.50$, and $h_1 = 3.10$.

$$PE = (5.00)(32.0)(6.50 - 3.10)$$

$$PE = (5.00)(32.0)(3.40)$$

$$PE = 544$$

**a.** Using $x = ab(z_1 + z_2)$, find $x$ when $a = 3.0$, $b = 4.0$, $z_1 = 3.8$, and $z_2 = 1.2$.
**b.** Using $PE = mg(h_2 - h_1)$, find $PE$ when $m = 7.00$, $g = 32.0$, $h_2 = 9.70$, and $h_1 = 8.20$.

**7**

The following example shows a formula that requires squaring a number.

6. a. 60
b. 336

**Example** Given $KE = \frac{1}{2}mv^2$, find $KE$ when $m = 4$ and $v = 36$.

$$KE = \frac{1}{2}(4)(36)^2$$

$$KE = \frac{1}{2}(4)(36)(36)$$

$$KE = 2{,}592$$

**a.** Using $P = I^2R$, find $P$ when $I = 10.0$ and $R = 7.00$

**b.** Using $E = \frac{mv^2}{2} + mgh$, find $E$ when $m = 4.00$, $v = 8.00$, $g = 2.00$, and $h = 5.00$.

**8**

Answers to frame 7.

**a.** $P = 700$ **b.** $E = 168$

# Exercise Set, Sections 4-2–4-3

## Formula Evaluation

Evaluate the following formulas (a calculator may be used).

1. Given K = °C + 273, find K when °C = 40.
2. Given °F = 1.8°C + 32, find °F when °C = 20.
3. Given °F = 1.8°C + 32, find °F when °C = 24.5.
4. Given °F = 1.8°C + 32, find °F when °C = −15.5.
5. Given $y = mx + b$, find $y$ when $m = 0.50$, $x = 4.0$, and $b = 3.2$.
6. Given $y = mx + b$, find $y$ when $m = 0.60$, $x = 2.1$, and $b = -4.0$.
7. Given $°C = \frac{(°F - 32)}{1.8}$, find °C when °F = 68.
8. Given $°C = \frac{(°F - 32)}{1.8}$, find °C when °F = 14.
9. Given $°C = \frac{(°F - 32)}{1.8}$, find °C when °F = 50.
10. Given $°C = \frac{(°F - 32)}{1.8}$, find °C when °F = −22.
11. Given $V_1 = \frac{T_1V_2}{T_2}$, find $V_1$ when $T_1 = 200$, $T_2 = 100$, and $V_2 = 40$.
12. Given $P_1 = \frac{P_2V_2}{V_1}$, find $P_1$ when $P_2 = 4$, $V_1 = 8$, and $V_2 = 10$.
13. Given $V_1 = \frac{T_1V_2}{T_2}$, find $V_1$ when $T_1 = 300$, $T_2 = 200$, and $V_2 = 120$.
14. Given $P_1 = \frac{P_2V_2}{V_1}$, find $P_1$ when $P_2 = 15$, $V_1 = 3.00$, and $V_2 = 30$.

**15.** Given $P_1 = \frac{P_2 V_2}{V_1}$, find $P_1$ when $P_2 = 250$, $V_1 = 15.0$, and $V_2 = 50.0$.

**16.** Given $V_1 = \frac{T_1 V_2}{T_2}$, find $V_1$ when $T_1 = 7.50$, $T_2 = 10.0$, and $V_2 = 25.0$.

**17.** Given $a = \frac{A}{bc}$, find $a$ when $A = 72$, $b = 3.2$, and $c = 4.5$.

**18.** Given $V = \frac{nRT}{P}$, find $V$ when $n = 2.00$, $R = 0.0800$, $T = 273$, and $P = 1.00$.

**19.** Given $PE = mg(h_2 - h_1)$, find $PE$ when $m = 7.20$, $g = 30.0$, $h_2 = 4.70$, and $h_1 = 1.70$.

**20.** Given $KE = \frac{1}{2}mv^2$, find $KE$ when $m = 7.20$ and $v = 2.10$.

**21.** Given $P = I^2R$, find $P$ when $I = 1.5$ and $R = 4.2$.

**22.** Given $E = \frac{1}{2}mv^2 + mgh$, find $E$ when $m = 4.20$, $v = 7.10$, $g = 32.0$, and $h = 9.10$.

**23.** Given $F = ma$, find $F$ when $m = 5.28$ and $a = 14.0$.

**24.** Given $B = \frac{mi}{6.28d}$, find $B$ when $m = 21.0$, $i = 31.4$, and $d = 7.00$.

**25.** Given $A = \frac{1}{2}bh$, find $A$ when $b = 54.8$ and $h = 13.0$.

**26.** Given $HP = \frac{nd^2}{2.5}$, find $HP$ when $n = 4.00$ and $d = 8.00$.

**27.** Given $P = \frac{2500A}{2000}$, find $P$ when $A = 28$.

**28.** Given $V = \frac{W_0 - W_d}{W_0}$, find $V$ when $W_o = 4.2$ and $W_d = 2.4$.

**29.** Given $S = \frac{W_a}{W_a - W_w}$, find $S$ when $W_a = 0.52$ and $W_w = 0.48$.

**30.** Given $S = \frac{3p\ell}{2bd^2}$, find $S$ when $b = 5.0$, $d = 2.5$, $p = 17.0$, and $\ell = 6.2$.

## Supplementary Exercise Set, Sections 4-2–4-3

Evaluate the following formulas. A calculator may be used.

**1.** Given $P = \frac{F}{A}$, find $P$ when $F = 45$ and $A = 15$.

**2.** Given $T = \frac{\ell}{A}$, find $T$ when $\ell = 25.0$ and $A = 1.25$.

**3.** Given $S = \frac{W_a}{W_a - W_w}$, find $S$ when $W_a = 6.8$ and $W_w = 7.2$.

**4.** Given $V = \frac{W_0 - W_d}{W_0}$, find $V$ when $W_0 = 18.2$ and $W_d = 16.8$.

**5.** Given $HP = \frac{nd^2}{2.5}$, find $HP$ when $n = 8.0$ and $d = 15.0$.

**6.** Given K = °C + 273, find K when °C = −20.

**7.** Given $Bhp = \frac{DPN}{792{,}000}$, find $Bhp$ when $D = 500$, $P = 2{,}000$, and $N = 80$.

**8.** Given $V_1 = \frac{T_1 V_2}{T_2}$, find $V_1$ when $T_1 = 52$, $T_2 = 2.6$, and $V_2 = 39$.

**9.** Given $W = 0.036sv$, find $W$ when $s = 400$ and $v = 350$.

**10.** Given $F = ma$, find $F$ when $m = 35.2$ and $a = 28.4$.

**11.** Given $P = I^2R$, find $P$ when $I = 7.5$ and $R = 14.0$.

**12.** Given $P_1 = \frac{P_2 V_2}{V_1}$, find $P_1$ when $V_1 = 15$, $V_2 = 18$, and $P_2 = 985$.

# Section 4-4 Formula Rearrangement

**1**

The following type of formula is used in science to calculate the concentration of a solution.

$$C_1V_1 = C_2V_2$$

This type of formula first has to be rearranged before it can be evaluated. Notice that there are four variables: $C_1$, $C_2$, $V_1$, and $V_2$. Each of these variables can be solved for by using formula rearrangement. Formula rearrangement uses the same principles as equation solving. This is illustrated below.

**Example 1** *Equation solving:* Solve for $X$.

$$5X = 20$$

$$\left(\frac{1}{5}\right)\left(5X\right) = \left(\frac{1}{5}\right)\left(20\right)$$

$$\left(\frac{5}{5}\right)\left(X\right) = \frac{20}{5}$$

$$X = 4$$

**Example 2** *Formula rearrangement:* Solve for $V_1$.

$$C_1V_1 = C_2V_2$$

$$\left(\frac{1}{C_1}\right)\left(C_1V_1\right) = \left(\frac{1}{C_1}\right)\left(C_2V_2\right)$$

$$\left(\frac{C_1}{C_1}\right)\left(V_1\right) = \frac{C_2V_2}{C_1}$$

$$V_1 = \frac{C_2V_2}{C_1}$$

Multiply both sides of the equation by the reciprocal of the coefficient of the letter being solved for.

Given $C_1V_1 = C_2V_2$, solve for:

**a.** $C_1$

**b.** $C_2$

## 2

Answers to frame 1.

**a.**

$$C_1V_1 = C_2V_2$$

$$\left(\frac{1}{V_1}\right)\left(C_1V_1\right) = \left(\frac{1}{V_1}\right)\left(C_2V_2\right)$$

$$\left(\frac{V_1}{V_1}\right)\left(C_1\right) = \frac{C_2V_2}{V_1}$$

$$C_1 = \frac{C_2V_2}{V_1}$$

**b.**

$$C_1V_1 = C_2V_2$$

$$\left(\frac{1}{V_2}\right)\left(C_1V_1\right) = \left(C_2V_2\right)\left(\frac{1}{V_2}\right)$$

$$\frac{C_1V_1}{V_2} = \left(C_2\right)\left(\frac{V_2}{V_2}\right)$$

$$\frac{C_1V_1}{V_2} = C_2$$

## 3

Given the formula $P_1V_1 = P_2V_2$, solve for:

**a.** $P_1$ **b.** $P_2$

**c.** $V_1$ **d.** $V_2$

## 4

Formulas of this type are also very common in science. Solve for $T_2$ given

$$\frac{V_1}{T_1} = \frac{V_2}{T_2}$$

The first step in solving for any one of the variables involves applying the cross-multiplication shortcut to clear the fraction as follows:

Solve for $T_2$: $\dfrac{V_1}{T_1} = \dfrac{V_2}{T_2}$

$$\frac{V_1}{T_1} \times \frac{V_2}{T_2}$$

$V_1T_2 = V_2T_1$ ← The formula is now of the type solved in the previous frame and can be solved accordingly.

$$\left(\frac{1}{V_1}\right)\left(V_1T_2\right) = \left(\frac{1}{V_1}\right)\left(V_2T_1\right)$$

$$\left(\frac{V_1}{V_1}\right)T_2 = \frac{V_2T_1}{V_1}$$

$$T_2 = \frac{V_2T_1}{V_1}$$

**a.** Given $\dfrac{V_1}{T_1} = \dfrac{V_2}{T_2}$, solve for $V_1$. **b.** Given $\dfrac{V_1}{T_1} = \dfrac{V_2}{T_2}$, solve for $V_2$.

## 5

Given $\dfrac{P_1V_1}{T_1} = \dfrac{P_2V_2}{T_2}$, solve for:

**a.** $P_1$ **b.** $V_1$

**c.** $T_2$ **d.** $P_2$

3. **a.** $P_1 = \dfrac{P_2V_2}{V_1}$

**b.** $P_2 = \dfrac{P_1V_1}{V_2}$

**c.** $V_1 = \dfrac{P_2V_2}{P_1}$

**d.** $V_2 = \dfrac{P_1V_1}{P_2}$

4. **a.** $V_1 = \dfrac{T_1V_2}{T_2}$

**b.** $V_2 = \dfrac{T_2V_1}{T_1}$

## 6

Answers to frame 5.

a. $P_1 = \frac{P_2V_2T_1}{V_1T_2}$

b. $V_1 = \frac{P_2V_2T_1}{P_1T_2}$

c. $T_2 = \frac{P_2V_2T_1}{P_1V_1}$

d. $P_2 = \frac{P_1V_1T_2}{T_1V_2}$

## 7

This type of formula can also be rearranged using the techniques of equation solving.

**Example 1** *Equation solving:*

Solve for $Y$.

$$5Y + 3 = 18$$
$$5Y + 3 + (-3) = 18 + (-3)$$
$$5Y + 0 = 15$$
$$5Y = 15$$
$$\left(\frac{1}{5}\right)5Y = \left(\frac{1}{5}\right)15$$
$$\frac{5}{5}Y = \frac{15}{5}$$
$$Y = 3$$

**Example 2** *Formula rearrangement:*

Solve for $x$.

$$mx + b = Y$$
$$mx + b + (-b) = Y + (-b)$$
$$mx + 0 = Y - b$$
$$mx = Y - b$$
$$\frac{1}{m}mx = \frac{1}{m}(Y - b)$$
$$\frac{m}{m}x = \frac{Y - b}{m}$$
$$x = \frac{Y - b}{m}$$

a. Given $AH + W = K$, solve for $H$.

b. Given $aw - r = z$, solve for $w$.

## 8

Formulas of the following type can be rearranged as shown.

**Example 1** *Equation solving:*

Solve for $h$.

$$5h + 7 = 2h + 22$$
$$5h + 7 + (-2h) = 2h + 22 + (-2h)$$
$$3h + 7 = 22$$
$$3h + 7 + (-7) = 22 + (-7)$$
$$3h + 0 = 15$$
$$\frac{1}{3}(3h) = \frac{1}{3}(15)$$
$$\frac{3}{3}h = \frac{15}{3}$$
$$h = 5$$

**Example 2** *Formula rearrangement:*

Solve for $w$.

$$aw + h = bw + k$$
$$aw + h + (-bw) = bw + k + (-bw)$$
$$(a - b)w + h = k$$
$$(a - b)w + h + (-h) = k + (-h)$$
$$(a - b)w + 0 = k - h$$
$$(a - b)w = k - h$$
$$\frac{1}{a - b}(a - b)w = \frac{1}{a - b}(k - h)$$
$$\frac{a - b}{a - b}w = \frac{k - h}{a - b}$$
$$w = \frac{k - h}{a - b}$$

7. a. $H = \frac{K - W}{A}$

b. $w = \frac{z + r}{a}$

**a.** Given $am + b = cm + d$, solve for $m$.

**b.** Given $ax + y = hx - z$, solve for $x$.

## 9

Many formula rearrangements require simplifying by "clearing the fraction" as follows:

8. a. $m = \dfrac{d - b}{a - c}$

b. $x = \dfrac{-z - y}{a - h}$

**Example** Solve for $m$.

$$KE = \frac{mv^2}{2} \qquad \text{original equation}$$

$$2(KE) = 2\frac{mv^2}{2} \qquad \text{simplify (clear the fraction by multiplying both sides by the denominator)}$$

$$2KE = \frac{2}{2}(mv^2)$$

$$2KE = mv^2 \qquad \frac{n}{n} = 1$$

$$\frac{1}{v^2}(2KE) = \frac{1}{v^2}(mv^2) \qquad \text{multiplication axiom}$$

$$\frac{2KE}{v^2} = \frac{v^2}{v^2}m \qquad \frac{n}{n} = 1$$

$$\frac{2KE}{v^2} = m$$

**a.** Given $b = \dfrac{AH}{5}$, solve for $H$.

**b.** Given $h = \dfrac{RK}{m}$, solve for $R$.

## 10

Answers to frame 9.

**a.** $H = \dfrac{5b}{A}$

**b.** $R = \dfrac{hm}{K}$

# Exercise Set, Section 4-4

## Formula Rearrangement

Solve the following formulas for the indicated variable.

1. Given $P_1V_1 = P_2V_2$, find $P_1$.
2. Given $P_1V_1 = P_2V_2$, find $V_2$.
3. Given $\dfrac{V_1}{T_1} = \dfrac{V_2}{T_2}$, find $V_2$.
4. Given $\dfrac{V_1}{T_1} = \dfrac{V_2}{T_2}$, find $T_1$.
5. Given $\dfrac{P_1V_1}{T_1} = \dfrac{P_2V_2}{T_2}$, find $P_2$.
6. Given $\dfrac{P_1V_1}{T_1} = \dfrac{P_2V_2}{T_2}$, find $T_1$.
7. Given $PV = nRT$, find $n$.
8. Given $PV = nRT$, find $T$.

**9.** Given $Y = mx + b$, find $x$.

**10.** Given $kh - r = m$, find $h$.

**11.** Given $af + h = bf - k$, find $f$.

**12.** Given $hx + y = kx + z$, find $x$.

**13.** Given $\frac{47h}{k} = b$, find $h$.

**14.** Given $w = \frac{Rk}{n}$, find $R$.

**15.** Given $z = \frac{x - m}{s}$, find $x$.

## Supplementary Exercise Set, Section 4-4

Solve the following formulas for the indicated variable.

**1.** Given $P_1V_1 = P_2V_2$, find $P_2$.

**2.** Given $\frac{V_1}{T_1} = \frac{V_2}{T_2}$, find $T_2$.

**3.** Given $\frac{P_1V_1}{T_1} = \frac{P_2V_2}{T_2}$, find $V_2$.

**4.** Given $PV = nRT$, find $R$.

**5.** Given $V = \frac{s}{t}$, find $s$.

**6.** Given $y = mx + b$, find $m$.

**7.** Given $I = BRT$, find $T$.

**8.** Given $E = IR$, find $I$.

**9.** Given $Z = \frac{x - m}{s}$, find $m$.

**10.** Given $E = IR$, find $R$.

# Section 4-5 Strategies for Rearranging Formulas

**1**

An efficient strategy for rearranging formulas involves these steps.

*Step 1.* Simplify each side of the formula. (This process will include clearing the fractions by multiplying both sides by any denominator.)
*Step 2.* Use the addition axiom to get the "solved-for" variables together on one side.
*Step 3.* Use the addition axiom to get the terms containing the other variables together on the other side.
*Step 4.* Factor out the solved-for variable if it occurs in more than one term.
*Step 5.* Use the multiplication axiom to get a coefficient of 1 for the solved-for variable.

Each of these steps can be used to rearrange any formula so that any solved-for variable occurs alone on one side of the formula.

**2**

The strategy in the previous frame is used to rearrange the following formula.

**Example** Solve for $m$.

$$E = \frac{mv^2}{2} + mgh$$

$$2(E) = 2\frac{mv^2}{2} + 2(mgh)$$

*Step 1.* Simplify. (This process can be done by clearing the fraction: multiply each term by the denominator, 2.)

$$2E = mv^2 + 2mgh$$

*Steps 2 and 3.* The solved-for variable, $m$, is on one side; therefore, the addition axiom is not used at this time.

$$2E = m(v^2 + 2gh)$$

*Step 4.* Factor out the solved-for variable ($m$).

$$\frac{1}{v^2 + 2gh}(2E) = \frac{1}{v^2 + 2gh}m(v^2 + 2gh)$$

*Step 5.* Use the multiplication axiom to get a coefficient of 1 for $m$. This means multiplying both sides by $\frac{1}{(v^2 + 2gh)}$.

$$\frac{2E}{v^2 + 2gh} = \frac{v^2 + 2gh}{v^2 + 2gh}m$$

$$\frac{2E}{v^2 + 2gh} = m$$

**Example** Solve for $V_2$.

$$a = \frac{V_2 - V_1}{t_2 - t_1}$$

$$(t_2 - t_1)(a) = (t_2 - t_1)\frac{V_2 - V_1}{t_2 - t_1}$$

*Step 1.* Simplify by clearing the fraction (multiply each term by $t_2 - t_1$).

$$a(t_2 - t_1) = \frac{t_2 - t_1}{t_2 - t_1}(V_2 - V_1)$$

$$a(t_2 - t_1) = V_2 - V_1$$

$$a(t_2 - t_1) + V_1 = V_2 - V_1 + V_1$$

*Step 2.* Use the addition axiom to get the solved-for variable alone on one side. (Add $V_1$ to both sides.)

$$a(t_2 - t_1) + V_1 = V_2$$

Simplify.

Given the formula

$$\frac{E_n - E_f}{E_f} = V$$

**a.** Solve for $E_n$.

**b.** Solve for $E_f$.

**3**

Rearrange these formulas, and solve for the required variable.

**a.** Given $V = V_o + at$, solve for $t$.

**b.** Given $E = \frac{100P}{I}$, solve for $I$.

**c.** Given $K = 0.5mv^2$, solve for $m$.

**d.** Given $P = \frac{\pi D}{N}$, solve for $D$.

**2. a.** $E_n = VE_f + E_f$
$E_n = E_f(V + 1)$

**b.** $E_f = \frac{E_n}{V + 1}$

**4**

Answers to frame 3.

**a.** $V - V_o = at,\ t = \frac{V - V_o}{a}$

**b.** $IE = 100P,\ I = \frac{100P}{E}$

**c.** $\frac{1}{0.5v^2} K = \frac{1}{0.5v^2} m;\ m = \frac{K}{0.5v^2}$

**d.** $D = \frac{PN}{\pi}$

# Exercise Set, Section 4-5

## Strategies for Rearranging Formulas

Solve the following formulas for the indicated variables.

**1.** Given $P = I^2R$, find $R$.

**2.** Given $n = \frac{T_2 - T_1}{T_1}$, find $T_2$.

**3.** Given $F_c = \frac{Wv^2}{gr}$, find $g$.

**4.** Given $T = \frac{4\pi^2 m}{K}$, find $m$.

**5.** Given $A = \pi r^2 + 2\pi rh$, find $h$.

**6.** Given $F = \frac{KQ_1Q_2}{d_2}$, find $K$.

**7.** Given $P = \frac{WG}{550t}$, find $G$.

**8.** Given $V = f\lambda$, find $\lambda$.

**9.** Given $B = \frac{mi}{2\pi d}$, find $i$.

**10.** Given $P = 2L + 2W$, find $W$.

# Supplementary Exercise Set, Section 4-5

Solve the following formulas for the indicated variable.

**1.** Given $E = \frac{mv^2}{2} + mgh$, find $h$.

**2.** Given $V = V_o rat$, find $a$.

**3.** Given $E = \frac{100P}{I}$, find $P$.

**4.** Given $P = \frac{\pi D}{N}$, find $N$.

**5.** Given $F_c = \dfrac{Wv^2}{gr}$, find $r$.

**6.** Given $T = \dfrac{4\pi^2 m}{K}$, find $K$.

**7.** Given $A = \dfrac{bh}{2}$, find $b$.

**8.** Given $P = 2L + 2W$, find $L$.

**9.** Given $MA = \dfrac{F_e}{F_s}$, find $F_s$.

**10.** Given $V = lwh$, find $w$.

# Section 4-6 Applied Problems

## 1

Charles' Law, describing the effect of temperature on gases, is an excellent example of an applied situation that requires formula rearrangement, formula evaluation, and dimensional analysis.

Charles' Law states that the volume of a gas is directly proportional to the temperature at constant pressure. This means that, when the temperature increases, the volume increases. This is best expressed with the following formula:

$$\frac{V_1}{T_1} = \frac{V_2}{T_2}$$

where:

$T_1$ = initial temperature of the gas (in degrees Kelvin, or K)
$V_1$ = initial volume of the gas (in milliliters or liters)
$T_2$ = new temperature of the gas (K)
$V_2$ = new volume of the gas (in milliliters or liters)

In the formula for Charles' Law, the new volume is labeled ____________.

## 2

Formula rearrangement can be used to solve the Charles' Law formula for any of the four variables. Generally, the formula is solved for the new volume. This is shown in problem **a.** Use the same procedure to solve for $T_2$ in problem **b.**

**1.** $V_2$

**a.** Solve for $V_2$.

$$\frac{V_1}{T_1} = \frac{V_2}{T_2}$$

$$\frac{V_1}{T_1} \times \frac{V_2}{T_2}$$

$$V_2 T_1 = V_1 T_2$$

$$\left(\frac{1}{T_1}\right)\left(V_2 T_1\right) = \left(\frac{1}{T_1}\right)\left(V_1 T_2\right)$$

$$V_2 = \frac{V_1 T_2}{T_1}$$

**b.** Solve for $T_2$.

$$\frac{V_1}{T_1} = \frac{V_2}{T_2}$$

## 3

Applied problems using Charles' Law can be solved using the five steps of problem solving discussed earlier.

2. b. $T_2 = \frac{V_2 T_1}{V_1}$

*Step 1.* Determine what is being asked for.
*Step 2.* Determine what information is already known.
*Step 3.* Find a mathematical model that describes the relationship.
*Step 4.* Substitute the data into the model.
*Step 5.* Do the calculations. (A calculator may be used.)

## 4

The five steps of problem solving can be used to solve the following Charles' Law problem.

At 300 K, a gas has a volume of 200 liters. What is the volume of this gas at 400 K?

*Step 1.* Determine what is being asked for.

$$V_2 = ?$$

*Step 2.* Determine what information is already known.

$$T_1 = 300 \text{ K}$$
$$T_2 = 400 \text{ K}$$
$$V_1 = 200 \ \ell$$

*Step 3.* Find a mathematical model that describes the relationship. (Solve Charles' Law for $V_2$.)

$$\frac{V_1}{T_1} = \frac{V_2}{T_2}$$

$$\frac{V_1}{T_1} \times \frac{V_2}{T_2}$$

$$T_1 V_2 = V_1 T_2$$

$$V_2 = \frac{V_1 T_2}{T_1}$$

*Step 4.* Substitute the data into the model.

$$V_2 = \frac{(200 \ \ell)(400 \text{ K})}{(300 \text{ K})}$$

*Step 5.* Do the calculations. (Note the use of dimensional analysis.)

$$V_2 = \frac{(200 \ \ell)(\overset{4}{\cancel{400}} \cancel{\text{K}})}{(\underset{3}{\cancel{300}} \cancel{\text{K}})} = \frac{(200 \ \ell)(4)}{(3)} = 266.66 \ \ell$$

$V_2 = 267 \ \ell$ (rounded to the correct number of significant digits)

Use the five steps of problem solving to solve the following Charles' Law problem.

At 300 K, a gas has a volume of 600 milliliters (m*l*). What is the volume of this gas at 380 K?

*Step 1.*

*Step 2.*

*Step 3.*

*Step 4.*

*Step 5.*

**5**

---

Answers to frame 4.

*Step 1.* $V_2 = ?$

*Step 2.* $T_1 = 300$ K
$T_2 = 380$ K
$V_1 = 600$ m*l*

*Step 3.* $V_2 = \dfrac{V_1 T_2}{T_1}$

*Step 4.* $V_2 = \dfrac{(600 \text{ m}l)(380 \text{ K})}{300 \text{ K}}$

*Step 5.* $V_2 = 760$ m*l*

**6**

---

Use the five steps of problem solving to solve this applied problem.

At 500 K, a gas has a volume of 200 milliliters. What is the volume of this gas at 250 K?

*Step 1.*

*Step 2.*

*Step 3.*

*Step 4.*

*Step 5.*

**7**

---

Answers to frame 6.

*Step 1.* $V_2 = ?$

*Step 2.* $V_1 = 200$ m*l*
$T_1 = 500$ K
$T_2 = 250$ K

*Step 3.* $V_2 = \dfrac{V_1 T_2}{T_1}$

*Step 4.* $V_2 = \dfrac{(200 \text{ m}l)(250 \text{ K})}{(500 \text{ K})}$

*Step 5.* $V_2 = 100$ m*l*

## 8

Fluid–power systems are widely used in industry, science, and technology. The relationship between pressure, force, and area is expressed by the formula $P = \frac{F}{A}$, where $P$ is pressure, $F$ is force, and $A$ is area. An illustration of a machine used to apply pressure to a fluid is shown in Figure 4-1.

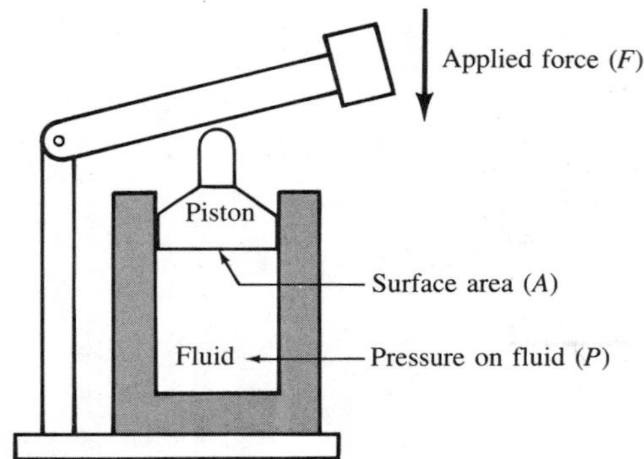


**Figure 4-1**

The five steps of problem solving can be used to solve problems regarding the pressure in a fluid–power system.

What pressure is developed in a cylinder when a force of 50.0 lb is applied to a piston with a surface area of 4.00 in.$^2$?

*Step 1.* Determine what is being asked for.

$$P = ?$$

*Step 2.* Determine what information is already known.

$$F = 50.0 \text{ lb}$$
$$A = 4.00 \text{ in.}^2$$

*Step 3.* Find a mathematical model that describes the relationship.

$$P = \frac{F}{A}$$

*Step 4.* Substitute the data into the model.

$$P = \frac{50.0 \text{ lb}}{4.00 \text{ in.}^2}$$

*Step 5.* Do the calculations.

$$P = \frac{50.0 \text{ lb}}{4.00 \text{ in.}^2} = 12.5 \text{ lb/in.}^2$$

When dealing with fluid power, plastics, and other industrial applications, pounds per square inch (lb/in.$^2$) is written as psi. Therefore, the answer can be written as $P = 12.5$ psi.

**a.** Find the pressure developed in a cylinder when a force of 500 lb is applied to a piston with a surface area of 0.400 $in.^2$.

**b.** What pressure is developed in a cylinder when a force of 225 lb is applied to a piston with a surface area of 7.5 $in.^2$?

**c.** What pressure is developed in a container if a piston with an area of 2.00 $in.^2$ has a pressure of 65.0 lb applied to it?

**d.** What is the pressure in a cylinder that has a force of 25.2 lb applied to a piston with an area of 3.4 $in.^2$?

**8. a.** 1,250 psi
**b.** 30 psi
**c.** 32.5 psi
**d.** 7.4 psi

## 9

More than one formula is needed to solve some problems. An example is given below.

What pressure is needed to mold a plastic part 10 inches long by 7 inches wide? The formula for determining molding pressure is

$$P = \frac{(2{,}500 \text{ lb/in.}^2)A}{2{,}000 \text{ lb}}$$

The pressure ($P$) will be in tons, and the area ($A$) will be in square inches.

*Step 1.* Determine what is being asked for.

$$P = ?$$

*Step 2.* What information is known?

Length ($\ell$) = 10 inches

Width (w) = 7.0 inches

*Step 3.* Use the mathematical model given for the pressure.

$$P = \frac{(2{,}500 \text{ lb/in.}^2)A}{2{,}000 \text{ lb}}$$

*Step 4.* Substitute the information into the model. Notice that the quantity $A$ is not given directly and must first be calculated using the information given in step 2.

$$A = \ell \times w$$

$$A = 10 \text{ in.} \times 7.0 \text{ in.} = 70 \text{ in.}^2$$

$$P = \frac{(2{,}500 \text{ lb/in.}^2)(70 \text{ in.}^2)}{2{,}000 \text{ lb/ton}}$$

*Step 5.* Do the calculation.

$$P = 88 \text{ tons}$$

Solve these problems.

**a.** Find the pressure needed to mold a plastic part 32 inches by 4.0 inches.

**b.** What pressure would be needed to mold a part with an area of 200 in.$^2$?

**c.** What pressure would be needed to mold a circular part with a radius of 4.00 inches? The formula for the area of a circle is $A = (3.14)(r)(r)$.

**d.** Find the pressure needed to mold a part 2.5 inches by 20.0 inches.

## 10

Large and small numbers often occur in the same formula. If these numbers are entered carefully into the calculator, they should present no difficulty.

9. **a.** 160 tons
**b.** 250 tons
**c.** 62.8 tons
**d.** 62.5 tons

The amount of longitudinal (length) shrinkage in a weld is quite small and is determined by the formula

$$\Delta L = \frac{(0.12 \text{ in.})(I)(L)}{(100{,}000A)(t)}$$

where $I$ = electrical current in amperes (A)
$L$ = length of weld in inches
$t$ = thickness of metal plate in inches
$\Delta L$ = longitudinal shrinkage

*Note:* The symbol $\Delta$ is the Greek letter delta and is used to indicate change. Used here, $\Delta L$, it means change in length.

Find the longitudinal shrinkage when plate thickness is 1/4 inch and the current in amperes (A) is 200. The length of the weld is 28 inches.

*Step 1.* What is asked for?

$$\Delta L = ?$$

*Step 2.* What is known?

$I$ = 200 A

$L$ = 28 in.

$t$ = 1/4 in. or 0.25 in.

*Step 3.* Use the mathematical model.

$$\Delta L = \frac{(0.12 \text{ in.})(I)(L)}{(100{,}000 \text{ A})(t)}$$

*Step 4.* Substitute into the model.

$$\Delta L = \frac{(0.12 \text{ in.})(200 \text{ A})(28 \text{ in.})}{(100{,}000 \text{ A})(0.25 \text{ in.})}$$

*Step 5.* Do the calculation.

$$\Delta L = 0.03 \text{ in.}$$

Find the longitudinal shrinkage of a weld when the plate thickness is 1/2 inch and the current and length are the same as in the previous example. Does the shrinkage become more or less if the plate is thicker?

**11**

Answers to frame 10.

$\Delta L = 0.01$ in.; the shrinkage becomes less as the plate becomes thicker.

# Exercise Set, Section 4-6

## Applied Problems

1. At 600 K, a gas has a volume of 300 milliliters. What is its volume at 400 K?
2. At 245 K, a gas has a volume of 35 liters. What is its volume at 300 K?
3. At 273 K, a gas has a volume of 500 milliliters. What is its volume at 173 K?
4. At 273 K, a gas has a volume of 10 liters. What is its volume at 546 K?
5. At 900 K, a gas has a volume of 600 milliliters. What is its volume at 600 K?
6. At 500 K, a gas has a volume of 3.50 liters. What is its volume at 750 K?
7. In a fluid–power system, what pressure is developed in a cylinder when a force of 325 lb is applied to a piston with an area of 15.5 in.$^2$?
8. What pressure is needed to mold a plastic part that is 15.0 in. by 18.0 in.?
9. Find the longitudinal shrinkage of a 40-in.-long weld when the plate is 0.50 in. thick and the welding current is 220 A.
10. What pressure is developed in a container if a piston with an area of 3.0 in.$^2$ has a pressure of 150 lb applied to it?

# Supplementary Exercise Set, Section 4-6

1. At 380 K, a gas has a volume of 200 milliliters. What is its volume at 420 K?
2. At 273 K, a gas has a volume of 42 liters. What is its volume at 240 K?

3. Find the longitudinal shrinkage of a butt weld when the plate thickness is 0.25 inch and the weld is 25 inches long. The current is 220 amperes.

4. What pressure is needed to mold a plastic part that is 28.0 inches by 32.0 inches?

5. In a fluid–power system, what pressure is developed in a cylinder when a force of 450 lbs is applied to a piston that has an area of 35.0 in.$^2$?

## Summary

1. Fractional equations are solved by using the multiplication axiom: both sides are multiplied by the coefficient of the $x$ term.
2. The cross-multiplication shortcut is used to solve proportions.
3. Formulas are mathematical shorthand for describing a relationship between numbers.
4. Formulas often have to be rearranged before they can be solved. Formula rearrangement uses the same principles as equation solving.
5. Dimensions as well as numbers must be evaluated. The evaluation of dimensions is usually performed by the reducing shortcut.

# Chapter 4 Self-Test

## Fractional Equations and Formulas

Solve the following fractional equations.

1. $\frac{x}{4} = 7$

2. $\frac{3a}{5} = 15$

3. $\frac{2x}{3} = 27$

4. $\frac{1}{y} = \frac{9}{10}$

5. $\frac{3}{4} = \frac{k}{16}$

6. $\frac{13}{7} = \frac{3}{m}$

7. $\frac{4.3}{x} = \frac{3}{4}$

8. $\frac{6.3}{4.2} = \frac{2.1}{t}$

9. $\frac{380}{150} = \frac{a}{75}$

Evaluate the following formulas.

10. Given K = °C + 273, find K when °C = 28.

11. Given °F = 1.8 °C + 32, find °F when °C = −125.

12. Given $y = mx + b$, find $y$ when $m = 0.80$, $x = 3.3$, and $b = -2.0$.

13. Given $°C = \frac{(°F - 32)}{1.8}$, find °C when °F = 68.

14. Given $V_1 = \frac{T_1 V_2}{T_2}$, find $V_1$ when $T_1 = 200$, $T_2 = 400$, and $V_2 = 80$.

15. Given $a = \frac{A}{bc}$, find $a$ when $A = 38$, $b = 4.6$, and $c = 3.7$.

**16.** Given $V = \frac{nRT}{P}$, find $V$ when $n = 3.00$, $R = 0.080$, $T = 300$, and $P = 3.00$.

**17.** Given $P = I^2R$, find $P$ when $I = 10.5$ and $R = 3.00$.

**18.** Given $KE = \frac{1}{2}mv^2$, find $KE$ when $m = 6.40$ and $v = 1.80$.

**19.** Given $B = \frac{mi}{6.28d}$, find $B$ when $m = 18.0$, $i = 41.4$, and $d = 6.00$.

**20.** Given $PE = mg(h_2 - h_1)$, find $PE$ when $m = 6.20$, $g = 20.0$, $h_2 = 13.60$, and $h_1 = 6.70$.

**21.** Given $E = \frac{1}{2}mv^2 + mgh$, find $E$ when $m = 3.20$, $v = 7.10$, $g = 32.0$, and $h = 15.0$.

Solve each of the following problems for the indicated variable.

**22.** Given $\frac{V_1}{T_1} = \frac{V_2}{T_2}$, find $V_2$.

**23.** Given $\frac{P_1V_1}{T_1} = \frac{P_2V_2}{T_2}$, find $P_2$.

**24.** Given $Kh - 3 = 2$, find $h$.

**25.** Given $Y = mx + b$, find $x$.

**26.** Given $xy + h = by - k$, find $y$.

**27.** Given $z = \frac{hk}{m}$, find $h$.

**28.** Given $P = \frac{a - m}{s}$, find $a$.

**29.** Given $w = \frac{24K}{z}$, find $K$.

**30.** Given $T = \frac{2\pi^2 m}{h}$, find $h$.

**31.** Given $P = 2L + 2w$, find $L$.

Use dimensional analysis and the five steps of problem solving to solve these problems. Charles' Law, $\frac{V_1}{T_1} = \frac{V_2}{T_2}$, is needed in problems 32 and 33.

**32.** At 300 K, a gas has a volume of 200 mℓ. What is its volume at 450 K?

**33.** At 200 K, a gas has a volume of 100 mℓ. At what temperature would the gas have a volume of 300 mℓ?

**34.** Examine the compression molding formula closely.

$$P = \frac{(2{,}500 \text{ psi})\,A}{2{,}000 \text{ lb}}$$

Can this formula be written in a simpler form? If it can, rewrite it.

**35.** What pressure would be needed to mold a part with an area of 356 in.$^2$?

# CHAPTER 5

# Ratios, Proportions, and Inverse Variations

Ratios and proportions are among the most useful of the mathematical tools used in the sciences. A large variety of measurement problems can be solved using these techniques. This chapter will discuss ratios, proportions, inverse variations, and their applications.

## Section 5-1 Ratios

### 1

A ratio is a comparison of two numbers using the process of division.

$\frac{4}{5}$ is a comparison of 4 to 5.

$\frac{9}{2}$ is a comparison of 9 to 2.

**a.** Write a comparison of 3 to 2.
**b.** Write a comparison of 1 to 5.
**c.** Write a comparison of 7 to 1.

### 2

The ratios in the previous frame were written as fractions, but there are several other ways to write ratios. For example, the ratio 2 to 5 can be written in any of four ways:

$\frac{2}{5}$ (or 2/5) $\quad 2 \div 5 \quad 2:5 \quad 0.4$

*Note:* $\frac{2}{5} \longrightarrow 5\,\overline{)2.0}$ = 0.4

Write the ratio 4 to 7 four ways.

**a.** **b.**

**c.** **d.**

**1. a.** $\frac{3}{2}$
**b.** $\frac{1}{5}$
**c.** $\frac{7}{1}$

### 3

Any comparison can be expressed as a ratio.

There are 19 employees in the print shop and 35 employees in the bindery of a printing company. The ratio of employees in the two units can be expressed as 19/35 or 19:35.

**2. a.** $\frac{4}{7}$
**b.** $4 \div 7$
**c.** 4:7
**d.** 0.57 (rounded)

Similarly, the ratio of 7 oz to 3 oz can be expressed as 7/3 or 7 : 3.

Express the following as ratios.

**a.** 45 centimeters to 11 centimeters **b.** 8 liters to 5 liters

**c.** 15 inches to 7 inches **d.** 1 gram to 10 grams

## 4

Ratios should always be reduced to lowest terms. This is done by dividing the numerator and denominator by the same number. The ratio 10/15 can be reduced to 2/3 as follows:

$$\frac{10}{15} = \frac{10 \div 5}{15 \div 5} = \frac{2}{3}$$

or

$$\frac{10}{15} = \frac{\overset{2}{\cancel{10}}}{\underset{3}{\cancel{15}}} = \frac{2}{3}$$

Reduce the following ratios to lowest terms.

**a.** $\frac{18}{9}$ **b.** $\frac{27}{81}$ **c.** $\frac{12}{144}$ **d.** $\frac{6}{8}$

## 5

Express the following as ratios in lowest terms.

**a.** 80 milliliters to 40 milliliters **b.** \$1,500 to \$300

**c.** 14 inches to 8 inches **d.** 18 feet to 81 feet

## 6

The principles of dimensional analysis are applied to the ratios below.

$$\frac{45 \text{ feet}}{15 \text{ feet}} = \frac{45 \cancel{\text{feet}}}{15 \cancel{\text{feet}}} = \frac{\overset{3}{\cancel{45}}}{\underset{1}{\cancel{15}}} = \frac{3}{1}$$

$$\frac{39 \text{ liters}}{78 \text{ liters}} = \frac{39 \cancel{\text{liters}}}{78 \cancel{\text{liters}}} = \frac{\overset{1}{\cancel{39}}}{\underset{2}{\cancel{78}}} = \frac{1}{2}$$

*Note:* In each case, the dimensions reduced and the ratios became numbers with no dimensions.

---

**3. a.** $\frac{45}{11}$ or 45 : 11

**b.** $\frac{8}{5}$ or 8 : 5

**c.** $\frac{15}{7}$ or 15 : 7

**d.** $\frac{1}{10}$ or 1 : 10

**4. a.** $\frac{2}{1}$

**b.** $\frac{1}{3}$

**c.** $\frac{1}{12}$

**d.** $\frac{3}{4}$

**5. a.** $\frac{2}{1}\left(\text{from } \frac{80}{40}\right)$

**b.** $\frac{5}{1}\left(\text{from } \frac{1{,}500}{300}\right)$

**c.** $\frac{7}{4}\left(\text{from } \frac{14}{8}\right)$

**d.** $\frac{2}{9}\left(\text{from } \frac{18}{81}\right)$

Use the principles of dimensional analysis to write the ratios below.

**a.** 15 pt to 3 pt

**b.** 25 qt to 80 qt

**c.** 6 in. to 10 in.

**d.** 18 capsules to 27 capsules

## 7

In many cases it is important to express a ratio with a denominator of 1. This is done by dividing the denominator into the numerator. The following ratios are expressed with a denominator of 1.

$$\frac{15}{3} = \frac{15 \div 3}{3 \div 3} = \frac{5}{1}$$

$$\frac{25}{5} = \frac{25 \div 5}{5 \div 5} = \frac{5}{1}$$

Express the following ratios with a denominator of 1.

**a.** $\frac{21}{3}$ **b.** $\frac{30}{5}$ **c.** $\frac{35}{7}$ **d.** $\frac{44}{11}$

**6. a.** $\frac{15\text{ pt}}{3\text{ pt}} = \frac{5}{1}$
**b.** $\frac{25\text{ qt}}{80\text{ qt}} = \frac{5}{16}$
**c.** $\frac{6\text{ in.}}{10\text{ in.}} = \frac{3}{5}$
**d.** $\frac{18\text{ cap.}}{27\text{ cap.}} = \frac{2}{3}$

## 8

The calculator can be helpful in expressing ratios with a denominator of 1.

$$\frac{25}{7} = \frac{25 \div 7}{7 \div 7} = \frac{25 \div 7}{1} = ?$$

The calculation 25 ÷ 7 can be performed on the calculator.

| *Enter* | *Press* | *Display* |
|---|---|---|
| 25 | [÷] | 25 |
| 7 | [=] | 3.5714285 |

The answer is 3.57 (rounded).

Therefore, $\frac{25}{7} = \frac{3.57}{1}$ (rounded).

Use the calculator to express the following ratios with denominators of 1.

**a.** $\frac{31}{8}$ **b.** $\frac{14}{17}$ **c.** $\frac{11}{9}$ **d.** $\frac{44}{5}$

**7. a.** $\frac{7}{1}$ or 7 : 1
**b.** $\frac{6}{1}$ or 6 : 1
**c.** $\frac{5}{1}$ or 5 : 1
**d.** $\frac{4}{1}$ or 4 : 1

## 9

**a.** Express the ratio of an alternator (1,075 rpm) to an idling engine (500 rpm).

**b.** Find the ratio of an 820 ohm resistor to a 500 ohm resistor.

**8. a.** 3.88
**b.** 0.82
**c.** 1.22
**d.** 8.8

## 10

Answers to frame 9.

**a.** 2.15 **b.** 1.64

# Section 5-2 Proportions

## 1

A proportion is a statement of equality between two ratios. The following is an example of a proportion:

$$\frac{5}{8} = \frac{N}{16}$$

This proportion can also be stated two other ways:

5 is to 8 as $N$ is to 16

$5:8 = N:16$

Express each of the following proportions in two other ways.

**a.** $\frac{1}{2} = \frac{2}{4}$ **b.** $\frac{a}{b} = \frac{c}{d}$

## 2

It is common in some sciences to see proportions expressed in the style below.

$$a:b = c:d$$

This proportion can also be written in fractional-equation form:

$$\frac{a}{b} = \frac{c}{d}$$

Write the following proportions in fractional-equation form.

**a.** $5:x = 3:4$ **b.** $9:N = 7:8$

**c.** $N:4 = 2:8$ **d.** $6:7 = 3:x$

**1. a.** 1 is to 2 as 2 is to 4; $1:2 = 2:4$
**b.** $a$ is to $b$ as $c$ is to $d$; $a:b = c:d$

## 3

When a proportion is expressed in fractional-equation form, it can be solved using the methods discussed in Chapter 4.

$$\frac{6}{x} = \frac{3}{4}$$

$$3x = (6)(4)$$

$$3x = 24$$

$$x = \frac{24}{3}$$

$$x = 8$$

Solve the following proportions using the methods from Chapter 4.

**a.** $\frac{9}{N} = \frac{3}{4}$ **b.** $\frac{2}{N} = \frac{1}{200}$

**2. a.** $\frac{5}{x} = \frac{3}{4}$
**b.** $\frac{9}{N} = \frac{7}{8}$
**c.** $\frac{N}{4} = \frac{2}{8}$
**d.** $\frac{6}{7} = \frac{3}{x}$

**4**

The following proportion is solved using the methods from Chapter 4, the techniques of dimensional analysis, and the calculator.

3. a. $N = 12$
b. $N = 400$

$$\frac{24.0\text{ g}}{9.20\text{ g}} = \frac{16.8\text{ g}}{N}$$

$$(24.0\text{ g})\,N = (16.8\text{ g})(9.20\text{ g})$$

$$N = \frac{(16.8\text{ g})(9.20\text{ g})}{24.0\text{ g}}$$

$$N = \frac{(16.8\text{ g})(9.20\text{ g})}{24.0\text{ g}}$$

| *Enter* | *Press* | *Display* |
|---|---|---|
| 16.8 | × | 16.8 |
| 9.20 | ÷ | 154.56 |
| 24.0 | = | 6.44 |

Therefore, $N = 6.44$ g.

**5**

Solve the following proportions.

a. $\dfrac{3.8\ \ell}{14\ \ell} = \dfrac{N}{25\ \ell}$ b. $\dfrac{x}{7.20\text{ g}} = \dfrac{3.00\text{ g}}{2.00\text{ g}}$

**6**

When a proportion is not expressed in fractional-equation form, it can be converted to fractional-equation form and solved using the methods from Chapter 4.

5. a. $N = 6.79\ \ell$ (rounded)
b. $x = 10.8$ g

$$N:3 = 2:6$$

$$\frac{N}{3} = \frac{2}{6} \longleftarrow \text{converted to fractional-equation form}$$

$$6N = (3)(2)$$

$$6N = 6$$

$$N = \frac{6}{6}$$

$$N = 1$$

Solve the following proportions by first converting to fractional-equation form, rounding to two decimal places. (A calculator may be helpful.)

a. $5:N = 7:9$ b. $3:4 = x:15$

**7**

Each part of a proportion has a special name. These names are shown below.

6. a. $N = 6.43$
b. $x = 11.25$

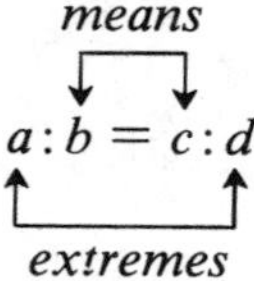

The first and last numbers are called the *extremes,* and the two middle numbers are called the *means.*

The proportion $a:b = c:d$ can be written as $a/b = c/d$. By cross-multiplication:

$$\frac{a}{b} \times \frac{c}{d}$$

The proportion $a:b = c:d$ can be written as $ad = bc$. Therefore, $a:b = c:d$ is the same as $ad = bc$.

Notice that $a$ and $d$ are the extremes and that $b$ and $c$ are the means. Therefore, the cross-multiplication shortcut can also be stated as "the product of the means equals the product of the extremes."

**8**

List the means and the extremes in the proportions below.

**a.** $2.8:N = 13.0:35.1$

means:

extremes:

**b.** $x:32 = 14:43$

means:

extremes:

**8. a.** means: $N$, 13.0
extremes: 2.8, 35.1
**b.** means: 32, 14
extremes: $x$, 43

**9**

The proportion below is solved using two methods.

*Product of Means Equals Product of Extremes*

$$3:N = 2:18$$

means ⟶ $2N = (3)(18)$ ⟵ extremes

$$2N = 54$$

$$N = \frac{54}{2}$$

$$N = 27$$

*Cross-Multiplication*

$$\frac{3}{N} = \frac{2}{18}$$

cross-multiplication ⟶ $2N = (3)(18)$

$$2N = 54$$

$$N = \frac{54}{2}$$

$$N = 27$$

Notice that the two methods are the same except for the way the proportion is initially set up. Therefore, the "product of the means equals the product of the extremes" rule works the same way as the cross-multiplication shortcut.

Solve the following proportions using either method.

**a.** $2:5 = N:30$

**b.** $1:500 = 3:N$

## 10

Solve the following proportions using a calculator.

a. $\frac{2.8}{15.1} = \frac{N}{10}$  b. $\frac{1}{200} = \frac{N}{500}$

c. $1:500 = 4.2:N$  d. $4.2:N = 3:80$

9. a. $N = 12$
b. $N = 1{,}500$

## 11

One proportion is solved below. Use the same method to solve the other proportion.

a. $\frac{\frac{1}{2}}{\frac{2}{3}} = \frac{N}{\frac{1}{4}}$  b. $\frac{\frac{1}{2}}{\frac{3}{4}} = \frac{N}{\frac{1}{3}}$

$$N \times \frac{2}{3} = \frac{1}{2} \times \frac{1}{4}$$

$$N \times \frac{2}{3} = \frac{1}{8}$$

$$N \times \frac{\frac{2}{3}}{\frac{2}{3}} = \frac{\frac{1}{8}}{\frac{2}{3}}$$

$$N = \frac{\frac{1}{8}}{\frac{2}{3}}$$

$$N = \left(\frac{1}{8}\right)\left(\frac{3}{2}\right)$$

$$N = \frac{3}{16}$$

10. a. $N = 1.85$ (rounded)
b. $N = 2.5$
c. $N = 2{,}100$
d. $N = 112$

## 12

When a proportion contains a whole number and a fraction, it is sometimes helpful to rewrite the whole number as a fraction (over 1). This procedure is shown in problem **a** below. Solve problem **b** the same way.

a. $\frac{N}{6} = \frac{\frac{1}{4}}{\frac{2}{3}}$  b. $\frac{N}{5} = \frac{\frac{1}{2}}{\frac{1}{4}}$

$$\frac{2}{3} \times N = \frac{6}{1} \times \frac{1}{4}$$

$$\frac{2}{3} \times N = \frac{6}{4}$$

$$\frac{2}{3} \times N = \frac{3}{2}$$

$$\frac{\frac{2}{3} \times N}{\frac{2}{3}} = \frac{\frac{3}{2}}{\frac{2}{3}}$$

$$N = \left(\frac{3}{2}\right)\left(\frac{3}{2}\right)$$

$$N = \frac{9}{4}$$

11. b. $N \times \frac{3}{4} = \frac{1}{2} \times \frac{1}{3}$

$$N \times \frac{3}{4} = \frac{1}{6}$$

$$N = \frac{\frac{1}{6}}{\frac{3}{4}}$$

$$N = \frac{1}{6} \times \frac{4}{3}$$

$$N = \frac{4}{18}$$

$$N = \frac{2}{9}$$

## 13

Solve the following proportions.

a. $\frac{N}{\frac{1}{2}} = \frac{\frac{3}{8}}{5}$ b. $\frac{2}{N} = \frac{\frac{3}{5}}{\frac{1}{4}}$

c. $\frac{\frac{3}{4}}{\frac{1}{8}} = \frac{N}{\frac{1}{3}}$ d. $\frac{\frac{1}{2}}{\frac{3}{4}} = \frac{5}{N}$

12. b. $\frac{1}{4} \times N = \frac{5}{1} \times \frac{1}{2}$

$\frac{1}{4} \times N = \frac{5}{2}$

$N = \frac{\frac{5}{2}}{\frac{1}{4}}$

$N = \frac{5}{2} \times \frac{4}{1}$

$N = \frac{20}{2}$

$N = 10$

## 14

Answers to frame 13.

a. $\frac{3}{80}$ b. $\frac{5}{6}$ c. 2 d. $\frac{15}{2}$

# Exercise Set, Sections 5-1–5-2

## Ratios

Express the following as ratios reduced to lowest terms.

1. 1 to 5
2. 2 to 11
3. 2 grams to 9 grams
4. 15 centimeters to 23 centimeters
5. 3 feet to 27 feet
6. 25 liters to 5 liters
7. 36 centimeters to 6 centimeters
8. 24 inches to 9 inches
9. 3 meters to 9 meters
10. 100 grams to 1,000 grams

Express the following ratios with a denominator of 1.

11. $\frac{42}{7}$
12. $\frac{38}{12}$
13. $\frac{4}{9}$
14. $\frac{3}{4}$
15. $\frac{2}{7}$

## Proportions

Write the following proportions in fractional-equation form.

16. $3:x = 6:7$
17. $5:8 = N:10$
18. $K:8 = 4:16$
19. $7:9 = 14:N$
20. $2:3 = x:9$
21. $7:N = 14:28$

Solve the following proportions for $N$.

**22.** $N:7 = 6:14$

**23.** $8:10 = 4:N$

**24.** $3:N = 6:18$

**25.** $5:8 = N:24$

**26.** $N:9 = 6:18$

**27.** $8:N = 16:32$

**28.** $\frac{2.5}{16} = \frac{N}{8}$

**29.** $\frac{4.2}{N} = \frac{2.1}{100}$

**30.** $\frac{6.8}{50} = \frac{13.6}{N}$

**31.** $\frac{\frac{1}{4}}{\frac{3}{4}} = \frac{10}{N}$

**32.** $\frac{\frac{1}{3}}{\frac{2}{3}} = \frac{N}{9}$

**33.** $\frac{\frac{1}{4}}{\frac{5}{8}} = \frac{N}{\frac{1}{2}}$

**34.** $\frac{\frac{2}{3}}{\frac{3}{4}} = \frac{8}{N}$

**35.** $\frac{12.0\text{ g}}{4.6\text{ g}} = \frac{8.4\text{ g}}{N}$

**36.** $\frac{8.2\text{ oz}}{N} = \frac{3.7\text{ oz}}{6.4\text{ oz}}$

**37.** $\frac{N}{22.4\text{ g}} = \frac{15.3\text{ g}}{8.9\text{ g}}$

**38.** $\frac{1.9\ \ell}{7.0\ \ell} = \frac{N}{12.5\ \ell}$

**39.** $\frac{17.8\text{ g}}{N} = \frac{27.9\text{ g}}{5.3\text{ g}}$

**40.** $\frac{22.8\text{ g}}{11.4\text{ g}} = \frac{26.8\text{ g}}{N}$

## Supplementary Exercise Set, Sections 5-1–5-2

Express the following ratios in lowest terms.

**1.** 1 to 4

**2.** 3 to 17

**3.** 3 to 15

**4.** 2 inches to 7 inches

**5.** 5 grams to 19 grams

**6.** 23 centimeters to 25 centimeters

**7.** 3 liters to 18 liters

**8.** 35 meters to 7 meters

**9.** 1,000 grams to 10 grams

**10.** 12 meters to 48 meters

Express the following ratios with a denominator of 1.

**11.** $\frac{48}{2}$

**12.** $\frac{52}{15}$

**13.** $\frac{74}{30}$

**14.** $\frac{5}{7}$

**15.** $\frac{1}{4}$

Write the following proportions in fractional-equation form.

**16.** $4:X = 7:9$

**17.** $3:5 = N:8$

**18.** $N:72 = 4.8:12.0$

**19.** $15.0:N = 31.0:1$

**20.** $17:1 = N:2$

**21.** $315:32 = N:35$

Solve the following proportions for $N$.

**22.** $N:2 = 15:3$

**23.** $4:5 = 8:N$

**24.** $5:N = 10:17$

**25.** $\frac{3.5}{N} = \frac{70}{100}$

**26.** $\frac{N}{1.4} = \frac{12}{28}$

**27.** $\frac{N}{2.2} = \frac{1.7}{1}$

**28.** $\frac{\frac{1}{4}}{\frac{3}{8}} = \frac{\frac{1}{2}}{N}$

**29.** $\frac{\frac{5}{8}}{\frac{1}{2}} = \frac{N}{16}$

**30.** $\frac{N}{\frac{1}{4}} = \frac{\frac{7}{8}}{1}$

**31.** $\frac{14.0 \text{ g}}{3.8 \text{ g}} = \frac{12.0 \text{ g}}{N}$

**32.** $\frac{2.1\ \ell}{8.0\ \ell} = \frac{N}{12.0\ \ell}$

**33.** $\frac{14.1 \text{ oz}}{5.8 \text{ oz}} = \frac{4.0 \text{ oz}}{N}$

**34.** $\frac{3\frac{1}{2} \text{ g}}{2\frac{1}{2} \text{ g}} = \frac{N}{5.0 \text{ g}}$

**35.** $\frac{N}{\$712} = \frac{\$0.04}{\$1.00}$

# Section 5-3 Applications of Ratios and Proportions

**1**

The concept of ratio is used to describe various relationships. Once a ratio is determined, a proportion can be used to solve for other relationships. Some commonly used ratios will be examined first. For example, the *pitch* of a roof is defined as the *rise of the roof* (vertical distance) compared to the *run of the roof* (horizontal distance), as shown in Figure 5-1.

$$\text{Pitch} = \frac{\text{rise}}{\text{run}}$$

$$\text{Pitch} = \frac{10 \text{ ft}}{4 \text{ ft}}$$

$$\text{Pitch} = \frac{1}{4}$$

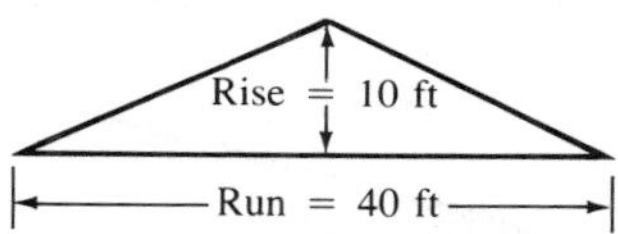


**Figure 5-1**

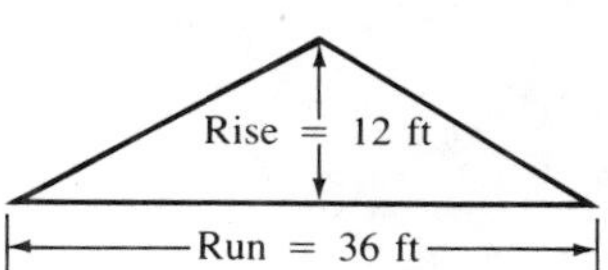


**Figure 5-2**

What is the pitch of the roof in Figure 5-2?

**1.** $\text{Pitch} = \frac{\text{rise}}{\text{run}}$

$\text{Pitch} = \frac{12 \text{ ft}}{36 \text{ ft}}$

$\text{Pitch} = \frac{1}{3}$

**2**

The volume in an automobile-engine cylinder before compression of the gases compared to the volume after compression of the gases is called the *compression ratio.* Figure 5-3 illustrates this ratio.

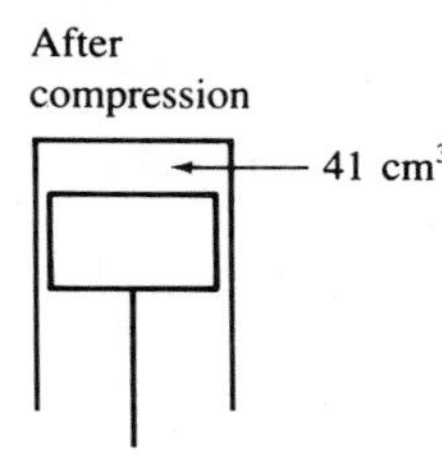


**Figure 5-3**

$$\text{Compression ratio} = \frac{\text{volume before compression}}{\text{volume after compression}}$$

$$\text{Compression ratio} = \frac{375 \text{ cm}^3 \div 41}{41 \text{ cm}^3 \div 41}$$

$$\text{Compression ratio} = \frac{9.1}{1}$$

What is the compression ratio of the cylinders shown in Figure 5-4?

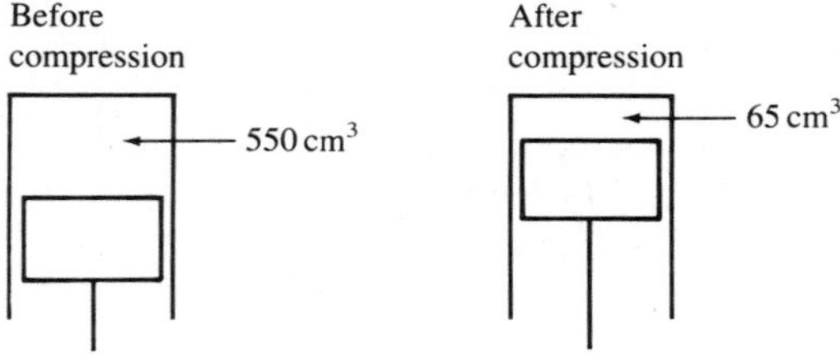


**Figure 5-4**

**3**

Answer to frame 2.

$$\text{Compression ratio} = \frac{550 \text{ cm}^3 \div 65}{65 \text{ cm}^3 \div 65} = \frac{8.5}{1}$$

**4**

The gear ratio of a bicycle is a comparison of the number of teeth on the front gear compared to the number of teeth on the back gear.

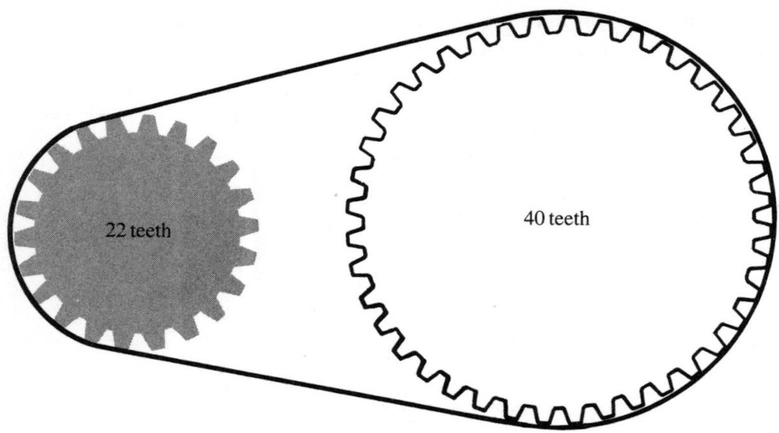


$$\text{gear ratio} = \frac{40 \text{ teeth} \div 22}{22 \text{ teeth} \div 22}$$

$$\text{gear ratio} = \frac{1.8}{1}$$

What is the gear ratio of these gears?

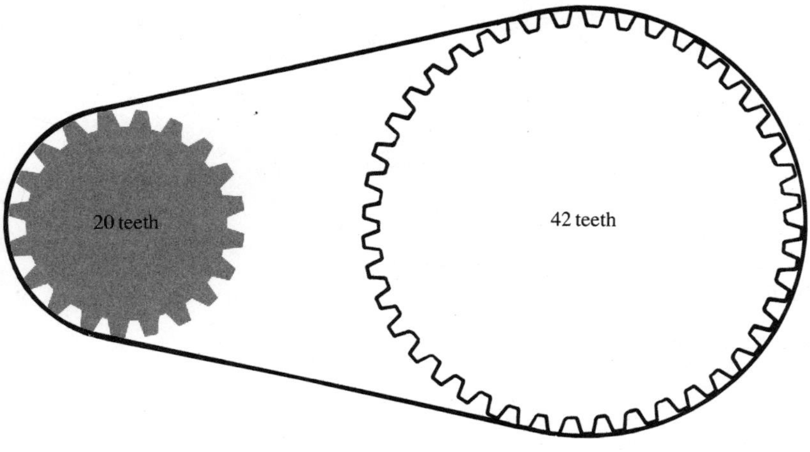


**5**

Answer to frame 4.

$$\frac{42\text{ teeth} \div 20}{20\text{ teeth} \div 20} = \frac{2.1}{1}$$

**6**

The five-step procedure is used to solve the following proportion problem.

The clutch linkage on a truck has an overall advantage of 25 : 1. How much force must the driver apply to release the clutch if the pressure plate has a force of 675 lb?

*Step 1.* Determine what is being asked for.

$F_d$ = force applied by driver

*Step 2.* Determine what information is already known.

Ratio of advantage of clutch linkage = 25 : 1
Force of pressure plate ($F_p$) = 675 lb

*Step 3.* Find a mathematical model that describes the relationship.
25 : 1 = force of pressure plate : force of driver or

$$\frac{25}{1} = \frac{F_p}{F_d}$$

*Step 4.* Substitute the data into the model.

$$\frac{25}{1} = \frac{675\text{ lb}}{F_d}$$

*Step 5.* Do the calculations.

$$25\ F_d = (1)(675)$$

$$25\ F_d = 675\ \text{lb}$$

$$F_d = \frac{675\ \text{lb}}{25}$$

$$F_d = 27\ \text{lb}$$

**7**

Use the five-step procedure to solve this problem.

If it takes 20 days to assemble 3 milling machines, how long should it take to assemble 11 milling machines?

*Step 1.* Determine what is being asked for.

*Step 2.* Determine what information is already known.

*Step 3.* Find a mathematical model that describes the relationship.

*Step 4.* Substitute the data into the model.

*Step 5.* Do the calculations.

**8**

Answers to frame 7.

*Step 1.* Time to assemble 11 milling machines = $T$ (days)
*Step 2.* 20 days to assemble 3 machines = 20 : 3
Number of machines wanted = 11
*Step 3.* 20 : 3 = time to assemble machines : number of machines
*Step 4.* $\dfrac{20\ \text{days}}{3\ \text{machines}} = \dfrac{T\ \text{days}}{11\ \text{machines}}$
*Step 5.* ($T$ days)(3 machines) = (20 days)(11 machines)
$T$ = 73 days (rounded)

**9**

Use the five-step procedure to solve this problem.

A building assessed at $75,000 is billed $1,200 for property tax. The following year it is reassessed at $100,000. What will the new property tax be?

*Step 1.*

*Step 2.*

*Step 3.*

*Step 4.*

*Step 5.*

**10**

Answers to frame 9.

*Step 1.* New property tax = $N$
*Step 2.* Ratio of old value to old tax = \$75,000 : \$1,200

New value = \$100,000

*Step 3.* \$75,000 : \$1,200 = \$100,000 : $N$

*Step 4.* $\frac{\$75{,}000}{\$1{,}200} = \frac{\$100{,}000}{N}$

$N = \$1{,}600$

**11**

100 mℓ of gas are at a temperature of 290 K. What is the new volume when the temperature is increased to 350 K?

*Note:* Use the proportion $\frac{V_1}{V_2} = \frac{T_1}{T_2}$.

*Step 1.*

*Step 2.*

*Step 3.*

*Step 4.*

*Step 5.*

**12**

Answers to frame 11.

*Step 1.* New volume = $V_2$
*Step 2.* Original volume = 100 mℓ
Original temperature = 290 K
New temperature = 350 K

*Step 3.* $\frac{V_1}{V_2} = \frac{T_1}{T_2}$

*Step 4.* $\frac{100\text{ m}\ell}{V_2} = \frac{290\text{ K}}{350\text{ K}}$

*Step 5.* $V_2 = 121$ mℓ

# Section 5-4 Direct and Inverse Variations

**1**

As stated earlier, a proportion is a statement of equality between two ratios. The three proportions below have the same ratio on the right side.

$$\frac{Y_1}{X_1} = \frac{20}{5}$$

$$\frac{Y}{10} = \frac{20}{5}$$

$$\frac{10}{X} = \frac{20}{5}$$

Whenever a situation such as this occurs, it is helpful to divide the ratio on the right side and obtain the following proportions:

$$\frac{Y_1}{X_1} = 4$$

$$\frac{Y}{10} = 4$$

$$\frac{10}{X} = 4$$

These proportions can be written as:

$$\frac{Y_1}{X_1} = k$$

$$\frac{Y}{10} = k$$

$$\frac{10}{X} = k$$

where $k = 4$, and the $k$ is called the *constant of proportionality*.

**2**

---

A proportion of the type in the previous frame is called a *direct variation*. Direct variations are usually written as:

$$\frac{Y}{X} = k$$

Proportions usually are written as:

$$\frac{Y}{X} = \frac{n}{m}$$

In the proportion below the numbers on the right are divided, and the result is an equivalent direct variation.

Proportion: $\dfrac{Y}{X} = \dfrac{7.5}{2.5}$

Direct variation: $\dfrac{Y}{X} = 3$

Perform the division below, and write the equivalent direct variation.

$$\frac{Y}{X} = \frac{8.61}{2.05}$$

**3**

---

Convert these proportions to direct variations. 2. $\dfrac{Y}{X} = 4.2$

**a.** $\dfrac{Y}{X} = \dfrac{0.175}{2.50}$ **b.** $\dfrac{Y}{X} = \dfrac{1{,}000}{40}$

c. $\frac{Y}{X} = \frac{1}{200}$ d. $\frac{Y}{X} = \frac{0.04}{0.15}$

## 4

The direct variation $Y/X = k$ states that $Y$ is proportional to $X$ and that $k$ is the constant of proportionality. Direct-variation problems can be solved using formula rearrangement techniques from Chapter 4. Using direct variation below, we find $Y$ when $X$ is 7.

3. **a.** 0.07
**b.** 25
**c.** 0.005
**d.** 0.27 (rounded)

$$\frac{Y}{X} = 0.35$$

$$\frac{Y}{7} = 0.35$$

$$7\left(\frac{Y}{7}\right) = (0.35)(7)$$

$$Y = 2.45$$

In each of these direct variations, find $Y$ when $X = 0.8$.

a. $\frac{Y}{X} = 5.1$ b. $\frac{Y}{X} = 0.03$

## 5

An extra step is required when $Y$ is given and $X$ is the unknown. Using direct variation below, we find $X$ when $Y$ is 2.31.

4. **a.** 4.08
**b.** 0.024

$$\frac{Y}{X} = 5.8$$

$$\frac{2.31}{X} = 5.8$$

$$X\left(\frac{2.31}{X}\right) = (5.8)X$$

$$2.31 = 5.8X$$

$$\left(\frac{1}{5.8}\right)(2.31) = (5.8X)\left(\frac{1}{5.8}\right)$$

$$\frac{2.31}{5.8} = X$$

$$0.40 = X$$

In each of these direct variations, find $X$ when $Y = 5$.

a. $\frac{Y}{X} = 4.5$ b. $\frac{Y}{X} = 0.25$

**6**

When one number varies directly as the reciprocal of another, the two numbers are said to be inversely proportional.

5. **a.** 1.11 (rounded)
**b.** 20

$$\frac{Y}{\left(\frac{1}{X}\right)} = k$$

*Note:* $\frac{1}{X}$ is the reciprocal of $X$.

The variation above is usually written in the simpler form below.

Inverse Variation: $XY = k$

The important thing to note about an inverse variation is that, when $X$ gets larger, $Y$ will get smaller.

In the inverse variation $XY = k$, what happens to $X$ if $Y$ gets larger?

____________________

**7**

Boyle's Gas Law is an example of an inverse variation. Boyle's Law is expressed by the following equation:

6. $X$ becomes smaller.

$$P_1V_1 = P_2V_2$$

where

$P_1$ = initial pressure on the gas

$V_1$ = initial volume of the gas

$P_2$ = new pressure on the gas

$V_2$ = new volume of the gas

Boyle's Law can be written as the following inverse variation:

$$P_2V_2 = k \text{ (when } k = P_1V_1)$$

According to Boyle's Gas Law, when pressure increases, the volume

____________________.

**8**

Boyle's Law is used to solve the following problem.

7. decreases

300 mℓ of gas is under a pressure of 4.00 atm. What will be the volume of the gas at 3.00 atm?

*Step 1.* Determine what is being asked for.

New volume = $V_2$

*Step 2.* Determine what information is already known.

$P_1$ (initial pressure) = 4.00 atm

$V_1$ (initial volume) = 300 mℓ

$P_2$ (new pressure) = 3.00 atm

*Step 3.* Find a mathematical model that describes the relationship.

$$P_1V_1 = P_2V_2$$

*Step 4.* Substitute the data into the model.

$$(4.00 \text{ atm})(300 \text{ m}\ell) = (3.00 \text{ atm})\, V_2$$

*Step 5.* Do the calculations.

$$V_2 = \frac{(4.00 \text{ atm})(300 \text{ m}\ell)}{(3.00 \text{ atm})}$$

$$V_2 = 400 \text{ m}\ell$$

**9**

Use Boyle's Law to solve the following problem.

If 200 mℓ of gas is under a pressure of 6.00 atm, what will be the volume of the gas at a pressure of 8.00 atm?

*Step 1.*

*Step 2.*

*Step 3.*

*Step 4.*

*Step 5.*

**10**

Answers to frame 9.

*Step 1.* New volume = $V_2$

*Step 2.* $P_1 = 6.00$ atm

$V_1 = 200$ mℓ

$P_2 = 8.00$ atm

*Step 3.* $P_1V_1 = P_2V_2$

*Step 4.* $(6.00 \text{ atm})(200 \text{ m}\ell) = (8.00 \text{ atm})\, V_2$

*Step 5.* $V_2 = 150$ mℓ

**11**

The rotation of gears illustrates the concept of inverse ratios. In Figure 5-5, gear *B* has only half as many teeth, and will rotate twice as fast, as gear *A*. If gear *A* rotates 25 revolutions per minute, how fast will gear *B* rotate? The proportion is set up as follows:

$$\frac{\text{Teeth in gear } A}{\text{Teeth in gear } B} = \frac{\text{Rotation gear } B}{\text{Rotation gear } A}$$

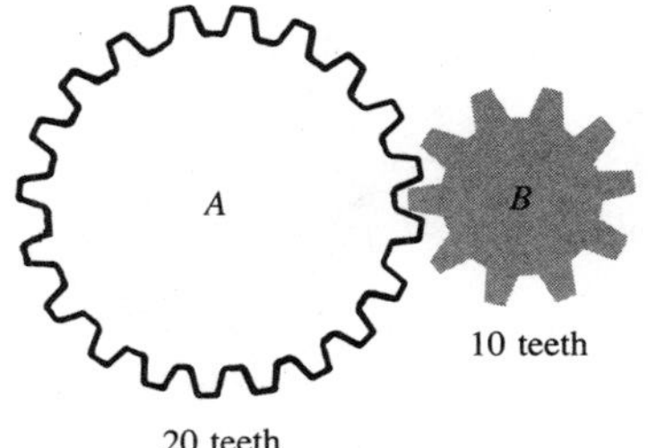


**Figure 5-5**

Note that the $B$ term appears in the denominator in the first ratio and in the numerator in the second ratio.

$$\text{Rotation gear } B = \frac{(\text{Teeth gear } A)(\text{Rotation gear } A)}{(\text{Teeth gear } B)}$$

$$\text{Rotation gear } B = \frac{(20 \text{ teeth})(25 \text{ rev per min})}{(10 \text{ teeth})}$$

$$\text{Rotation gear } B = 50 \text{ rev per min}$$

In Figure 5-6, if gear $A$ rotates 60 revolutions per minute, how fast will gear $B$ rotate?

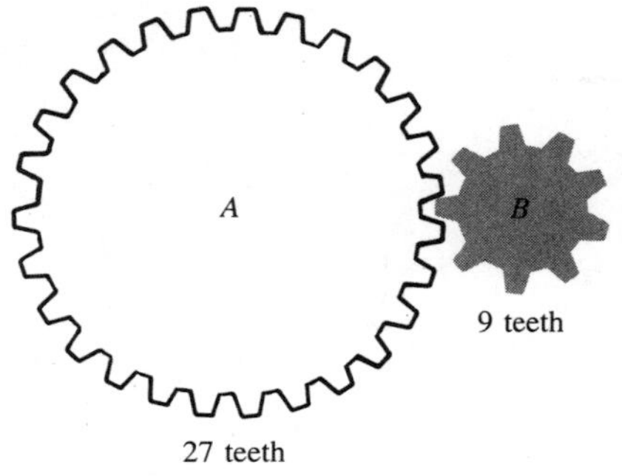


**Figure 5-6**

**12**

Answers to frame 11.

$$\frac{\text{Teeth } A}{\text{Teeth } B} = \frac{\text{Rotation } B}{\text{Rotation } A}; \text{ Rotation } B = \frac{(\text{Teeth } A)(\text{Rotation } A)}{(\text{Teeth } B)}$$

$$\text{Rotation } B = \frac{(27 \text{ teeth})(60 \text{ rev per min})}{(9 \text{ teeth})}; \text{ Rotation } B = 180 \text{ rev/min}$$

# Exercise Set, Sections 5-3–5-4

## Applications of Ratios and Proportions

Find the ratios using the information from the illustrations.

1. Find the pitch of the roof in Figure 5-7.

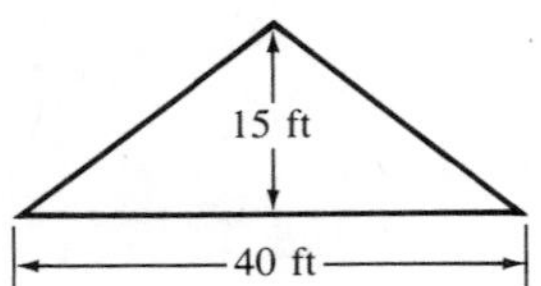


**Figure 5-7**

2. Find the gear ratio in Figure 5-8.

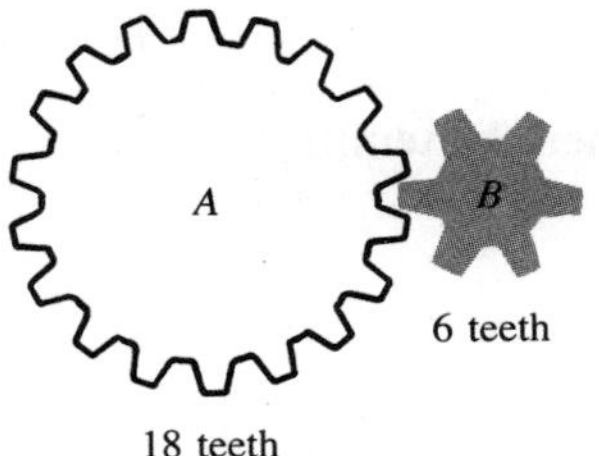


**Figure 5-8**

3. Find the compression ratio of the cylinders in Figure 5-9.

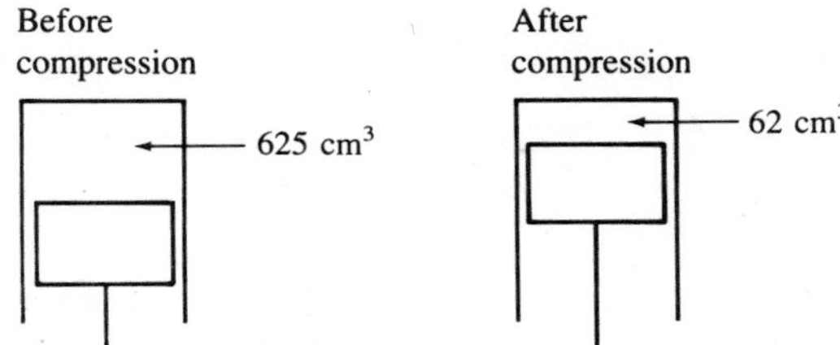


**Figure 5-9**

4. The ratio of two pulleys is the ratio of the pulley diameters. Find the pulley ratio of the pulleys in Figure 5-10.

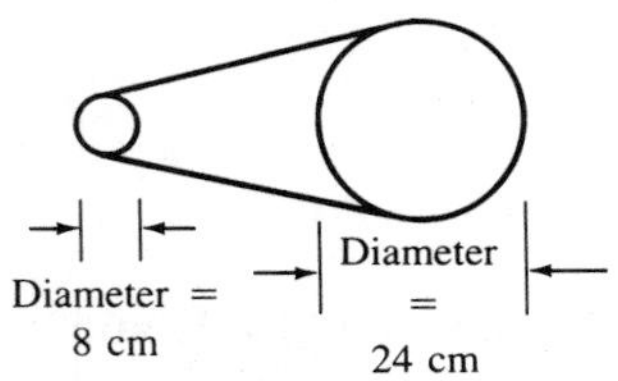


**Figure 5-10**

Use the five-step procedure to solve the following problems.

5. Given a power-to-weight ratio of 0.03 : 1, what is the horsepower of a 2,000 lb car?

6. Twenty linear feet of stock is needed to make the blanks for 50 bolts. How many feet of stock will be needed for 745 bolts?

7. The pitch of a roof is the ratio of rise (vertical distance) to run (horizontal distance). If the pitch of a roof is 3 : 8 and the run is 24 ft, find the rise.

8. A gas is under a pressure of 650 mm of mercury at 310 K. What is the new temperature when the pressure is increased to 800 mm of mercury?

*Note:* $\dfrac{P_1}{P_2} = \dfrac{T_1}{T_2}$

**9.** Each patient is required to take six capsules of a drug each day. If there are 14 patients, how many capsules are required each day?

**10.** If 200 m$\ell$ of a gas are at a temperature of 300 K, what is the new volume of the gas when the temperature is increased to 400 K?

*Note:* $\frac{V_1}{V_2} = \frac{T_1}{T_2}$

**11.** A machine can do a milling operation in 1/4 hour. How many times can that operation be done in 6 hours?

**12.** If each light fixture costs \$6.50, what will 144 fixtures cost?

## Direct and Inverse Variations

In each of these direct variations, find $Y$ when $X = 0.6$.

**13.** $\frac{Y}{X} = 0.12$ **14.** $\frac{Y}{X} = 5.8$ **15.** $\frac{X}{Y} = 0.12$ **16.** $\frac{X}{Y} = 5.8$

**17.** $\frac{Y}{X} = 0.68$ **18.** $\frac{X}{Y} = 1.34$ **19.** $\frac{Y}{X} = 0.07$ **20.** $\frac{X}{Y} = 7.9$

Use Boyle's Law to solve the following problem.

**21.** If 250 m$\ell$ of gas are under a pressure of 5.00 atm, what will be the volume of the gas at a pressure of 7.00 atm?

**22.** How many revolutions will gear $B$ make when gear $A$ makes 20 revolutions per minute? This is an inverse relationship.

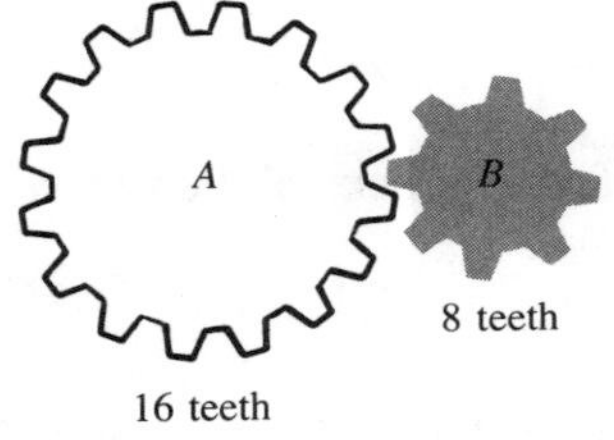


**23.** Under zoning laws, four houses can be built on 2 acres of land. How many houses can be built on 17 acres of land?

**24.** On a blueprint drawing, 1/4 in. equals 1 ft. A steel beam is 8 ft long. How long will the drawing be?

**25.** If a car burned 25.5 gallons of gasoline in 510 miles, how many gallons would it burn in 765 miles?

# Supplementary Exercise Set, Sections 5-3–5-4

Solve the following problems.

1. Find the ratio of the two pulley diameters.

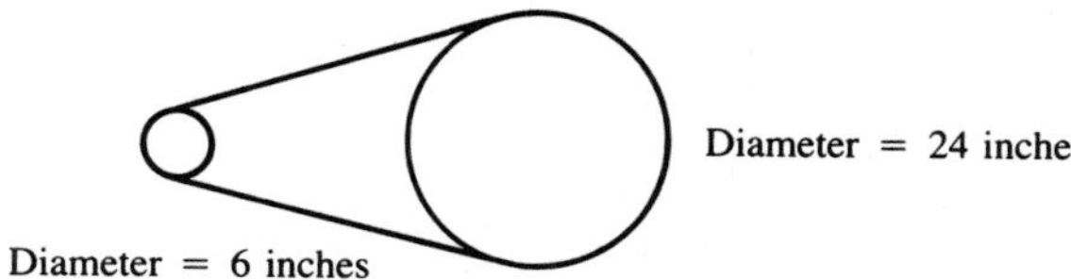


2. Find the pitch of the roof.

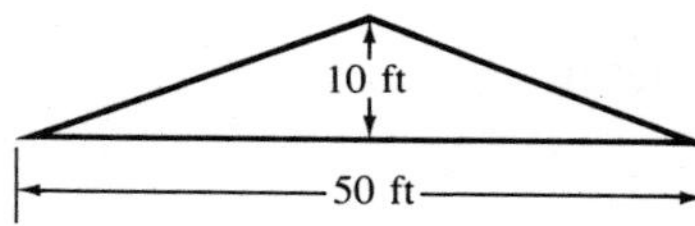


3. A race car with a power-to-weight ratio of 0.5 : 1 weighs 1,500 lb. What is its horsepower?

4. The pitch of a roof is 2 : 9. What is the rise if the run is 27 feet?

5. A gas is under a pressure of 830 mm of mercury at 305 K. What is the new temperature when the pressure is decreased to 730 mm of mercury?

6. If 10 milliliters of gas are at a temperature of 295 K, what is the new volume when the temperature is increased to 340 K?

7. The resistance of a wire is proportional to its length. If a copper wire 500 ft long has a resistance of 1.40 ohms, what will be the resistance of copper wire 300 ft long?

In each of these direct variations, find $Y$ when $X = 0.8$.

8. $\frac{Y}{X} = 5.2$

9. $\frac{Y}{X} = 0.03$

10. $\frac{X}{Y} = 0.04$

11. $\frac{X}{Y} = 8.4$

Use Boyle's Law to solve the following problem.

12. If 340 m$\ell$ of gas are under a pressure of 4.00 atm, what will be the volume of the gas at a pressure of 5.00 atm?

**13.** How many revolutions will gear $B$ make when gear $A$ makes 40 revolutions per minute?

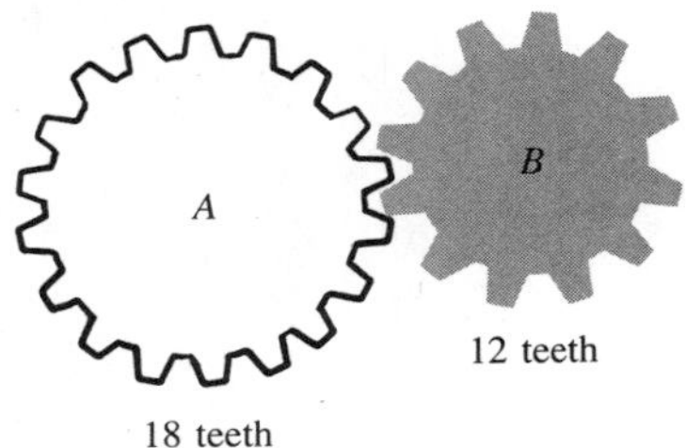


# Summary

1. A ratio is a comparison of two numbers.
2. Ratios are always reduced to the lowest terms.
3. To obtain a ratio with a denominator of 1, divide the numerator by the denominator.
4. A proportion is a statement of equality between two ratios.
5. Proportions can be written in fractional form or as means and extremes.
6. Cross-multiplication of proportions can be stated as "the product of the means equals the product of the extremes."
7. Use the five-step procedure to solve proportion problems.

# Chapter 5 Self-Test

## Ratios, Proportions and Inverse Variations

Express the following as ratios reduced to lowest terms.

**1.** 1 to 8 **2.** 2 g to 7 g **3.** 36 in. to 9 in. **4.** 20 ℓ to 5 ℓ

Express the following ratios with a denominator of 1.

**5.** $\frac{36}{9}$ **6.** $\frac{32}{11}$ **7.** $\frac{4}{5}$

Solve the following.

**8.** Express the ratio of engine speed (3,000 rpm) to final drive speed (900 rpm).

Write the following proportions in fractional-equation form.

**9.** $4:x = 8:12$ **10.** $5:9 = Y:10$

Solve the following proportions for $N$.

**11.** $N:9 = 4:6$ **12.** $\frac{2.5}{12} = \frac{N}{8}$ **13.** $\frac{\frac{1}{4}}{\frac{7}{8}} = \frac{\frac{1}{2}}{N}$ **14.** $\frac{4.2\text{ g}}{16.0\text{ g}} = \frac{8.4\text{ g}}{N}$

Use the five-step procedure to solve the following problems.

**15.** 3 cubic feet of sand is needed to make 10 cubic feet of concrete. How much sand should be ordered to make 150 cubic feet of concrete?

**16.** If a company makes a profit of $1,200 on a sale of $8,200, what profit can be expected on a sale of $27,800?

In each of these variations, find $Y$ when $x = 0.4$.

**17.** $\frac{Y}{x} = 1.2$

**18.** $\frac{x}{Y} = 5.4$

**19.** $xY = 16.4$

Use Boyle's Law ($P_1V_1 = P_2V_2$) to solve the following problem.

**20.** If 200 m$\ell$ of a gas are under a pressure of 5.00 atm, what will be the volume of the gas at a pressure of 9.00 atm?

# CHAPTER 6

# Percents

Percentage calculations are used in all aspects of science and technology. Expressing measurements as percentages is convenient, and it reduces the incidence of error. In this chapter you will learn the concept of percent, how to convert among fractions, decimals, and percents, and how to solve problems involving percents.

## Section 6-1 Converting Percents to Fractions and Decimals

**1**

The word *percent* means "per one hundred" and is written as %. Therefore, any percent can be written as a fraction with a denominator of 100.

For example, 21% means "21 per one hundred" and can be written as $\frac{21}{100}$.

35% means "35 per one hundred" and can be written as ____________.

**2**

Convert the following percents to fractions with a denominator of 100.

**1.** $\frac{35}{100}$

**a.** 45% = **b.** 1% = **c.** 15% = **d.** 95% =

**3**

Notice that, in the example below, the fraction can be reduced to lowest terms.

**2. a.** $\frac{45}{100}$

$$40\% = \frac{40}{100} = \frac{4}{10} = \frac{2}{5}$$

**b.** $\frac{1}{100}$

**c.** $\frac{15}{100}$

Change the following percents to fractions, and reduce the fractions to lowest terms.

**d.** $\frac{95}{100}$

**a.** $30\% = \frac{30}{100} =$ **b.** $10\% = \frac{10}{100} =$ **c.** 80% =

**d.** 75% = **e.** 25% = **f.** 50% =

## 4

Certain common percents, such as $33\frac{1}{3}\%$ and $66\frac{2}{3}\%$, require a lengthier technique when converting to a fraction. This technique is illustrated below.

$$33\tfrac{1}{3}\% = \frac{33\frac{1}{3}}{100}$$

In order to complete this conversion, $33\frac{1}{3}$ must be changed to an improper fraction: 100/3.

$$\frac{\frac{100}{3}}{100} = \left(\frac{100}{3}\right) \times \left(\frac{1}{100}\right) = \left(\frac{\overset{1}{\cancel{100}}}{3}\right) \times \left(\frac{1}{\underset{1}{\cancel{100}}}\right) = \frac{1}{3}$$

Use the same steps to convert $66\frac{2}{3}\%$ to a fraction.

$$66\tfrac{2}{3}\% = \frac{66\frac{2}{3}}{100} = \frac{\frac{200}{3}}{100} = \left(\frac{200}{3}\right) \times \left(\frac{1}{100}\right) =$$

## 5

Convert the following percents to fractions.

**a.** $37\frac{1}{2}\% = \dfrac{37\frac{1}{2}}{100} = \dfrac{\frac{75}{2}}{100} = \left(\dfrac{75}{2}\right) \times \left(\dfrac{1}{100}\right) =$

**b.** $12\frac{1}{2}\% = \dfrac{12\frac{1}{2}}{100} = \dfrac{\frac{25}{2}}{100} = (\quad) \times (\quad) =$

**c.** $62\frac{1}{2}\% = \dfrac{62\frac{1}{2}}{100} = \dfrac{\quad}{100} = (\quad) \times (\quad) =$

## 6

The conversions in the two previous frames are time consuming and can be avoided if memorized. The following tables list common mixed-number percents and their equivalent fractions.

**Table 6-1**

| |
|---|
| $33\frac{1}{3}\% = \frac{1}{3}$ |
| $66\frac{2}{3}\% = \frac{2}{3}$ |

**Table 6-2**

| |
|---|
| $12\frac{1}{2}\% = \frac{1}{8}$ |
| $37\frac{1}{2}\% = \frac{3}{8}$ |
| $62\frac{1}{2}\% = \frac{5}{8}$ |
| $87\frac{1}{2}\% = \frac{7}{8}$ |

A more complete table of conversions will appear later in this section.

**3. a.** $\frac{3}{10}$

**b.** $\frac{1}{10}$

**c.** $\frac{80}{100} = \frac{8}{10} = \frac{4}{5}$

**d.** $\frac{75}{100} = \frac{3}{4}$

**e.** $\frac{25}{100} = \frac{1}{4}$

**f.** $\frac{50}{100} = \frac{1}{2}$

**4.** $\frac{2}{3}$

**5. a.** $\frac{3}{8}$

**b.** $\frac{25}{2} \times \frac{1}{100} = \frac{1}{8}$

**c.** $\dfrac{\frac{125}{2}}{100} = \left(\dfrac{125}{2}\right) \times \left(\dfrac{1}{100}\right) = \dfrac{5}{8}$

## 7

As noted in Chapter 2, a fraction can be converted to a decimal by dividing the denominator into the numerator. Therefore, any fraction with a denominator of 100 can be converted to a decimal by dividing the numerator by 100. This is illustrated below.

$$\frac{37}{100} = 100\overline{)37} = 0.37$$

$$\frac{61}{100} = 100\overline{)61} = 0.61$$

Complete the following conversion.

$\frac{83}{100} = 100\overline{)83} =$ ________

## 8

7. 0.83

Any division by 100 can be performed by shifting the decimal point two places to the left. This is called the *decimal-point-shift* shortcut.

$$\frac{63}{100} = \underset{\smile}{63.} = 0.63$$

Complete the following conversions.

a. $\frac{47}{100} = \underset{\smile}{47.} =$ ________ b. $\frac{29}{100} =$ ________ $=$ ________

## 9

8. a. 0.47
b. $\underset{\smile}{29.} = 0.29$

In some cases it is necessary to write a 0 when converting from a fraction to a decimal.

$$\frac{5}{100} = \underset{\smile}{05.} = 0.05$$

Complete the following conversions.

a. $\frac{3}{100} = \underset{\smile}{03.} =$ ________ b. $\frac{7}{100} =$ ________ $=$ ________

## 10

9. a. 0.03
b. $\underset{\smile}{07.} = 0.07$

The decimal-point-shift shortcut can be used when converting a percent to a decimal. This process is illustrated below.

$$31\% = \frac{31}{100} = \underset{\smile}{31.} = 0.31$$

Therefore, 31% = 0.31.

Convert the following percents to decimals.

a. 43% = b. 91% =

c. 15% = d. 9% =

## 11

The procedure in the previous frame can be stated as follows: when converting from a percent to a decimal, drop the percent sign and move the decimal point two places to the left.

Convert the following percents to decimals.

**a.** 54% = **b.** 83% =

**c.** 12% = **d.** 4% =

10. **a.** 0.43
**b.** 0.91
**c.** 0.15
**d.** 0.09

## 12

The same procedure applies to percents that contain decimals. For example:

$$25.8\% = 25.8 = 0.258$$

Convert the following percents to decimals.

**a.** 15.2% = **b.** 62.5% =

**c.** 12.5% = **d.** 6.25% =

11. **a.** 0.54
**b.** 0.83
**c.** 0.12
**d.** 0.04

## 13

The following table shows some commonly used percents and their equivalent fractions and decimals.

12. **a.** 0.152
**b.** 0.625
**c.** 0.125
**d.** 0.0625

**Table 6-3**

| *Common percents* | *Equivalent fractions* | *Equivalent decimals* |
|---|---|---|
| $12\frac{1}{2}\%$ | $\frac{1}{8}$ | 0.125 |
| 25% | $\frac{1}{4}$ | 0.250 |
| $33\frac{1}{3}\%$ | $\frac{1}{3}$ | 0.333 |
| $37\frac{1}{2}\%$ | $\frac{3}{8}$ | 0.375 |
| 50% | $\frac{1}{2}$ | 0.500 |
| $62\frac{1}{2}\%$ | $\frac{5}{8}$ | 0.625 |
| $66\frac{2}{3}\%$ | $\frac{2}{3}$ | 0.667 |
| 75% | $\frac{3}{4}$ | 0.750 |
| $87\frac{1}{2}\%$ | $\frac{7}{8}$ | 0.875 |
| 100% | 1 | 1.000 |

## 14

Because *percent* means "per hundred," any percent that is a multiple of 100 is equivalent to a whole number. This is shown below.

$$600\% = \frac{600}{100} = \frac{6}{1} = 6$$

Convert the following percents to their whole number equivalents.

**a.** 100% = **b.** 800% =

**c.** 400% = **d.** 2,500% =

## 15

Percents such as 250% can be converted to mixed-number equivalents.

$$250\% = 200\% + 50\% = 2 + \frac{1}{2} = 2\frac{1}{2}$$

Convert the following percents to their mixed-number equivalents.

**a.** 525% = **b.** $133\frac{1}{3}\%$ =

**c.** 275% = **d.** $366\frac{2}{3}\%$ =

**14. a.** 1
**b.** 8
**c.** 4
**d.** 25

## 16

Most often, large percents are converted to decimal-number equivalents. This is done using the decimal-point shift. For example, to convert 348% to an equivalent decimal:

$$348\% = 348. = 3.48$$

Convert the following percents to their decimal-number equivalents.

**a.** 495% = **b.** 103% =

**c.** 1,942% = **d.** 888% =

**15. a.** $5\frac{1}{4}$
**b.** $1\frac{1}{3}$
**c.** $2\frac{3}{4}$
**d.** $3\frac{2}{3}$

## 17

Percents smaller than 1% are converted to decimals using the same techniques that are used to convert other percents. The most common conversion technique is again the decimal-point shift. For example, to convert 0.75% to an equivalent decimal:

$$0.75\% = 00.75 = 0.0075$$

Convert the following percents to their decimal-number equivalents.

**a.** 0.2% = **b.** 0.08% =

**c.** 0.35% = **d.** 0.001% =

**16. a.** 4.95
**b.** 1.03
**c.** 19.42
**d.** 8.88

**18**

Answers to frame 17.

**a.** 0.002 **b.** 0.0008 **c.** 0.0035 **d.** 0.00001

# Exercise Set, Section 6-1

## Converting Percents to Fractions and Decimals

Convert the following percents to fractions.

**1.** 25% = **2.** 51% = **3.** $33\frac{1}{3}\%$ = **4.** 70% =

**5.** $62\frac{1}{2}\%$ = **6.** 40% = **7.** 82% = **8.** 8.5% =

Convert the following percents to whole numbers or mixed numbers.

**9.** 300% = **10.** 250% = **11.** 175% = **12.** 1,500% =

**13.** $666\frac{2}{3}\%$ = **14.** 140% = **15.** 1,250% =

Convert the following percents to decimals.

**16.** 85% = **17.** 52% = **18.** 9% = **19.** 41% =

**20.** 15% = **21.** 1% = **22.** 91% = **23.** 38% =

**24.** 42% = **25.** 11% = **26.** 0.65% = **27.** 0.03% =

**28.** 0.75% = **29.** 0.99% = **30.** 2.4% = **31.** 7% =

**32.** 1.06% = **33.** 9.01% =

# Supplementary Exercise Set, Section 6-1

Convert the following percents to fractions.

**1.** 75% **2.** 43% **3.** $66\frac{2}{3}\%$ **4.** 80% **5.** $87\frac{1}{2}\%$

Convert the following percents to whole numbers or mixed numbers.

**6.** 400% **7.** 320% **8.** $162\frac{1}{2}\%$ **9.** 3,500% **10.** $333\frac{1}{3}\%$

Convert the following percents to decimals.

**11.** 65% **12.** 48% **13.** 7% **14.** 57% **15.** 18%

**16.** 3% **17.** 89% **18.** 31% **19.** 0.63% **20.** 0.04%

**21.** 0.0085% **22.** 9.5% **23.** 0.75% **24.** 0.005% **25.** 0.89%

# Section 6-2 Converting Fractions and Decimals to Percents

**1**

The previous section showed how percents can be converted to fractions and decimals. This section will show how fractions and decimals can be converted to percents.

Any fraction with a denominator of 100 can easily be converted to a percent, because percent means "per 100." Therefore:

$$\frac{27}{100} = 27 \text{ "per 100"} = 27\%$$

Complete the following conversions.

**a.** $\frac{69}{100} =$ **b.** $\frac{41}{100} =$

**c.** $\frac{22}{100} =$ **d.** $\frac{73}{100} =$

**2**

Convert each of the following fractions to percents.

**a.** $\frac{2}{100} =$ **b.** $\frac{25}{100} =$

**c.** $\frac{75}{100} =$ **d.** $\frac{99}{100} =$

**1. a.** 69% **b.** 41% **c.** 22% **d.** 73%

**3**

The fractions in the previous frame also can be written in decimal form, as shown below.

$$\frac{2}{100} = 0.02 = 2\%$$

$$\frac{25}{100} = 0.25 = 25\%$$

**2. a.** 2% **b.** 25% **c.** 75% **d.** 99%

Complete the following.

**a.** $\frac{75}{100} = 0.75 =$ **b.** $\frac{99}{100} = 0.99 =$

**4**

A decimal can be converted to a percent by shifting the decimal point two places to the right. Therefore:

3. a. 75%
   b. 99%

0.27 = .27 = 27%

0.08 = .08 = 8%

Complete these conversions.

**a.** 0.21 = **b.** 0.58 =

**c.** 0.03 = **d.** 0.71 =

**5**

The decimal-point-shift conversion also works with decimals that contain more than two digits. This is shown below.

4. a. 21%
   b. 58%
   c. 3%
   d. 71%

0.385 = .385 = 38.5%

0.7412 = .7412 = 74.12%

Complete these conversions.

**a.** 0.428 = **b.** 0.654 =

**c.** 0.9541 = **d.** 0.1112 =

**6**

The decimal-point-shift conversion also may be used with decimals containing one digit.

5. a. 42.8%
   b. 65.4%
   c. 95.41%
   d. 11.12%

0.4 = .40 = 40%

0.1 = .10 = 10%

Complete these conversions.

**a.** 0.3 = **b.** 0.9 =

**c.** 0.2 = **d.** 0.5 =

**7**

Fractions can be converted to percents by first converting to a decimal and then using the decimal-point-shift conversion. This is shown below.

6. a. 30%
   b. 90%
   c. 20%
   d. 50%

$\frac{2}{5} = 5\overline{)2.0}$ (quotient 0.4)  0.4 = .40 = 40%

$\frac{3}{4} = 4\overline{)3.0}$ (quotient 0.75)  0.75 = .75 = 75%

Complete these conversions.

**a.** $\frac{1}{4} =$ **b.** $\frac{3}{5} =$

**c.** $\frac{1}{10} =$ **d.** $\frac{1}{2} =$

## 8

In some cases, the conversion will have a remainder. The remainder can be written as a fraction as follows:

7. **a.** 25% **b.** 60% **c.** 10% **d.** 50%

$$\frac{1}{3} = 3\overline{)1.00}^{\,0.33\frac{1}{3}} \qquad .33\frac{1}{3} = 33\frac{1}{3}\%$$

$$\frac{7}{8} = 8\overline{)7.00}^{\,0.87\frac{1}{2}} \qquad .87\frac{1}{2} = 87\frac{1}{2}\%$$

```
   0.87½
8 )7.00
   6 4
     60
     56
      4
```

$$\left[\textit{Note: } \frac{4}{8} = \frac{1}{2}\right]$$

Complete these conversions.

**a.** $\frac{2}{3} =$ **b.** $\frac{5}{8} =$

## 9

In many industrial applications percents are reported several places to the right of the decimal. Therefore, percents such as 31.5%, 52.55%, and 5.01% are common.

8. **a.** $66\frac{2}{3}\%$ **b.** $62\frac{1}{2}\%$

In all additions and subtractions of percents, it is necessary to line up the decimal points. Notice that, when the three percents below are added, the decimal points are written directly beneath one another.

```
   31.5 %
   52.55%
 +  5.01%
 --------
   89.06%   or   89.1%
```

Perform the following additions and subtractions.

**a.** 25.1% + 5.49% = **b.** 75.81% − 14.4% =

## 10

Additions and subtractions of percents can be performed on a calculator. The numbers—but not the percent sign—should be entered in the calculator. This procedure is shown below.

9. **a.** 30.6% **b.** 61.4%

Add: 47.2% + 5.35% + 92.1%

| *Enter* | *Press* | *Display* |
|---|---|---|
| 47.2 | [+] | 47.2 |
| 5.35 | [+] | 52.55 |
| 92.1 | [+] | 144.65 |

The answer is 144.65% or 144.7%.

Use a calculator to perform these additions and subtractions.

**a.** 58.0% + 5.2% + 9.1% =
**b.** 41.5% − 3.22% =
**c.** 57.28% − 5.2% =

**11**

Answers to frame 10.

**a.** 72.3% **b.** 38.3% **c.** 52.1%

# Exercise Set, Section 6-2

## Converting Fractions and Decimals to Percents

Convert the following fractions and decimals to percents.

**1.** $\frac{1}{100} =$ **2.** $\frac{28}{100} =$ **3.** $\frac{1}{4} =$ **4.** $\frac{3}{3} =$ **5.** $\frac{2}{3} =$

**6.** $\frac{5}{50} =$ **7.** $\frac{3}{4} =$ **8.** $\frac{1}{3} =$ **9.** $\frac{7}{8} =$ **10.** $\frac{37}{100} =$

**11.** 0.4 = **12.** 0.78 = **13.** 0.972 = **14.** 0.1991 =

**15.** 0.04 = **16.** 0.18 = **17.** 0.01 = **18.** 0.23 =

Perform the following additions and subtractions.

**19.** 55.2% + 20.1% = **20.** 14.8% − 4.5% = **21.** 25.01% + 22.7% + 35.80% =

**22.** 16.3% − 2.8% = **23.** 1.1% + 0.8% + 4.2% = **24.** 400% − 125% =

**25.** 0.8% + 1.9% + 25.0% = **26.** 0.05% − 0.01% = **27.** 95.6% + 0.5% + 4.5% =

**28.** 600.8% − 10.05% = **29.** 85.9% + 0.02% + 6.3% = **30.** 0.9% − 0.01% =

# Supplementary Exercise Set, Section 6-2

Convert the following fractions and decimals to percents.

**1.** $\frac{3}{100}$ **2.** $\frac{34}{100}$ **3.** $\frac{3}{4}$ **4.** $\frac{4}{4}$ **5.** $\frac{1}{3}$

**6.** 0.3 **7.** 0.92 **8.** 0.675 **9.** 0.2114 **10.** 0.03

Perform the following additions and subtractions.

**11.** 46.2% + 30.4% =

**12.** 15.7% − 5.4% =

**13.** 24.01% + 22.6% + 33.70% =

**14.** 18.3% − 3.4% =

**15.** 2.9% + 0.7% + 3.3% =

**16.** 500% − 175% =

**17.** 0.9% + 2.1% + 36.6% =

**18.** 0.07% − 0.01% =

**19.** 85.2% + 14.5% + 0.5% =

**20.** 700.8% − 11.04% =

# Section 6-3 Solving Problems Involving Percents

## 1

Problems involving percents can be solved using the following information.

Percent means "per hundred"; therefore, any percent can be written as a ratio with a denominator of 100.

**Example** 50% can be written as $\frac{50}{100}$

A proportion is a statement of equality between two ratios.

**Examples** $\frac{50}{100} = \frac{N}{42}$ $\frac{N}{100} = \frac{21}{42}$ $\frac{50}{100} = \frac{21}{N}$

Write the following percents as ratios with denominators of 100.

**a.** 65% = **b.** 31.2% = **c.** 254% = **d.** 0.09% =

## 2

There are three types of percentage problems. The first type is shown below.

What is 50% of 86?

The percentage problem above can be written as a proportion with the ratio 50/100 equal to the ratio PART/TOTAL.

$$\frac{50}{100} = \frac{\text{PART}}{\text{TOTAL}}$$

In percentage problems, the number representing the total is the number preceded in the word problem by the word *of.* In the example above, 86 is the total. Therefore, this proportion can be written as:

$$\frac{50}{100} = \frac{\text{PART}}{86}$$

**1. a.** $\frac{65}{100}$ **b.** $\frac{31.2}{100}$ **c.** $\frac{254}{100}$ **d.** $\frac{0.09}{100}$

This proportion can then be rewritten

$$\frac{50}{100} = \frac{N}{86}$$

Therefore, the problem "what is 50% of 86?" can be written as the proportion:

$$\frac{50}{100} = \frac{N}{86} \leftarrow \text{the "of" number}$$

**3**

The percentage problem below is rewritten as a proportion.

80% of 40 is what number?

80% can be written as $\frac{80}{100}$.

40 is the total. (It is the "of" number.) The proportion is $\frac{80}{100} = \frac{N}{40}$.

Rewrite the following percentage problems as proportions.

**a.** Find 40% of 20.

**b.** What is 15% of 200?

**4**

Rewrite the following percentage problems as proportions.

**a.** 12.2% of 208 is what number?

**b.** Find 0.7% of 200.

**c.** What is 250% of 50?

**d.** 8% of 1,000 is what number?

**5**

After a percentage problem is rewritten as a proportion, the proportion can be solved for $N$ using the techniques from Chapter 5. This procedure is shown below.

12.2% of 208 is what number?

$$\frac{12.2}{100} = \frac{N}{208}$$

$$\frac{12.2}{100} \times \frac{N}{208}$$

$$100N = (12.2)(208)$$

$$N = \frac{(12.2)(208)}{100}$$

$$N = 25.376 \quad \text{(calculator answer)}$$

Therefore, 12.2% of 208 is 25.4.

Find the following.

**a.** Find 0.70% of 200.

$$\frac{0.70}{100} = \frac{N}{200}$$

**b.** What is 250% of 50.0?

$$\frac{250}{100} = \frac{N}{50.0}$$

**3. a.** $\frac{40}{100} = \frac{N}{20}$

**b.** $\frac{15}{100} = \frac{N}{200}$

**4. a.** $\frac{12.2}{100} = \frac{N}{208}$

**b.** $\frac{0.7}{100} = \frac{N}{200}$

**c.** $\frac{250}{100} = \frac{N}{50}$

**d.** $\frac{8}{100} = \frac{N}{1,000}$

**c.** 8.0% of 1,000 is what number?

$$\frac{8.0}{100} = \frac{N}{1{,}000}$$

## 6

The calculator can be used to perform the calculations to solve percentage problems.

Find 9.00% of 750.

$$\frac{9.00}{100} = \frac{N}{750}$$

| *Enter* | *Press* | *Display* |
|---|---|---|
| 9.00 | ⊠ ($\times$) | 9 |
| 750 | ($\div$) | 6750 |
| 100 | (=) | 67.5 |

Therefore, $N = 67.5$.

Use a calculator to find the following.

**a.** What is 0.25% of 37?

$$\frac{0.25}{100} = \frac{N}{37}$$

**b.** Find 110% of 60.

$$\frac{110}{100} = \frac{N}{60}$$

5. **a.** $N = 1.4$
**b.** $N = 125$
**c.** $N = 80$

## 7

Rewrite the following percentage problems as proportions and solve.

**a.** Find 5.0% of 86.

**b.** What is 47% of 65?

**c.** 150% of 44 is what number?

**d.** Find 0.070% of 5,000.

6. **a.** $N = 0.093$
**b.** $N = 66$

## 8

An example of the second type of percentage problem is shown below.

What percent of 300 is 180?

The percent is unknown and can be written as $N\%$, or $N/100$. The ratio $N/100$ is equal to $180/300$. Therefore, the percentage problem above can be written as the following proportion:

$$\frac{N}{100} = \frac{180}{300} \leftarrow \text{the number after "of"}$$

Rewrite the following percentage problems as proportions.

**a.** 28 is what percent of 50?

**b.** What percent of 3.50 is 2.1?

**c.** What percent of 7 is 35?

**d.** 500 is what percent of 50?

7. **a.** $\frac{5.0}{100} = \frac{N}{86}$
$N = 4.3$
**b.** $\frac{47}{100} = \frac{N}{65}$
$N = 31$
**c.** $\frac{150}{100} = \frac{N}{44}$
$N = 66$
**d.** $\frac{0.070}{100} = \frac{N}{5{,}000}$
$N = 3.5$

## 9

The percentage problem below is rewritten as a proportion and solved using the techniques from Chapter 5.

What percent of 700 is 280?

$$\frac{N}{100} = \frac{280}{700}$$

$$\frac{N}{100} \times \frac{280}{700}$$

$$700N = (100)(280)$$

$$N = \frac{(100)(280)}{(700)}$$

$$N = 40.0$$

Therefore, $\frac{N}{100} = \frac{40}{100} = 40.0\%$.

Find the following.

**a.** What percent of 350 is 70?

**b.** What percent of 400 is 600?

**c.** What percent of 500 is 2?

**8. a.** $\frac{N}{100} = \frac{28}{50}$

**b.** $\frac{N}{100} = \frac{2.1}{3.50}$

**c.** $\frac{N}{100} = \frac{35}{7}$

**d.** $\frac{N}{100} = \frac{500}{50}$

## 10

The calculator can be used as follows:

What percent of 241 is 106?

$$\frac{N}{100} = \frac{106}{241} \qquad N = \frac{(106)(100)}{(241)}$$

| *Enter* | *Press* | *Display* |
|---|---|---|
| 106 | [×] | 106 |
| 100 | [÷] | 10600 |
| 241 | [=] | 43.983402 |

Therefore, $N = 44.0\%$.

Use a calculator to find the following.

**a.** What percent of 45.2 is 15.8? ____________

**b.** What percent of 207.0 is 31.2? ____________

**9. a.** 20%
**b.** 150%
**c.** 0.4%

## 11

Rewrite the following percentage problems as proportions and solve.

**a.** Find 12% of 45.

**b.** What percent of 18 is 1.0?

**c.** What is 38% of 200?

**d.** What percent of 200 is 76?

**10. a.** 35.0%
**b.** 15.1%

## 12

An example of the third type of percentage problem is shown below.

75 is 15% of what number?

This can be written as a proportion:

$$\frac{15}{100} = \frac{75}{N} \leftarrow \text{"of what number"}$$

Rewrite the following percentage problems as proportions. (You do not need to solve the problems.)

**a.** 70% of what number is 55? **b.** 45 is 50% of what number?

**c.** 3.56 is 41.2% of what number? **d.** 100 is 0.02% of what number?

**11. a.** $\frac{12}{100} = \frac{N}{45}$

$N = 5.4\%$

**b.** $\frac{N}{100} = \frac{1.0}{18}$

$N = 5.6\%$

**c.** $\frac{38}{100} = \frac{N}{200}$

$N = 76\%$

**d.** $\frac{N}{100} = \frac{76}{200}$

$N = 38\%$

## 13

The percentage problem below is rewritten as a proportion and solved using the techniques from Chapter 5.

32 is 8% of what number?

$$\frac{8}{100} = \frac{32}{N}$$

$$\frac{8}{100} \times \frac{32}{N}$$

$$8N = (100)(32)$$

$$N = \frac{(100)(32)}{(8)}$$

$$N = 400$$

Therefore, 32 is 8% of 400.

Find the following.

**a.** 12 is 3% of what number? **b.** 75 is 150% of what number?

**c.** 11 is 0.25% of what number?

**12. a.** $\frac{70}{100} = \frac{55}{N}$

**b.** $\frac{50}{100} = \frac{45}{N}$

**c.** $\frac{41.2}{100} = \frac{3.56}{N}$

**d.** $\frac{0.02}{100} = \frac{100}{N}$

## 14

The calculator can be used as follows:

5.2 is 15% of what number?

$$\frac{15}{100} = \frac{5.2}{N} \qquad N = \frac{(100)(5.2)}{(15)}$$

| *Enter* | *Press* | *Display* |
|---|---|---|
| 100 | ⊠ | 100 |
| 5.2 | ÷ | 520 |
| 15 | = | 34.666666 |

Therefore, $N = 35$.

**13. a.** 400
**b.** 50
**c.** 4,400

Use a calculator to find the following.

**a.** 35.2 is 89% of what number? **b.** 387 is 155% of what number?

## 15

Rewrite the following percentage problems as proportions and solve.

**14. a.** 40
**b.** 250

**a.** Find 15% of 65. **b.** What percent of 56 is 2.0?

**c.** 8.5 is 12% of what number? **d.** 60 is 110% of what number?

**e.** What is 20% of 280? **f.** 0.290 is 0.0200% of what number?

## 16

Answers to frame 15.

**a.** $\frac{15}{100} = \frac{N}{65}$ $N = 9.8$ **b.** $\frac{N}{100} = \frac{2.0}{56}$ $N = 3.6$

**c.** $\frac{12}{100} = \frac{8.5}{N}$ $N = 71$ **d.** $\frac{110}{100} = \frac{60}{N}$ $N = 55$

**e.** $\frac{20}{100} = \frac{N}{280}$ $N = 56$ **f.** $\frac{0.0200}{100} = \frac{0.290}{N}$ $N = 1{,}450$

# Exercise Set, Section 6-3

## Solving Problems Involving Percents

Solve the following problems involving percents.

**1.** What is 60% of 115? **2.** What is 240% of 35? **3.** What is 45% of 98?

**4.** What is 115% of 82? **5.** What is 82% of 3.0? **6.** What is 128% of 110?

**7.** What percent of 150 is 3? **8.** What percent of 60 is 40? **9.** What percent of 99 is 11?

**10.** What percent of 80 is 120? **11.** What percent of 50 is 75? **12.** What percent of 150 is 200?

**13.** 58 is 70% of what number? **14.** 217 is 40.0% of what number? **15.** 0.50 is 10% of what number?

**16.** 2.80 is 200% of what number? **17.** 1.4 is 50% of what number? **18.** 500 is 400% of what number?

**19.** What percent of 1.5 is 0.30? **20.** What is 180% of 22?

## Supplementary Exercise Set, Section 6-3

Solve the following problems involving percents.

**1.** What number is 40.0% of 280?

**2.** What number is 135% of 42?

**3.** What percent of 140 is 4.0?

**4.** What number is 0.06% of 5,000?

**5.** What percent of 30 is 40?

**6.** 49 is 50% of what number?

**7.** What number is 11.3% of 52?

**8.** What percent of 400 is 1.0?

**9.** 235 is 130% of what number?

**10.** 0.09 is 10% of what number?

## Section 6-4 Applied Problems

**1**

The following problem is solved using the five-step procedure for problem solving. If 12.0% of the cost of an \$82,000 building is for carpentry work, how much was spent for carpentry?

*Step 1.* Determine what is being asked for.

cost of carpentry = $C$

*Step 2.* Determine what information is already known.

total cost of building = \$82,000

$$\%\text{ cost of carpentry} = \frac{12.0}{100}$$

*Step 3.* Find a mathematical model that describes the relationship.

$$\frac{12.0}{100} = \frac{\text{cost of carpentry}}{\text{total cost of building}}$$

*Step 4.* Substitute the data into the model.

$$\frac{12.0}{100} = \frac{C}{\$82{,}000}$$

*Step 5.* Do the calculations.

$$C = \frac{(\$82{,}000)(12.0)}{(100)}$$

$$C = \$9{,}840$$

**2**

Use the five-step procedure to solve this problem.

If 80% of the 3,000 machinists in a metropolitan area belong to a union, how many of the machinists belong to a union?

*Step 1.* Determine what is being asked for.
*Step 2.* Determine what information is already known.
*Step 3.* Find a mathematical model that describes the relationship.
*Step 4.* Substitute the data into the model.
*Step 5.* Do the calculations.

**3**

Answers to frame 2.

*Step 1.* Machinists who belong to a union $= M$

*Step 2.* Percent who belong to a union $= \frac{80}{100}$

total number of machinists = 3,000

*Step 3.* $\frac{80}{100} = \frac{\text{percent who belong to a union}}{\text{total number of machinists}}$

*Step 4.* $\frac{80}{100} = \frac{M}{3{,}000}$

*Step 5.* $M = 2{,}400$ machinists

**4**

The following applied problem is solved using the five-step procedure.

If 7.2 g of a 20 g solution are alcohol, what percent of the solution is alcohol?

*Step 1.* Determine what is being asked for.

$$\text{percent of solution that is alcohol} = \frac{N}{100}$$

*Step 2.* Determine what information is already known.

total solution = 20 g

alcohol in solution = 7.2 g

*Step 3.* Find a mathematical model that describes the relationship.

$$\frac{N}{100} = \frac{\text{alcohol in solution}}{\text{total solution}}$$

*Step 4.* Substitute the data into the model.

$$\frac{N}{100} = \frac{7.2\text{ g}}{20\text{ g}}$$

*Step 5.* Do the calculations.

$$N = \frac{(7.2)(100)}{(20)}$$

$$N = 36; \qquad \frac{36}{100} = 36\%$$

Therefore, 36% of the solution is alcohol.

**5**

Use the five-step procedure to solve this applied problem.

If 9.9 g of a 45 g solution is alcohol, what percent of the solution is alcohol?

*Step 1.* Determine what is being asked for.
*Step 2.* Determine what information is already known.
*Step 3.* Find a mathematical model that describes the relationship.
*Step 4.* Substitute the data into the model.
*Step 5.* Do the calculations.

## 6

Answers to frame 5.

*Step 1.* percent of solution that is alcohol $= \frac{N}{100}$

*Step 2.* total solution = 45 g
alcohol in solution = 9.9 g

*Step 3.* $\frac{N}{100} = \frac{\text{alcohol in solution}}{\text{total solution}}$

*Step 4.* $\frac{N}{100} = \frac{9.9}{45}$

*Step 5.* $N = 22; \frac{22}{100} = 22\%$

## 7

This problem is solved using the five-step procedure.

In a study of mathematics skills needed for technology, 139 questionnaires were returned. This figure represented 60% of the questionnaires sent out. How many questionnaires were sent out?

*Step 1.* Determine what is being asked for.

number of questionnaires sent out $= N$

*Step 2.* Determine what information is already known.

number of questionnaires returned = 139

percent of questionnaires returned $= \frac{60}{100}$

*Step 3.* Find the mathematical model that describes the relationship.

$$\frac{60}{100} = \frac{\text{number of questionnaires returned}}{\text{number of questionnaires}}$$

*Step 4.* Substitute the data into the model.

$$\frac{60}{100} = \frac{139}{N}$$

*Step 5.* Do the calculations.

$$N = \frac{(139)(100)}{(60)}$$

$N = 231.67$; $N = 232$ (Round to a whole number. These are exact numbers.)

Therefore, 232 questionnaires were sent out.

## 8

Use the five-step procedure to solve this applied problem.

A certain mineral comprises 0.900% of the weight of an adult's body. If a person's body contains 1.44 pounds of the mineral, how much does this person weigh?

*Step 1.* Determine what is being asked for.
*Step 2.* Determine what information is already known.
*Step 3.* Find a mathematical model that describes the relationship.
*Step 4.* Substitute the data into the model.
*Step 5.* Do the calculations.

## 9

Answers to frame 8.

*Step 1.* total weight of person = $N$
*Step 2.* weight of mineral = 1.44 lb

$$\text{percent that is the mineral} = \frac{0.900}{100}$$

*Step 3.* $\frac{0.900}{100} = \frac{\text{weight of the mineral}}{\text{total weight of the person}}$

*Step 4.* $\frac{0.900}{100} = \frac{1.44}{N}$

*Step 5.* $N = 160$ lb

## 10

As computers become integrated into the manufacturing process and a new generation of factory automation occurs, it will be important for effective workers to have a basic knowledge of electronics. Percentage problems using electrical measurements are illustrated below.

In an electrical circuit, resistance to the flow of electrical current, $R$, is found by the formula $R = E/I$, where $E$ is the voltage and $I$ is the current. Use the formula to complete Table 6.4. (Round to the nearest whole number.)

**Table 6-4** Measured voltage across and current through a 1,000-ohm resistor

| *Voltage* $E$ | *Current* $I$ | *Calculated resistance* $R = E/I$ |
|---|---|---|
| 2.5 | 0.0022 | **a.** |
| 5.0 | 0.0051 | **b.** |
| 7.5 | 0.0077 | **c.** |
| 10.0 | 0.0100 | **d.** |
| 20.0 | 0.0170 | **e.** |

## 11

Each calculated resistance in Table 6-4 should theoretically have been 1,000 ohms. However, in only one case was the calculated resistance actually 1,000 ohms. The concept of *percent error* is used to describe the difference between the calculated resistance and the theoretical resistance in the other cases.

**10. a.** 1,136
**b.** 980
**c.** 974
**d.** 1,000
**e.** 1,176

To illustrate the concept of percent error, look at the calculated resistance for **a** in Table 6-4 ($R = 1{,}136$ ohms). This is larger than the theoretical value of 1,000 ohms. To calculate the percent error between 1,000 ohms and 1,136 ohms, use the following steps:

1. Subtract the calculated resistance value from the theoretical resistance value to find the difference between the two.

$$\text{difference} = \text{theoretical resistance} - \text{calculated resistance}$$

$$\text{difference} = 1{,}000 \text{ ohms} - 1{,}136 \text{ ohms}$$

$$\text{difference} = -136 \text{ ohms}$$

**2.** Set up the following proportion and solve for the percent error.

$$\frac{\text{difference}}{\text{theoretical resistance}} = \frac{\text{percent error}}{100\%}$$

$$\frac{-136 \text{ ohms}}{1{,}000 \text{ ohms}} = \frac{\text{percent error}}{100\%}$$

$$\text{percent error} = -13.6\%$$

The negative sign tells us that the calculated resistance value was larger than the theoretical resistance value. If the opposite is true, the percent error will have a positive sign. Use the same procedure to calculate the percent error for part **b** of Table 6-4.

## 12

Complete Table 6-5.

**Table 6-5** Percent error in a 1,000-ohm resistor

| *Voltage* $E$ | *Current* $I$ | *Calculated resistance* $R = E/I$ | *% error* |
|---|---|---|---|
| 2.5 | 0.0022 | 1,136 | −13.6% |
| 5.0 | 0.0051 | 980 | 2.0% |
| 7.5 | 0.0070 | 974 | **a.** |
| 10.0 | 0.0100 | 1,000 | **b.** |
| 20.0 | 0.0170 | 1,176 | **c.** |

**11.** % error =

$$\frac{(20 \text{ ohms})(100\%)}{(1{,}000 \text{ ohms})}$$

% error = 2.0%

## 13

A percent error can also be calculated by comparing the measured voltage to the theoretical voltage as shown.

$$\text{difference} = \text{theoretical} - \text{measured}$$

$$\frac{\text{difference}}{\text{theoretical}} = \frac{\text{percent error}}{100\%}$$

Use the proportion above to complete Table 6-6.

**Table 6-6** Measured and theoretical voltage across terminals

| *Terminals* | *Measured voltage* | *Theoretical voltage* | *% error* |
|---|---|---|---|
| A to B | 6.00 | 6.25 | **a.** |
| B to C | 2.75 | 2.19 | **b.** |
| C to D | 5.20 | 5.37 | **c.** |
| D to E | 7.00 | 7.31 | **d.** |
| E to F | 9.70 | 8.76 | **e.** |

**12. a.** 2.6%
**b.** 0.0%
**c.** −17.6%

## 14

Answers to frame 13.

**a.** 4.20% **b.** −25.6% **c.** 3.20% **d.** 4.20% **e.** −10.7%

# Exercise Set, Section 6-4

## Applied Problems

1. In a recent study, 58 mechanical technicians were sent a survey concerning the mathematics they need to do their jobs; 34 technicians completed the survey. What percent completed the survey? (Round to the nearest percent.)

2. A certain mineral makes up 0.70% of the weight of a human body. How many pounds of the mineral are in the body of a 65-lb child?

3. Brazing metal contains 15% zinc. How much zinc is in 42.00 kilograms of brazing steel?

4. Quality control standards allow 5% of the parts produced by a machine to be defective. If 160 parts are produced, how many defective parts can be expected?

5. A public health official reported that six out of ten people who had the flu recovered within two days. What percent recovered within two days?

6. A factory employing 450 workers recently laid off 20% of their workers. How many workers were laid off?

7. A firm purchases a microcomputer system for $6,800. Of that amount, $1,700 was for a printer. What percent of the cost was for the printer?

8. A new machine can produce a product in 65% of the time the old machine could produce it. If the old machine could produce it in 2.0 hr, how long will it take the new machine?

9. A malfunctioning machine produced 21 defective parts. If this represents 42% of the total number of parts that machine produced in an hour, how many parts did the machine produce in an hour?

10. A manufacturing process takes 180 hours to complete. If a technological advancement can reduce that time to 80.0% of 180 hours, what will the new time of the manufacturing process be?

11. Forty ounces of an alloy contain 2 ounces of lead. What percent of the alloy is lead?

12. There were 534 defective parts manufactured. If this represents 11% of all parts manufactured, how many total parts were manufactured? (Round to a whole number.)

13. Job application forms were completed correctly by 30 out of 36 people. What percent completed their forms correctly?

14. A lab technician was able to analyze 21 out of 28 samples. What percent of the samples did she analyze?

15. A forging weighed 2,500 kilograms. When a part was machined from the forging, 14% of the material was removed. How much material was removed?

16. What is the percent error when the measured voltage across the terminals is 2.5 volts and the theoretical voltage is 2.8 volts?

## Supplementary Exercise Set, Section 6-4

1. In a recent study, 70 technicians were surveyed concerning the mathematics they need to do their jobs; 39 completed the survey. What percent completed the survey? (Round to the nearest percent.)

2. In the survey above, 18 employers of technicians were also surveyed; 13 completed the survey. What percent completed the survey? (Round to the nearest percent.)

3. In a certain state, 65% of the biomedical technicians work in urban hospitals. If there are 140 biomedical technicians in the state, how many are working in urban hospitals? (Round to the nearest whole number.)

4. In a computer graphics firm, 12% of the 421 employees are over 35 years of age. How many are over 35? (Round to a whole number.)

5. A technician mixes a 30-gram solution that contains 9.2 grams of alcohol. What percent of the solution is alcohol?

6. If a technician wishes to mix a 40-gram solution that contains the same percentage of alcohol as the solution in problem 5, how much alcohol should be used?

7. A certain mineral comprises 0.700% of the weight of an adult's body. If a person contains 1.52 pounds of the mineral, how much does the person weigh?

8. A manufacturer in a small town spends 68% of its annual expenses for salaries. If salaries amount to $47,000, what are the total expenses for the manufacturer?

9. An electromechanical technician received a 15% pay raise. If she was making $21,000 a year, how much will her raise be? What will her new salary be?

10. One of the authors of this text weighs 170 lbs. He wants to lose 8.00% of his weight. How much should he lose? What will his new weight be?

11. Find the percent error when the measured voltage is 5.6 volts and the theoretical voltage is 5.2 volts.

12. Find the percent error when the measured resistance is 10,000 ohms and the calculated resistance is 9,090 ohms.

# Summary

1. A percent can be written as a fraction with a denominator of 100.
2. A percent can be written in decimal form by dividing the numerator by 100. This is most quickly accomplished by shifting the decimal point two places to the left.
3. A decimal can be converted to a percent by shifting the decimal point two places to the right and writing a % sign.
4. The decimal points must be lined up carefully when adding and subtracting percents.
5. Problems involving percents can be solved by setting the ratio %/100 equal to the ratio PART/TOTAL.

# Chapter 6 Self-Test

## Percents

Convert the following percents to fractions.

**1.** 42% =

**2.** $66\frac{2}{3}\% =$

**3.** 65% =

Convert the following percents to whole numbers or mixed numbers.

**4.** 400% =

**5.** 175% =

**6.** 1,500% =

Convert the following percents to decimals.

**7.** 95% =

**8.** 9% =

**9.** 0.25% =

Convert the following fractions and decimals to percents.

**10.** $\frac{2}{100} =$

**11.** $\frac{3}{10} =$

**12.** $\frac{1}{3} =$

**13.** 0.7 =

**14.** 0.467 =

**15.** 0.006 =

Perform the following additions and subtractions.

**16.** 38.6% + 20.2% =

**17.** 13.6% − 5.8% =

**18.** 1.9% + 0.3% + 6.4% =

**19.** 0.07% − 0.02% =

Solve the following problems involving percents.

**20.** What is 40% of 112?

**21.** What is 320% of 30?

**22.** What percent of 180 is 20?

**23.** What percent of 30 is 45?

**24.** 36 is 44% of what number?

**25.** 42 is 110% of what number?

Solve the following applied problems.

**26.** An alloy that weighs 210 kilograms is 80.0% nickel. How many kilograms of nickel are in the alloy?

**27.** A survey showed that 18 out of 42 people working in an office came down with the flu. What percent had the flu? (Round to the nearest percent.)

**28.** A casting is poured that shrinks 2.0% as it cools. If the length of the casting is 85.0 inches before it cools, what is its length after it cools?

**29.** In a certain computer firm, 18 out of 32 employees are female. What percent of the employees are female? (Round to the nearest percent.)

**30.** Six percent of the lumber delivered to a building site was unusable. If 220 pieces of lumber were delivered, how many were unusable? (Round to a whole number.)

**31.** Find the percent error when the measured resistance is 460 ohms and the calculated resistance is 500 ohms.

**32.** What is the percent error when the measured voltage across the terminals is 9.2 volts and the theoretical voltage is 10.0 volts?

## CHAPTER 7

# The Metric System of Measurement

The metric system of measurement is widely used in science and industry. Mastery of this system is a requirement for technically trained people. Chapter 7 will teach you to measure length, area, volume, weight, and temperature in the metric system. In addition, several measuring devices are examined.

## Section 7-1 Basic Organization of the Metric System

**1**

The metric system of measurement is superior to the English system because it is based on the number 10. This fact means that all multiples and subdivisions of the standard units are based on the number 10. The metric system is contrasted with the English system of measurement below.

| | *Subdivision* | *Standard unit of length* | *Multiple* |
|---|---|---|---|
| *English* | 12 inches | 1 foot | 3 ft = 1 yd |
| *Metric* | 10 decimeters | 1 meter | 10 m = 1 dkm |

Note that, in the English system, the number 12 is used for a subdivision and the number 3 for a multiple. The metric system uses the number 10 consistently. The number 10, of course, is easy to multiply and divide by.

The metric system is based on the number ______.

**2**

Multiples and subdivisions of standard units are used because the standard unit is often too big or too small to work with. A prefix before the name of a unit in the metric system is used to indicate what multiple or subdivision is wanted. The most common prefixes are listed below. (A more complete table is given as Table 1 in the Appendix.)

**1.** 10

| *Prefix* | *Meaning* | |
|---|---|---|
| milli | 1,000 of these in a standard unit | *Subdivisions* |
| centi | 100 of these in a standard unit | |
| deci | 10 of these in a standard unit | |
| deka | 10 times the standard unit | *Multiples* |
| hecto | 100 times the standard unit | |
| kilo | 1,000 times the standard unit | |

You must know these prefixes. They are used for all units in the metric system.

How many of these subdivisions are in a standard unit?

**a.** milli ________ **b.** deci ________ **c.** centi ________

**3**

How many of the standard units are there in each of these multiples?

**a.** deka ________ **b.** kilo ________ **c.** hecto ________

**2. a.** 1,000
**b.** 10
**c.** 100

**4**

**a.** If there are 1,000 subdivisions in the standard unit, the prefix is ________.

**b.** For a multiple containing 1,000 standard units, the prefix is ________.

**c.** If there are 100 subdivisions in the standard unit, the prefix is ________.

**3. a.** 10
**b.** 1,000
**c.** 100

**5**

Indicate whether the prefix represents a multiple or a subdivision of a standard unit.

**a.** deci ________

**b.** kilo ________

**c.** milli ________

**4. a.** milli
**b.** kilo
**c.** centi

**6**

Answers to frame 5.

**a.** subdivision **b.** multiple **c.** subdivision

# Section 7-2 Measurement of Length in the Metric System

**1**

The standard unit of length in the metric system is the meter. The meter is 39.4 inches long, or slightly longer than a yard in the English system of measurement.

Meter

Yard

**2**

The meter is useful for measuring certain objects, but to measure objects shorter or longer than the meter, we use a subdivision or multiple of the meter. The subdivision or multiple is obtained by dividing or multiplying the standard meter by 10, 100, or 1,000.

The table below contains the subdivisions and multiples of the standard meter. Note that the abbreviations are not capitalized.

| *Unit* | *Abbreviation* | *Subdivision or Multiple* |
|---|---|---|
| millimeter | mm | 1,000 of these in a meter |
| centimeter | cm | 100 of these in a meter |
| decimeter | dm | 10 of these in a meter |
| meter | m | standard unit |
| dekameter | dkm | 10 meters |
| hectometer | hm | 100 meters |
| kilometer | km | 1,000 meters |

The most common metric units of length are listed below. These are the only ones used in this book.

| *Unit* | *Abbreviation* | *Subdivision or Multiple* |
|---|---|---|
| millimeter | mm | 1,000 of these in a meter |
| centimeter | cm | 100 of these in a meter |
| meter | m | standard unit |
| kilometer | km | 1,000 meters |

Give the abbreviations of the following.

**a.** kilometer ________ **b.** centimeter ________

**c.** meter ________ **d.** millimeter ________

**3**

Give the name of the unit for each abbreviation.

**a.** cm ________ **b.** km ________

**c.** mm ________ **d.** m ________

**2. a.** km
**b.** cm
**c.** m
**d.** mm

**4**

**a.** In one meter there are ________ mm.

**b.** In one meter there are ________ cm.

**c.** In one kilometer there are ________ m.

**3. a.** centimeter
**b.** kilometer
**c.** millimeter
**d.** meter

**5**

It is easy to remember how many centimeters are in 1 m. It is more difficult, however, to know how many centimeters are in 3.86 m. To solve this problem, conversion factors and dimensional analysis are used. The conversion factor is set up in the following manner.

**4. a.** 1,000
**b.** 100
**c.** 1,000

**Example** Change 3.86 m to centimeters.

$$3.86 \text{ m} \times \frac{100 \text{ cm}}{1 \text{ m}}$$

unit desired on top or in numerator (100 cm)

unit to be changed at bottom or in denominator (1 m)

Which of these are set up correctly? (Circle your selections.)

**a.** $2.6\text{ km} \times \dfrac{1{,}000\text{ m}}{1\text{ km}}$

**b.** $10\text{ cm} \times \dfrac{10\text{ mm}}{1\text{ cm}}$

**c.** $182\text{ cm} \times \dfrac{100\text{ cm}}{1\text{ m}}$

**d.** $82\text{ mm} \times \dfrac{1{,}000\text{ mm}}{1\text{ m}}$

**6**

The following problem illustrates the conversion procedure step by step.

5. **a** and **b**

**Example** Change 3.86 m to centimeters.

*Step 1.* Set up the conversion-factor formula.

$$3.86\text{ m} \times \frac{\text{cm}}{\text{m}}$$

*Step 2.* Now determine how many of the smaller units are in the larger unit and insert the information into the conversion-factor formula.
smaller unit = cm
larger unit = m
There are 100 cm in 1 m.

$$3.86\text{ m} \times \frac{100\text{ cm}}{1\text{ m}}$$

*Step 3.* Solve the problem.

$$3.86\text{ }\cancel{\text{m}} \times \frac{100\text{ cm}}{1\text{ }\cancel{\text{m}}} = 386\text{ cm}$$

Notice that, if the problem has been set up correctly, the units to be changed will cancel. If the units to be changed do not cancel, the conversion factor has been set up incorrectly. The numbers used to convert between units in the metric system are exact numbers, and the answers are not rounded off.

**7**

Change 4,870 mm to meters.

*Step 1.* Set up the formula.

$$4{,}870\text{ mm} \times \frac{\text{m}}{\text{mm}}$$

*Step 2.* Determine how many of the smaller units are in the larger unit and insert the information in the formula.
smaller unit = mm
larger unit = m
There are 1,000 mm in one meter.

$$4{,}870\text{ mm} \times \frac{1\text{ m}}{1{,}000\text{ mm}}$$

*Step 3.* Solve the problem.

$$4{,}870\text{ }\cancel{\text{mm}} \times \frac{1\text{ m}}{1{,}000\text{ }\cancel{\text{mm}}} = 4.87\text{ m}$$

**8**

Try this problem using the three steps.

Change 8,690 m to kilometers.

*Step 1.*

*Step 2.*

*Step 3.*

**9**

Answers to frame 8.

**1.** $8{,}690 \text{ m} \times \dfrac{\text{km}}{\text{m}}$ **2.** $8{,}690 \text{ m} \times \dfrac{1 \text{ km}}{1{,}000 \text{ m}}$ **3.** $8{,}690 \cancel{\text{m}} \times \dfrac{1 \text{ km}}{1{,}000 \cancel{\text{m}}} = 8.69 \text{ km}$

**10**

Change 0.05 m to millimeters.

*Step 1.*

*Step 2.*

*Step 3.*

**11**

Answers to frame 10.

**1.** $0.05 \text{ m} \times \dfrac{\text{mm}}{\text{m}}$ **2.** $0.05 \text{ m} \times \dfrac{1{,}000 \text{ mm}}{1 \text{ m}}$ **3.** $0.05 \cancel{\text{m}} \times \dfrac{1{,}000 \text{ mm}}{1 \cancel{\text{m}}} = 50 \text{ mm}$

**12**

Conversion problems also can be solved using a calculator.

Change 0.2 m to centimeters.

*Step 1.* $0.2 \text{ m} \times \dfrac{\text{cm}}{\text{m}}$

*Step 2.* $0.2 \text{ m} \times \dfrac{100 \text{ cm}}{1 \text{ m}}$

*Step 3.* $0.2 \cancel{\text{m}} \times \dfrac{100 \text{ cm}}{1 \cancel{\text{m}}} =$

| *Enter* | *Press* | *Display* |
|---|---|---|
| 0.2 | [×] | 0.2 |
| 100 | [=] | 20 |

The answer is 20 cm.

**13**

Change 150 cm to meters.

*Step 1.* $150 \text{ cm} \times \dfrac{\text{m}}{\text{cm}}$

*Step 2.* $150 \text{ cm} \times \dfrac{1 \text{ m}}{100 \text{ cm}}$

*Step 3.* $150 \not{\text{cm}} \times \dfrac{1 \text{ m}}{100 \not{\text{cm}}} =$

| *Enter* | *Press* | *Display* |
|---|---|---|
| 150 | ÷ | 150 |
| 100 | = | 1.5 |

The answer is 1.5 m.

**14**

Solve these problems.

**a.** Change 1.516 m to millimeters.

**b.** Change 15 mm to meters.

**c.** Change 138 cm to meters.

**d.** Change 2.312 km to meters.

**e.** Change 1,650 m to kilometers.

**15**

Answers to frame 14.

**a.** 1,516 mm **b.** 0.015 m **c.** 1.38 m
**d.** 2,312 m **e.** 1.65 km

## Section 7-3 Measurement of Area in the Metric System

**1**

Area is derived from length measurements: area = length × width

> **Example** What is the area of a space 2 cm long by 3 cm wide?
> area = length × width
> area = 2 cm × 3 cm = 6 cm$^2$

**2**

The diagram below shows the relationship between a square centimeter and a square millimeter.

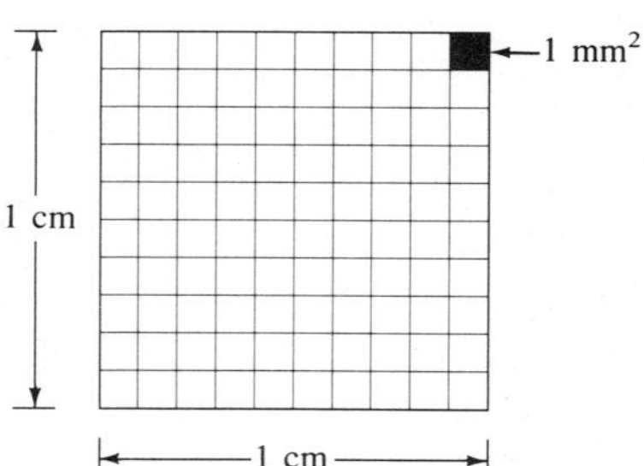

Thus, there are (10 mm × 10 mm), or 100 $mm^2$, in 1 $cm^2$. Notice that the numbers and units both are squared. All squared units in the metric system are related by multiples of 100 ($10^2$).

**a.** ______ $cm^2$ = 1 $dm^2$. **b.** 1 $cm^2$ = ______ $mm^2$.

**3**

What is the area of a rectangle 10.22 cm by 3.410 cm?

**2. a.** 100 (from 10 × 10)
**b.** 100 (from 10 × 10)

**4**

Answer to frame 3.

34.85 $cm^2$

# Exercise Set, Sections 7-1–7-3

## Basic Organization of the Metric System

How many of these subdivisions are in a standard unit?

**1.** centi ______ **2.** milli ______ **3.** deci ______

How many of the standard units are there in each of these multiples?

**4.** deka ______ **5.** kilo ______ **6.** hecto ______

## Measurement of Length in the Metric System

Complete the following equivalencies.

**7.** 1 km = ______ m **8.** 1 m = ______ mm **9.** 1 m = ______ cm **10.** 1 cm = ______ mm

Solve these problems using conversion factors and dimensional analysis.

**11.** 1.93 m = ______ mm **12.** 87 cm = ______ m **13.** 132 mm = ______ m

**14.** 0.86 m = ______ cm **15.** 0.67 m = ______ mm **16.** 976 cm = ______ m

**17.** 1,286 mm = ______ m **18.** 0.18 m = ______ cm **19.** 0.8 cm = ______ mm

**20.** 14 mm = ______ cm **21.** 1.6 cm = ______ mm **22.** 83 mm = ______ cm

## Measurement of Area in the Metric System

**23.** 1 $cm^2$ = ______ $mm^2$

**24.** A space 1.87 mm × 2.36 mm = ______ $mm^2$.

**25.** A rectangle 3.25 cm × 2.55 cm = ______ $cm^2$.

# Supplementary Exercise Set, Sections 7-1–7-3

Complete the following equivalencies.

**1.** 1 cm = ______ mm **2.** 1 km = ______ m **3.** 1 m = ______ cm **4.** 1 m = ______ mm

Solve these problems.

**5.** 2.76 m = ______ mm **6.** 193 cm = ______ m **7.** 426 mm = ______ m

**8.** 0.37 m = ______ cm **9.** 0.92 m = ______ mm **10.** 768 cm = ______ m

**11.** 1,467 mm = ______ m **12.** 0.14 m = ______ cm **13.** 0.7 cm = ______ mm

**14.** 18 mm = ______ cm **15.** 1.9 cm = ______ mm **16.** 89 mm = ______ cm

**17.** 1.09 m = ______ cm **18.** 0.08 km = ______ m

**19.** A rectangle 1.11 mm × 4.86 mm = ______ $mm^2$.

**20.** A square 1.46 cm × 1.46 cm = ______ $cm^2$.

# Section 7-4 Measurement of Volume in the Metric System

**1**

Volume measurements are derived from linear measurements. The volume of a box, for example, is found by multiplying the length times the width times the height of the box.

$$\text{volume of box} = l \times w \times h$$

If a box 10 cm on an edge is used, we have the following:

$$\text{volume} = 10\text{ cm} \times 10\text{ cm} \times 10\text{ cm}$$

$$\text{volume} = \left.\begin{matrix} 10 \times 10 \times 10 \\ \text{cm} \times \text{cm} \times \text{cm} \end{matrix}\right\} = 1{,}000\text{ cm}^3$$

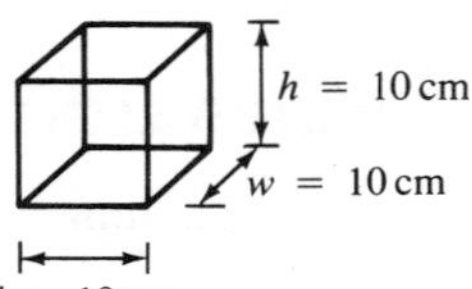


The answer is read as 1,000 cubic centimeters. Note that both the numbers and units are cubed.

To obtain volume, units of linear measurement must be ____________.

**2**

When converting linear measurements to volume measurements, we cube not only the numbers but also the ____________.

1. cubed

**3**

The volume contained in 1,000 $cm^3$ is the standard unit of volume in the metric system and is called the *liter*. The liter is very similar to the quart in the English system. The two most frequently used units are listed below.

2. units

| *Unit* | *Abbreviation* | *Subdivision* |
|---|---|---|
| milliliter | mℓ | 1,000 of these in a liter |
| liter | ℓ | standard unit |

Give the name or abbreviation of each of the following.

**a.** mℓ ________ **b.** ℓ ________

**c.** milliliter ________ **d.** liter ________

**4**

In 1 ℓ there are ________ mℓ.

3. **a.** milliliter
**b.** liter
**c.** mℓ
**d.** ℓ

**5**

A liter contains 1,000 $cm^3$. A liter also contains 1,000 mℓ. Using this information, we can show the relationship between the $cm^3$ and the mℓ.

4. 1,000

$$1{,}000 \text{ cm}^3 = 1\ \ell = 1{,}000 \text{ m}\ell$$

Since 1,000 $cm^3$ and 1,000 mℓ both equal 1 ℓ, they must also equal each other.

$$1{,}000 \text{ cm}^3 = 1{,}000 \text{ m}\ell \quad \text{or} \quad 1 \text{ cm}^3 = 1 \text{ m}\ell$$

If 1 $cm^3$ = 1 mℓ, then 6 $cm^3$ must equal ______ mℓ.

**6**

10 mℓ = ______ $cm^3$.

5. 6

**7**

Conversion factors and dimensional analysis can be used to change from one unit of volume to another.

6. 10

Change 3,289 mℓ to liters.

*Step 1.* $3{,}289 \text{ m}\ell \times \dfrac{\ell}{\text{m}\ell}$

*Step 2.* smaller unit = mℓ   There are 1,000 mℓ in 1 ℓ.
larger unit = ℓ

*Step 3.* $3{,}289 \cancel{\text{m}\ell} \times \dfrac{1\ \ell}{1{,}000 \cancel{\text{m}\ell}} = 3.289\ \ell$

Change 1,400 mℓ to liters.

*Step 1.*

*Step 2.*

*Step 3.*

**8**

---

Answers to frame 7.

1. $1{,}400 \text{ m}\ell \times \frac{\ell}{\text{m}\ell}$
2. smaller unit = mℓ
   larger unit = ℓ
   1ℓ = 1,000 mℓ
3. $1{,}400 \cancel{\text{m}\ell} \times \frac{1\ \ell}{1{,}000 \cancel{\text{m}\ell}} = 1.4\ \ell$

**9**

---

The calculator can be used to solve this type of problem.

**Example** Change 2.46 ℓ to milliliters.

*Step 1.* $2.46\ \ell \times \frac{\text{m}\ell}{\ell}$

*Step 2.* smaller unit = mℓ   In 1 ℓ there are 1,000 mℓ.
larger unit = ℓ

*Step 3.* $2.46 \cancel{\ell} \times \frac{1{,}000 \text{ m}\ell}{1 \cancel{\ell}} =$

| *Enter* | *Press* | *Display* |
|---|---|---|
| 2.46 | ⊠ (×) | 2.46 |
| 1,000 | (=) | 2,460 |

The answer is 2,460 mℓ.

Solve these problems.

a. Change 9,867 mℓ to liters.

b. Change 2.65 ℓ to milliliters.

c. Change 0.58 ℓ to milliliters.

d. Change 1,400 mℓ to liters.

**10**

---

Answers to frame 9.

a. 9.87 ℓ  b. 2,650 mℓ  c. 580 mℓ  d. 1.4 ℓ

# Section 7-5 Measurement of Weight in the Metric System

**1**

---

The standard unit of weight in the metric system is the kilogram. This unit is too large for many measurements, so subdivisions and multiples of the gram are used instead.

| *Unit* | *Abbreviation* | *Subdivisions and multiples* |
|---|---|---|
| milligram | mg | 1,000 of these in a gram |
| centigram | cg | 100 of these in a gram |
| gram | g | most common unit |
| kilogram | kg | 1,000 grams |

Give the abbreviations for the following units.

**a.** gram ________ **b.** milligram ________

**c.** kilogram ________ **d.** centigram ________

## 2

Name the unit indicated by each symbol.

**a.** mg ________ **b.** cg ________

**c.** kg ________ **d.** g ________

**1. a.** g
**b.** mg
**c.** kg
**d.** cg

## 3

How many of the following are contained in the given unit?

**a.** 1 g = ________ mg

**b.** 1 g = ________ cg

**c.** 1 kg = ________ g

**2. a.** milligram
**b.** centigram
**c.** kilogram
**d.** gram

## 4

Conversion factors and dimensional analysis are used to convert from one unit of weight to another.

Change 2,287 mg to grams.

*Step 1.* $2{,}287 \text{ mg} \times \dfrac{\text{g}}{\text{mg}}$

*Step 2.* smaller unit = mg   There are 1,000 mg in each gram.
larger unit = g

*Step 3.* $2{,}287 \cancel{\text{mg}} \times \dfrac{1 \text{ g}}{1{,}000 \cancel{\text{mg}}} = 2.287 \text{ mg}$

**3. a.** 1,000
**b.** 100
**c.** 1,000

## 5

The calculator can be used to solve the following problem.

Change 127 cg to grams.

*Step 1.* $127 \text{ cg} \times \dfrac{\text{g}}{\text{cg}}$

*Step 2.* smaller unit = cg   There are 100 cg in each gram.
larger unit = g

*Step 3.* $127 \cancel{\text{cg}} \times \dfrac{1 \text{ g}}{100 \cancel{\text{cg}}} =$

| *Enter* | *Press* | *Display* |
|---|---|---|
| 127 | ÷ | 127 |
| 100 | = | 1.27 |

The answer is 1.27 g.

Solve these problems.

**a.** Change 2.263 kg to grams.

**b.** Change 349 g to kilograms.

**c.** Change 4.265 g to milligrams.

**d.** Change 830 mg to grams.

**6**

Answers to frame 5.

**a.** 2,263 g **b.** 0.35 kg **c.** 4,265 mg **d.** 0.83 g

## Section 7-6 Conversion Between the English and Metric Systems

**1**

Converting between measurement systems is easy with the use of conversion factors and dimensional analysis. The relationships between the most commonly used units in each system are listed in Table 7-1.

**Table 7-1** English–metric conversion

| *Length* | | *Volume* | | *Weight* | |
|---|---|---|---|---|---|
| *English* | *Metric* | *English* | *Metric* | *English* | *Metric* |
| 1 in. = 2.54 cm | | 1 qt = 0.95 ℓ | | 1 lb = 454 g | |
| 1 mi = 1.6 km | | 1 fl oz* = 30 mℓ | | 2.2 lb = 1 kg | |
| 39.4 in. = 1 m | | | | 1 oz = 28 g | |

*Note that "fl oz" is a fluid ounce and "oz" is a weight ounce.

Complete the following equivalencies.

**a.** 1 mi = ________ km

**b.** 1 qt = ________ ℓ

**c.** 1 lb = ________ g

**d.** 1 oz = ________ g

**e.** 1 fl oz = ________ mℓ

**f.** 1 in. = ________ cm

**g.** 1 kg = ________ lb

**h.** 1 m = ________ in.

**2**

Conversion problems can be solved using Table 7-1 conversion factors and dimensional analysis. Round all answers to the hundredths place.

**1. a.** 1.6
**b.** 0.95
**c.** 454
**d.** 28
**e.** 30
**f.** 2.54
**g.** 2.2
**h.** 39.4

**Example 1** Change 3.2 in. to centimeters.

*Step 1.* $3.2 \text{ in.} \times \dfrac{\text{cm}}{\text{in.}}$

*Step 2.* 1 in. = 2.54 cm

*Step 3.* $3.2 \cancel{\text{in.}} \times \dfrac{2.54 \text{ cm}}{1 \cancel{\text{in.}}} = 8.13 \text{ cm}$

**Example 2** Change 40 mℓ to fluid ounces.

*Step 1.* $40 \text{ m}\ell \times \dfrac{\text{fl oz}}{\text{m}\ell}$

*Step 2.* 1 fl oz = 30 mℓ

*Step 3.* $40 \cancel{\text{m}\ell} \times \dfrac{1 \text{ fl oz}}{30 \cancel{\text{m}\ell}} = 1.33 \text{ fl oz}$

Make the following conversions.

**a.** 200 g = ________ lb  **b.** 4 liters = ________ qt

**c.** 50 mi = ________ km  **d.** 5 m = ________ in.

**e.** 3 in. = ________ cm  **f.** 4 lb = ________ kg

**g.** 50 g = ________ oz  **h.** 5 fl oz = ________ mℓ

2. **a.** 0.44
**b.** 4.21
**c.** 80
**d.** 197
**e.** 7.62
**f.** 1.82
**g.** 1.79
**h.** 150

## 3

Multiple conversion factors are needed to solve some problems. For example, to change 16 oz to kilograms, two methods can be used.

*Method 1.* Change ounces to pounds and then pounds to kilograms.

*Step 1.* $16 \text{ oz} \times \dfrac{\text{lb}}{\text{oz}} \times \dfrac{\text{kg}}{\text{lb}} =$

*Step 2.* 1 lb = 16 oz
1 kg = 2.2 lb

*Step 3.* $16 \cancel{\text{oz}} \times \dfrac{1 \cancel{\text{lb}}}{16 \cancel{\text{oz}}} \times \dfrac{1 \text{ kg}}{2.2 \cancel{\text{lb}}} = 0.45 \text{ kg}$

*Method 2.* Change ounces to grams and then grams to kilograms.

*Step 1.* $16 \text{ oz} \times \dfrac{\text{g}}{\text{oz}} \times \dfrac{\text{kg}}{\text{g}}$

*Step 2.* 1 oz = 28 g
1 kg = 1,000 g

*Step 3.* $16 \cancel{\text{oz}} \times \dfrac{28 \cancel{\text{g}}}{1 \cancel{\text{oz}}} \times \dfrac{1 \text{ kg}}{1{,}000 \cancel{\text{g}}} = 0.45 \text{ kg}$

Either method is correct, and you can decide which one you want to use.

Make these conversions using multiple conversion factors. (More than one method can be used to solve these problems and the answers may vary slightly because of rounding.)

**a.** 2 qt = ________ mℓ  **b.** 2 kg = ________ oz

**c.** 2.5 ℓ = ________ fl oz  **d.** 20 oz = ________ kg

**4**

Answers to frame 3.

a. $2\ \text{qt} \times \frac{0.95\ \ell}{1\ \text{qt}} \times \frac{1{,}000\ \text{m}\ell}{1\ \ell} = 1{,}900\ \text{m}\ell$  b. $2\ \text{kg} \times \frac{1{,}000\ \text{g}}{1\ \text{kg}} \times \frac{1\ \text{oz}}{28\ \text{g}} = 71.43\ \text{oz}$

c. $2.5\ \ell \times \frac{1{,}000\ \text{m}\ell}{1\ \ell} \times \frac{1\ \text{fl oz}}{30\ \text{m}\ell} = 83.33\ \text{fl oz}$  d. $20\ \text{oz} \times \frac{28\ \text{g}}{1\ \text{oz}} \times \frac{1\ \text{kg}}{1{,}000\ \text{g}} = 0.56\ \text{kg}$

# Exercise Set, Sections 7-4–7-6

## Measurement of Volume in the Metric System

Complete the following equivalencies.

**1.** 1 ℓ = ______ mℓ  **2.** 10 $\text{cm}^3$ = ______ mℓ  **3.** 1,000 mℓ = ______ ℓ  **4.** 1,000 mℓ = ______ $\text{cm}^3$

Use conversion factors and dimensional analysis to complete these equivalencies.

**5.** 697 mℓ = ________ ℓ  **6.** 2.6 ℓ = ________ mℓ  **7.** 827 $\text{cm}^3$ = ________ ℓ

**8.** 0.84 ℓ = ________ $\text{cm}^3$  **9.** 1,285 mℓ = ________ ℓ  **10.** 1.6 ℓ = ________ mℓ

**11.** 1,657 $\text{cm}^3$ = ________ ℓ  **12.** 0.25 ℓ = ________ $\text{cm}^3$

## Measurement of Weight in the Metric System

Complete the following statements.

**13.** 1 g = ______ mg  **14.** 1 g = ______ cg  **15.** 1 kg = ______ g  **16.** 1 cg = ______ mg

Use conversion factors and dimensional analysis to complete these equivalencies.

**17.** 2.5 kg = ________ g  **18.** 415 g = ________ kg  **19.** 0.15 kg = ________ g

**20.** 2,160 g = ________ kg  **21.** 8.9 kg = ________ g  **22.** 89 g = ________ kg

**23.** 2.87 g = ________ mg  **24.** 487 mg = ________ g

## Conversion Between the English and Metric Systems

Give the following English–metric relationships.

**25.** 1 qt = ______ ℓ

**26.** 1 m = ______ in.

**27.** 1 kg = ______ lb

**28.** 1 lb = ______ g

**29.** 1 oz = ______ g

**30.** 1 fl oz = ______ mℓ

**31.** 1 mi = ______ km

**32.** 1 in. = ______ cm

Make the following conversions.

**33.** 300 g = ________ lb

**34.** 5 lb = ________ kg

**35.** 80 mi = ________ km

**36.** 6 m = ________ in.

**37.** 4 in. = ________ cm

**38.** 3 ℓ = ________ qt

**39.** 30 oz = ________ g

**40.** 600 mℓ = ________ fl oz

**41.** 5.08 cm = ________ in.

**42.** 10 qt = ________ ℓ

**43.** 40 km = ________ mi

**44.** 300 mℓ = ________ fl oz

**45.** 10 lb = ________ g

**46.** 1,200 mℓ = ________ qt

**47.** 60 oz = ________ kg

**48.** 2 qt = ________ mℓ

**49.** 5 kg = ________ oz

**50.** 40 fl oz = ________ ℓ

## Supplementary Exercise Set, Sections 7-4–7-6

Solve these problems.

**1.** 557 mℓ = ________ ℓ

**2.** 0.67 ℓ = ________ mℓ

**3.** 2,867 mℓ = ________ ℓ

**4.** 1.9 ℓ = ________ mℓ

**5.** 2.7 kg = ________ g

**6.** 218 g = ________ kg

**7.** 0.95 g = ________ mg

**8.** 1,800 mg = ________ g

**9.** 2,670 g = ________ kg

**10.** 5.6 g = ________ mg

**11.** 200 g = ________ lb

**12.** 90 oz = ________ g

**13.** 150 mℓ = ________ fl oz

**14.** 45 oz = ________ g

**15.** 5 lb = ________ g

**16.** 1,400 mℓ = ________ qt

**17.** 8 kg = ________ oz

**18.** 2 ℓ = ________ fl oz

**19.** 2.2 in. = ________ mm

**20.** 0.21 qt = ________ mℓ

# Section 7-7 The Fahrenheit, Celsius, and Kelvin Temperature Systems

## 1

The Fahrenheit temperature scale is the system presently in use in the United States. The Celsius temperature scale is increasingly being used and will someday replace the Fahrenheit system. The Celsius system is compared with the Fahrenheit system below.

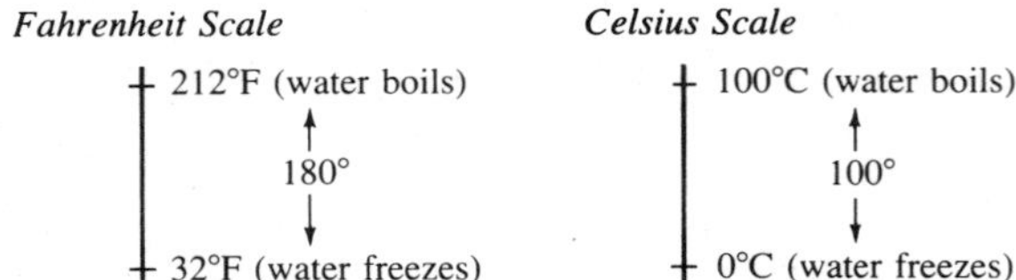


Notice that there are 100° between the freezing and boiling points of water on the Celsius scale, giving a number that is a multiple of 10.

## 2

The Kelvin temperature scale is used for calculations involving gases. The Kelvin scale is similar to the Celsius scale and is compared below.

*Celsius Scale*

100°C (water boils)

0°C (water freezes)

*Kelvin Scale*

373 K (water boils)

273 K (water freezes)

0 K (absolute zero)

The symbol for the Kelvin system is K (without a degree sign). The Kelvin system differs from the Celsius system by 273°. The zero point on the Kelvin scale is the point at which there is no heat left in a system; hence, the name ***absolute zero***.

## 3

Celsius temperatures are easily converted to Kelvin temperatures with the use of the following formula:

$$K = °C + 273$$

**Example** 15°C is what temperature on the Kelvin scale?

*Step 1.* K = °C + 273
*Step 2.* °C = 15
*Step 3.* K = 15 + 273
*Step 4.* K = 288

Change 40°C to the Kelvin temperature.

## 4

Formulas can be used to convert temperature readings from the Fahrenheit scale to the Celsius scale and from Celsius to Fahrenheit. The formula for converting Celsius to Fahrenheit is as follows:

**3.** K = °C + 273
°C = 40
K = 40 + 273
K = 313

$$°F = (1.8 \times °C) + 32$$

What is 30°C on the Fahrenheit scale?

Use the five steps to solve this problem.

*Step 1.* What is being asked for?
$°F = ?$
*Step 2.* What information is known?
$°C = 30$
*Step 3.* What mathematical formula should be used?
$°F = (1.8 \times °C) + 32$
*Step 4.* Substitute the information into the formula.
$°F = (1.8 \times 30) + 32$
*Step 5.* Do the calculations. (Do the operation in parentheses first.)
$°F = (1.8 \times 30) + 32$
$°F = 54 + 32 = 86°F$

**5**

What is 25°C on the Fahrenheit scale?

*Step 1.*
*Step 2.*
*Step 3.*
*Step 4.*
*Step 5.*

**6**

Answers to frame 5.

*Step 1.* $°F = ?$
*Step 2.* $°C = 25$
*Step 3.* $°F = (1.8 \times °C) + 32$
*Step 4.* $°F = (1.8 \times 25) + 32$
*Step 5.* $°F = 45 + 32 = 77°F$

**7**

Convert the following Celsius temperatures to Fahrenheit temperatures.

**a.** 5.0°C = __________ **b.** 27.0°C = __________

**8**

The formula for changing Fahrenheit to Celsius is as follows:

7. **a.** 41°F
**b.** 80.6°F

$$°C = \frac{(°F - 32)}{1.8}$$

Change 50°F to Celsius.

*Step 1.* $°C = ?$
*Step 2.* $°F = 50$

*Step 3.* $°C = \frac{(°F - 32)}{1.8}$

*Step 4.* $°C = \frac{(50 - 32)}{1.8}$ ← Do operation in parentheses first.

*Step 5.* $°C = \frac{18}{1.8} = 10°C$

Change 68° F to Celsius.

*Step 1.*
*Step 2.*
*Step 3.*
*Step 4.*
*Step 5.*

**9**

Answers to frame 8.

*Step 1.* °C = ?
*Step 2.* °F = 68

*Step 3.* $°C = \frac{(°F - 32)}{1.8}$

*Step 4.* $°C = \frac{68 - 32}{1.8}$

*Step 5.* $°C = \frac{36}{1.8} = 20°C$

**10**

Since negative (signed) numbers are sometimes involved in converting one temperature scale to another, the calculator is more convenient to use in this type of operation. Here are two examples.

Change −11.0°C to Fahrenheit.

*Step 1.* °F = ?
*Step 2.* °C = −11.0
*Step 3.* °F = (1.8 × °C) + 32
*Step 4.* °F = (1.8 × −11.0) + 32
*Step 5.*

| *Enter* | *Press* | *Display* |
|---|---|---|
| 1.8 | [×] | 1.8 |
| 11.0 | [+/−] [+] | −19.8 |
| 32 | [=] | 12.2 |

The answer is 12.2°F.

Change −11.0°F to Celsius.

*Step 1.* °C = ?
*Step 2.* °F = −11.0

*Step 3.* $°C = \frac{(°F - 32)}{1.8}$

*Step 4.* $°C = \frac{(-11.0 - 32)}{1.8}$

*Step 5.*

| *Enter* | *Press* | *Display* |
|---|---|---|
| 11.0 | [+/−] [−] | −11 |
| 32 | [=] [÷] | −43 |
| 1.8 | [=] | −23.888 |

The answer is −23.9.

Perform the indicated conversions.

**a.** $-10°C =$ ________ °F **b.** $-15°C =$ ________ °F

**c.** $27°F =$ ________ °C **d.** $-4.0°F =$ ________ °C

**11**

Answers to frame 10.

**a.** 14 **b.** 5.0 **c.** −2.8 **d.** −15.6

# Section 7-8 Changing Fractions of Inches to Millimeters

**1**

Tool sizes are manufactured in fractions of an inch in the English system of measurement and are manufactured in millimeters in the metric system. To convert from inches to millimeters, use the following conversion factor.

**Example** What size would a 9/16-inch wrench be in millimeters?

*Step 1.* $\frac{9}{16}\text{ in.} \times \frac{\text{mm}}{\text{in.}} =$

*Step 2.* $1\text{ in.} = 2.54\text{ cm} \times \frac{10\text{ mm}}{1\text{ cm}} = 25.4\text{ mm}$

(If you remember that 1 inch equals 25.4 mm, you need not do this step each time you make a conversion.)

*Step 3.* $\frac{9}{16}\cancel{\text{in.}} \times \frac{25.4\text{ mm}}{1\text{ in.}} = 14.3\text{ mm}$

Change these fractions of inches to millimeters.

**a.** 7/8″ = ______ mm **b.** 3/16″ = ______ mm

**c.** 9/32″ = ______ mm **d.** 39/64″ = ______ mm

**2**

Answers to frame 1.

**a.** 22.23 **b.** 4.75 **c.** 7.14 **d.** 15.48

# Exercise Set, Sections 7-7–7-8

## The Fahrenheit, Celsius, and Kelvin Temperature Systems

Make the following temperature conversions.

**1.** $-12°C =$ ______ K **2.** $200°C =$ ______ K **3.** $32°C =$ ______ °F

**4.** $10°F =$ ______ °C **5.** $-10°F =$ ______ °C **6.** $-15°C =$ ______ °F

7. $-40°C =$ ______ °F

8. $68°F =$ ______ °C

9. $45°F =$ ______ °C

10. $10°C =$ ______ °F

## Changing Fractions of Inches to Millimeters

Change these fractions to millimeters.

11. 3/8″ = ______ mm

12. 3/4″ = ______ mm

13. 5/8″ = ______ mm

14. 5/16″ = ______ mm

# Supplementary Exercise Set, Sections 7-7–7-8

Make the following temperature conversions.

1. $-5°C =$ ______ K

2. $25°C =$ ______ K

3. $40°C =$ ______ °F

4. $12°F =$ ______ °C

5. $-15°F =$ ______ °C

6. $-10°C =$ ______ °F

7. $-20°C =$ ______ °F

8. $72°F =$ ______ °C

9. $50°F =$ ______ °C

10. $20°C =$ ______ °F

Change these fractions to millimeters.

11. 1/4″ = ______ mm

12. 11/16″ = ______ mm

13. 13/32″ = ______ mm

14. 1/8″ = ______ mm

15. 9/64″ = ______ mm

# Section 7-9 Measuring Devices

**1**

Technical and scientific work often requires measurements where data in the form of numbers are read from scales. Several measuring devices are looked at in this section. The first is a simple ruler that measures in both the metric and English systems. A part of this ruler is shown.

The first step needed to read a ruler is to count the number of subdivisions in a main division.

**a.** How many subdivisions are in a one-inch main division? ______

**b.** How many subdivisions are in a one-centimeter main division? ______

**c.** Each subdivision of the 1-in. main division equals ______ of an inch.

**d.** Each subdivision of the 1-cm main division equals ______ of a centimeter.

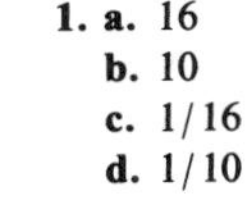
1. a. 16
b. 10
c. 1/16
d. 1/10

## 2

Use the ruler to answer questions **a** and **b**.

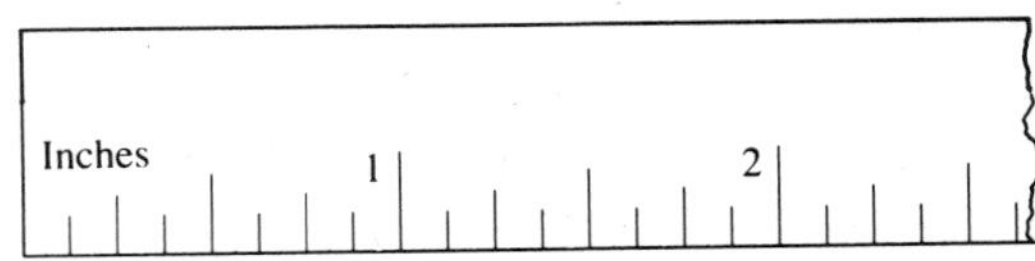


**a.** How many subdivisions are in a one-inch main division? ______

**b.** Each subdivision equals ______ of an inch.

2. a. 8
b. 1/8″

## 3

Place the ruler beside the object to be measured. The second step performed in reading the measuring device is to add the number of main divisions and subdivisions to equal the length of the object to be measured.

In the illustration above there is one main division, and each main division equals one inch. There are also three subdivisions, and each subdivision equals 1/16 of an inch.

$$1 \text{ in.} + \frac{3}{16} \text{ in.} = 1\frac{3}{16} \text{ in.}$$

Therefore, the object is $1\frac{3}{16}$ in. long.

Read the measurements on the ruler below in either inches or centimeters, as indicated. If a reading does not fall exactly on a mark, read the nearest scale division or subdivision.

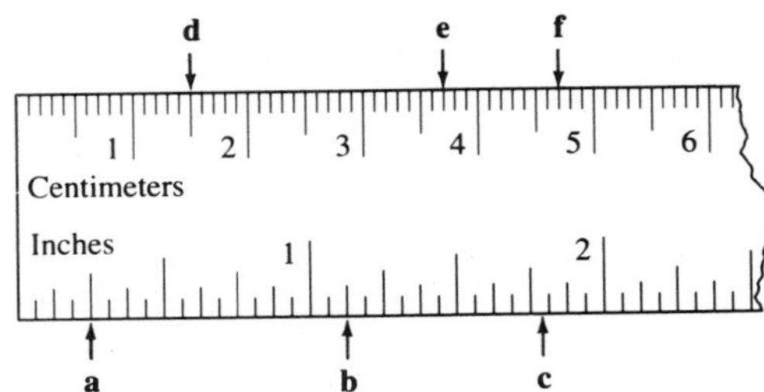


**a.** ______ in. **b.** ______ in. **c.** ______ in.

**d.** ______ cm **e.** ______ cm **f.** ______ cm

## 4

Some technical work requires reading the dial or scale on a meter. One example of such a meter is the tachometer of an automobile. To read a tachometer, multiply the reading on the dial by 100. The reading obtained is the revolutions per minute (rpm) at which the engine is rotating. Read the meter to the nearest scale division or subdivision.

3. **a.** 1/4
**b.** 1 2/16 or 1 1/8
**c.** 1 13/16
**d.** 1.5
**e.** 3.7
**f.** 4.7

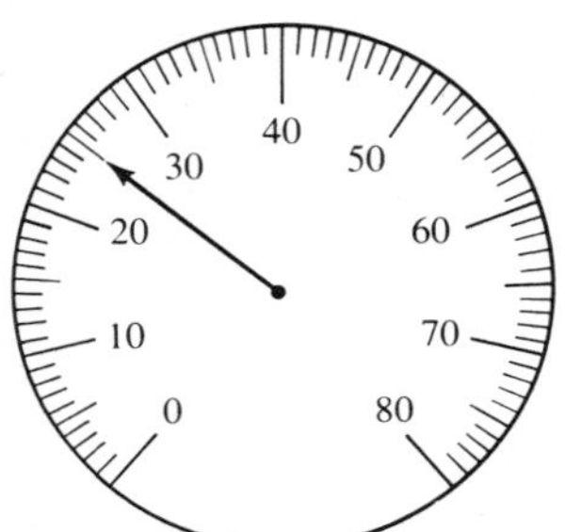


**a.** First notice how many subdivisions are between each main subdivision. There are ten subdivisions between 0 and 10.
**b.** Next notice that the needle points to 25. This reading of the meter includes 2.5 main divisions or 25 subdivisions on the scale.
**c.** Multiply the number of subdivisions (25) by 100; the tachometer reads 2,500 rpm.

## 5

Read these tachometers. Don't forget to multiply the dial reading by 100.

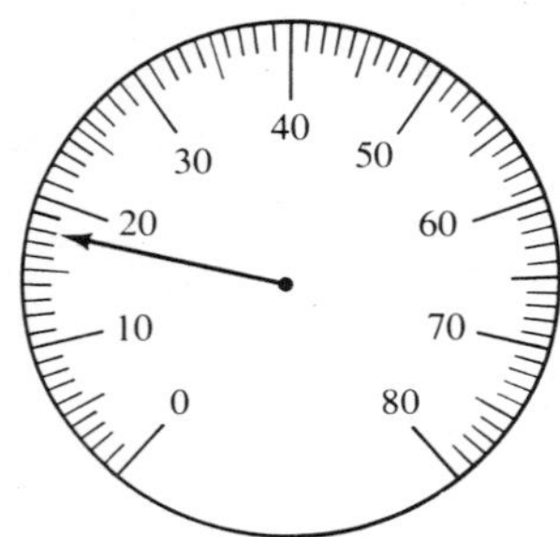


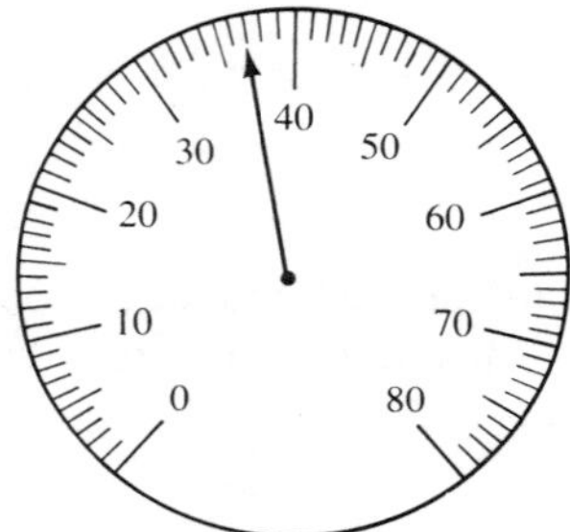


**a.** ______ **b.** ______

## 6

The vernier caliper is a precise instrument used to measure distances. It consists of an upper, main scale and a lower, sliding scale called a *vernier scale*. The vernier caliper shown below will measure to 0.01 centimeter. All measurements using this caliper must have two decimal places.

5. **a.** 1,800 rpm
**b.** 3,700 rpm

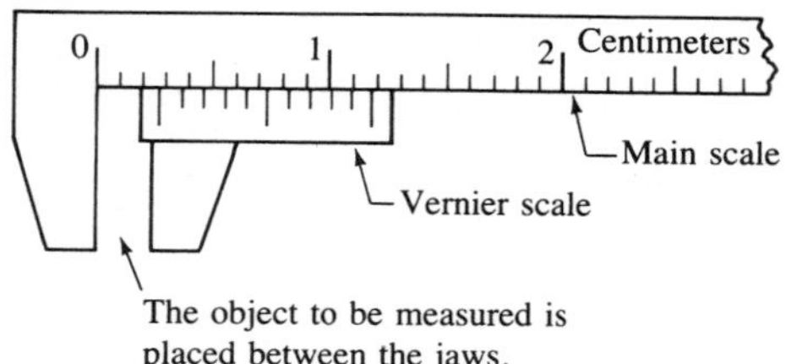

The main divisions on the upper scale are one centimeter (1 cm) apart. These main divisions are further divided into ten subdivisions equal to one tenth of a centimeter (0.1 cm).

The divisions on the lower, or vernier, scale tell how many hundredths of a centimeter are in the measurement. There are ten marks on the vernier scale, and each mark represents 1 one-hundredth of a centimeter (0.01 cm).

The steps in reading a vernier scale are illustrated below.

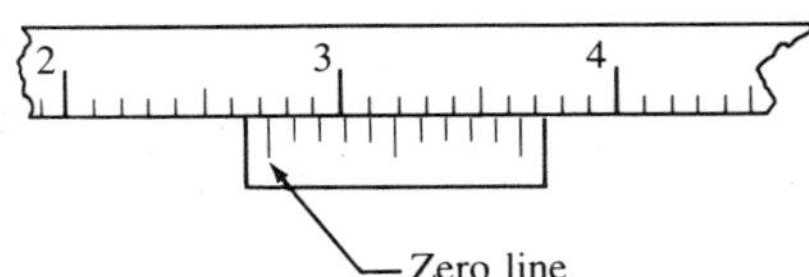


*Step 1.* The first mark, or zero line, on the lower scale is used to determine the readings on the upper scale. The main-division mark on the upper scale to the left of the zero mark on the lower scale is 2. Multiply this number by 1 cm.

$$2 \times 1 \text{ cm} = 2 \text{ cm}$$

*Step 2.* Next count the number of one-tenth-centimeter subdivision marks between the zero mark on the lower scale and the main division mark in step 1. You should count seven subdivision marks. Each subdivision mark equals 0.1 cm.

$$7 \times 0.1 \text{ cm} = 0.7 \text{ cm}$$

*Step 3.* The last reading is taken from the lower, or vernier, scale. Note which line on the lower scale coincides with a line on the upper scale. In this illustration, the fourth mark on the lower scale best matches a line on the upper scale. The marks on the lower scale equal 0.01 cm.

$$4 \times 0.01 \text{ cm} = 0.04 \text{ cm}$$

*Step 4.* Now add up the readings from steps 1, 2, and 3.

| | |
|---|---|
| Main division | 2 |
| Subdivisions | 0.7 |
| Lower scale | 0.04 |
| Reading | 2.74 cm |

Read the following scales.

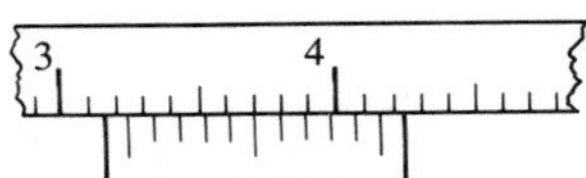


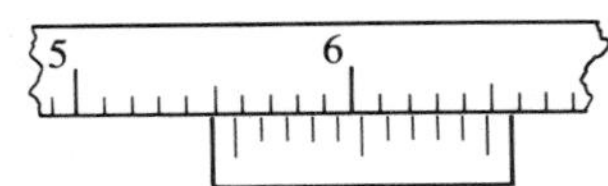


**a.** Main division
Subdivisions
Lower scale ____
Reading

**b.** Main division
Subdivisions
Lower scale ____
Reading

## 7

Here are two more examples of reading scales.

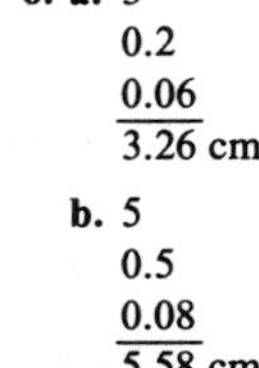
6. a. 3
0.2
0.06
3.26 cm
b. 5
0.5
0.08
5.58 cm

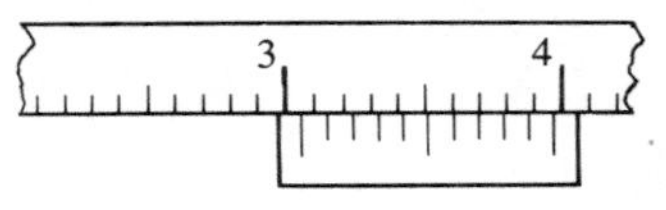


a.

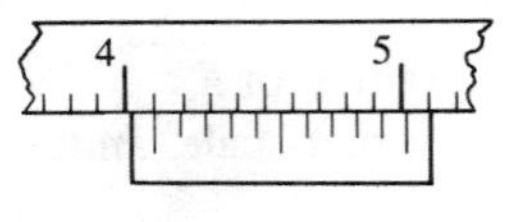


b.

In example **a**, there are no 0.1-cm subdivisions between the zero mark on the lower scale and the main division.

| | |
|---|---|
| Main division | 3 |
| Subdivisions | 0.0 |
| Lower scale | 0.07 |
| Reading | 3.07 |

In example **b**, the zero mark on the lower scale coincides with a mark on the upper scale. There will be no 0.01–cm readings.

| | |
|---|---|
| Main division | 4 |
| Subdivisions | 0.1 |
| Lower scale | 0.00 |
| Reading | 4.10 |

Read these scales.

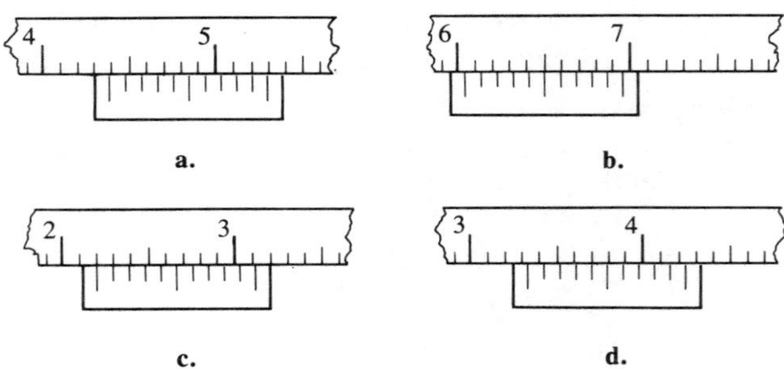


## 8

A vernier caliper allows us to make measurements to hundredths of a centimeter. For greater accuracy, a micrometer can be used to make measurements to hundredths of a millimeter.

7. a. 4.39
b. 6.05
c. 2.20
d. 3.35

The diagram below shows the main components of a metric micrometer. The object to be measured is placed between the *anvil* and the *spindle*. There is a scale on both the *sleeve* and the *thimble*.

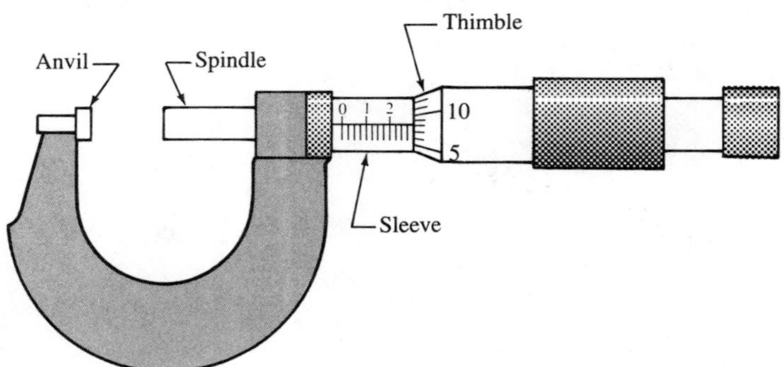


The diagram of the scales on the sleeve and thimble above will be used to illustrate how a measurement is obtained.

Main division
Horizontal line

*Step 1.* Read the main division number to the left of the thimble on the sleeve and multiply by 1 mm.

$$10 \times 1 \text{ mm} = 10 \text{ mm}$$

*Step 2.* Next count the number of marks above and below the horizontal line on the sleeve and between the main division number and the thimble. Multiply the number of marks by 0.5 mm.

$$2 \times 0.5 \text{ mm} = 1.0 \text{ mm}$$

*Step 3.* Read the number on the thimble that best lines up with the horizontal line on the sleeve. Multiply this reading by 0.01 mm.

$$17 \times 0.01 \text{ mm} = 0.17 \text{ mm}$$

*Step 4.* Add up the numbers from the three previous steps.

$$10 \text{ mm} + 1.0 \text{ mm} + 0.17 \text{ mm} = 11.17 \text{ mm}$$

Read these metric micrometer scales.

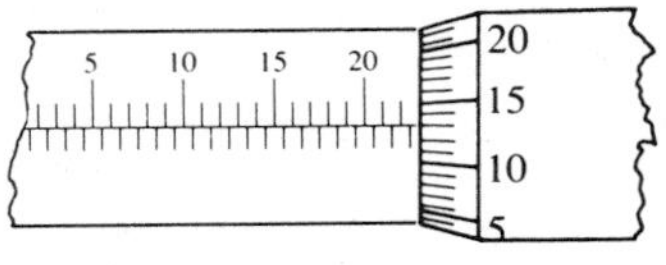

a.

*Step 1.*
*Step 2.*
*Step 3.* ______
Reading

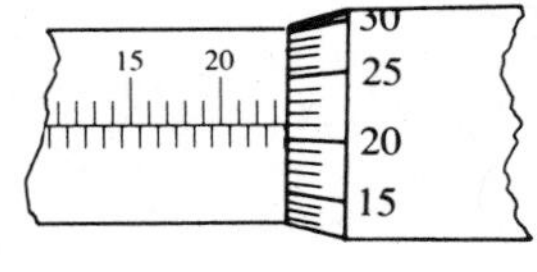

b.

*Step 1.*
*Step 2.*
*Step 3.* ______
Reading

**9**

Read these metric micrometer scales.

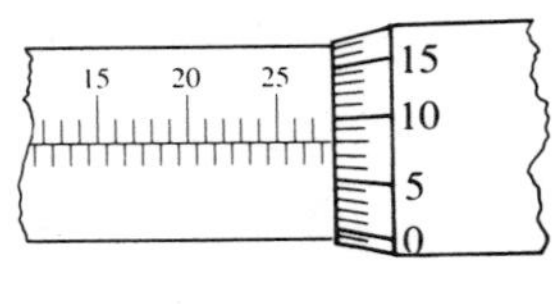

a.

b.

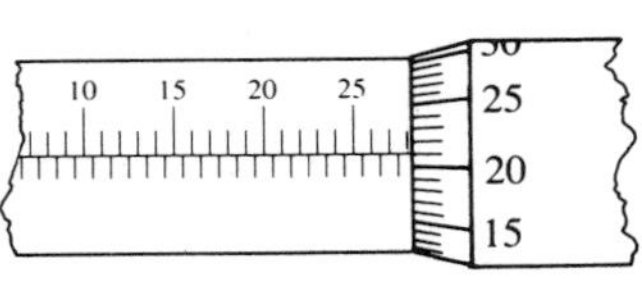

c.

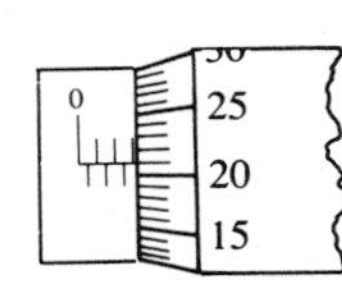

d.

**8. a.** 20
2.5
0.13
22.63 mm

**b.** 20
3.0
0.21
23.21 mm

**10**

Answers to frame 9.

**a.** 27.58 mm **b.** 11.72 mm **c.** 28.21 mm **d.** 3.21 mm

# Exercise Set, Section 7-9

## Measuring Devices

Read each measurement on the ruler below, in either inches or centimeters as indicated.

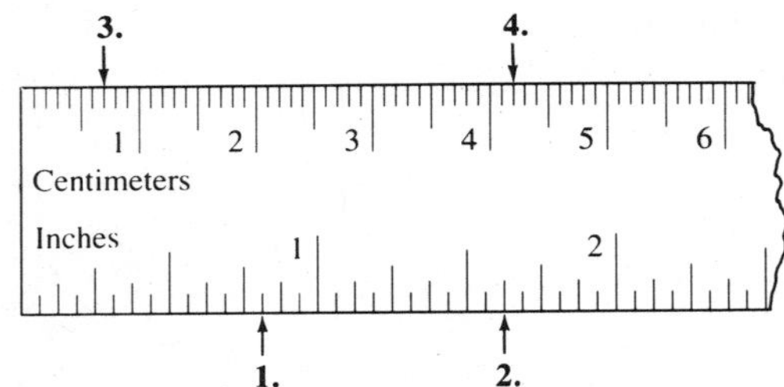


**1.** ________

**2.** ________

**3.** ________

**4.** ________

Read each tachometer.

**5.**

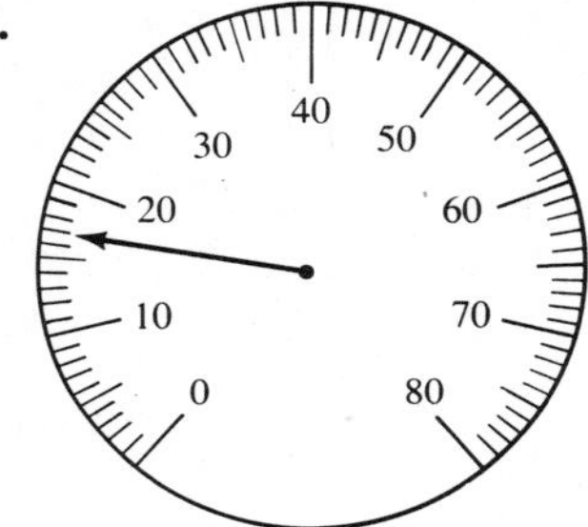


**6.**

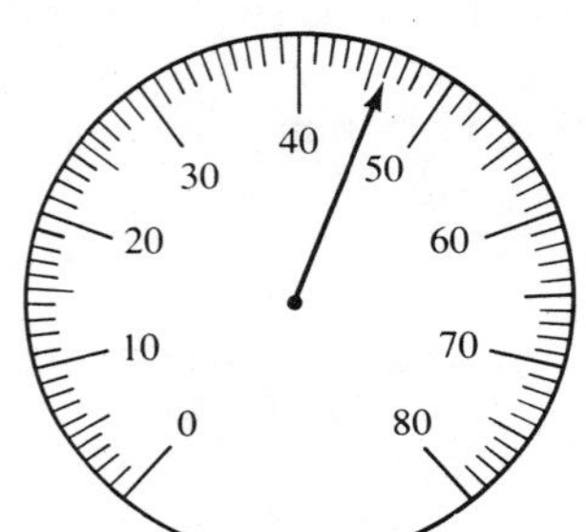


**5.** ________

**6.** ________

Read each vernier caliper scale.

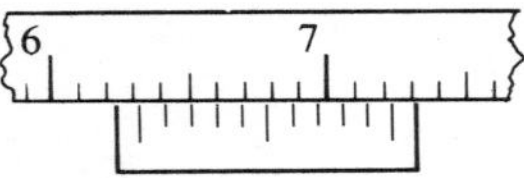


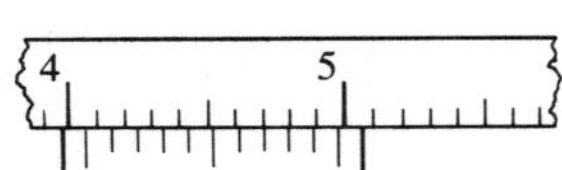


**7.** ________ cm

**8.** ________ cm

Read each metric micrometer scale.

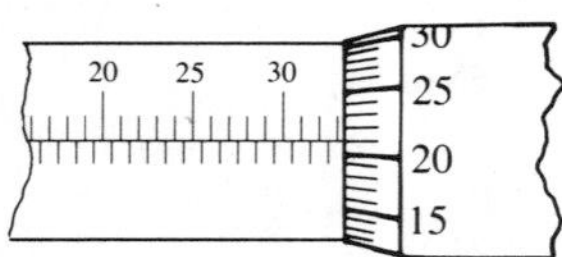


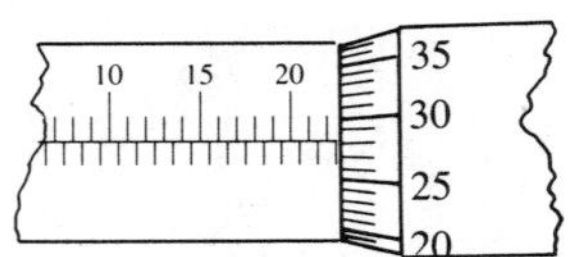


**9.** ________ mm

**10.** ________ mm

# Supplementary Exercise Set, Section 7-9

Read each measurement on the ruler, in either inches or centimeters as indicated.

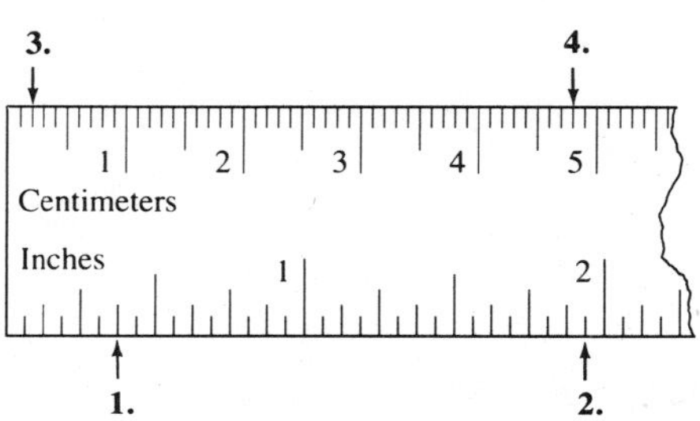


1. ________
2. ________
3. ________
4. ________

Read each tachometer.

5. 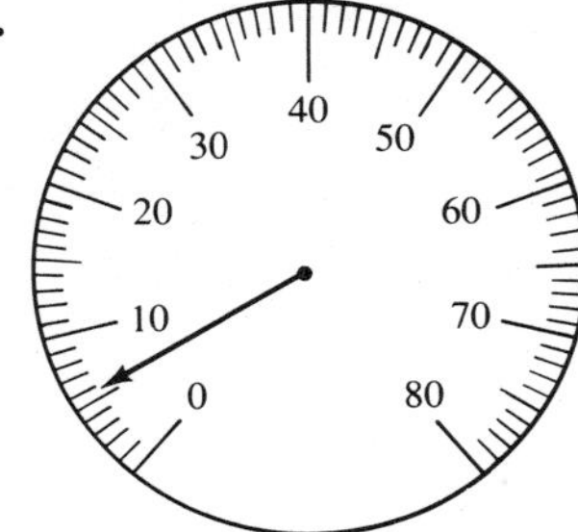


6. 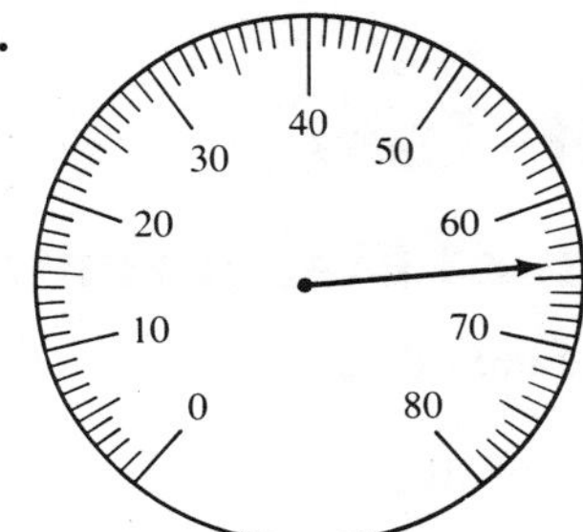


5. ________
6. ________

Read each vernier caliper scale.

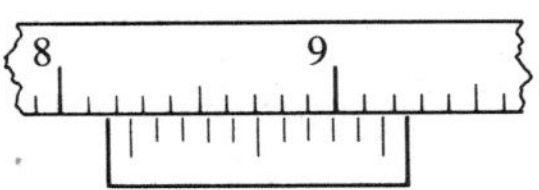


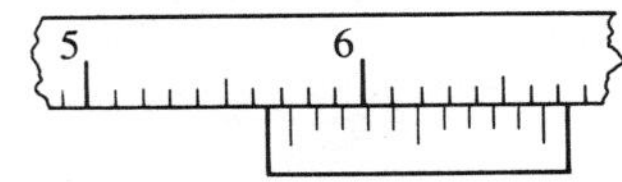


7. ________ cm

8. ________ cm

Read each metric micrometer scale.

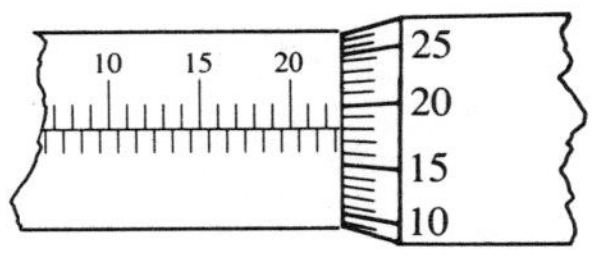


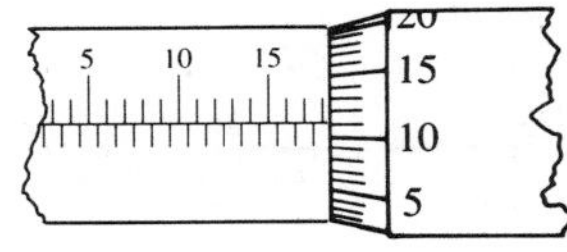


9. ________ mm

10. ________ mm

# Summary

1. The metric system is based on the number 10. Thus, all multiples and subdivisions of the standard units are based on the number 10.
2. Conversion factors and dimensional analysis are used to convert between multiples and subdivisions of a given unit.
3. The standard or most common units in the metric system are

   length: meter
   volume: liter
   weight: gram

4. 1 mℓ of water weighs approximately one gram.
5. The formulas for changing between Fahrenheit and Celsius temperatures are

$$°F = (1.8 \times °C) + 32$$

$$°C = \frac{(°F - 32)}{1.8}$$

6. The formula for changing Celsius to Kelvin is

$$K = °C + 273$$

7. Use the following conversion factor to change fractions of an inch to millimeters.

$$\text{fraction of inch} \times \frac{25.4 \text{ mm}}{1 \text{ in.}}$$

# Chapter 7 Self-Test

## The Metric System of Measurement

How many of these subdivisions are in a standard unit?

1. milli ________
2. centi ________
3. deci ________

How many of the standard units are in each of these multiples?

4. kilo ________
5. hecto ________
6. deka ________

Complete the following equivalencies.

7. 1 m = ________ mm
8. 1 km = ________ m

Change each of the following measurements to the desired unit.

9. 2.83 m = ________ mm
10. 92 cm = ________ m
11. 484 m = ________ km
12. 5.6 cm = ________ mm
13. 681 mℓ = ________ ℓ
14. 4.8 ℓ = ________ mℓ
15. 4.5 kg = ________ g
16. 82 g = ________ kg
17. 1 mi = ________ km
18. 1 oz = ________ g
19. 1 lb = ________ g
20. 1 qt = ________ ℓ
21. 1 fl oz = ________ mℓ
22. 1 in. = ________ cm
23. 6 ℓ = ________ qt
24. 2.5 lb = ________ kg
25. 120 g = ________ oz
26. 60 fl oz = ________ mℓ
27. 15 g of water = ________ mg of water

Perform the following temperature conversions.

**28.** $-18°C =$ ________ K

**29.** $68°F =$ ________ °C

**30.** $18°C =$ ________ °F

**31.** $-18°F =$ ________ °C

Change these fractions of inches to millimeters.

**32.** 15/16″ = ________ mm

**33.** 11/32″ = ________ mm

Read these scales.

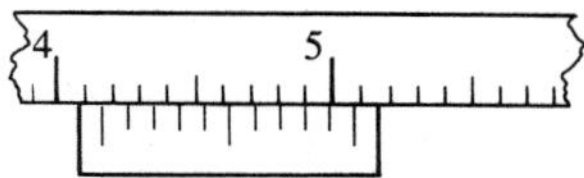


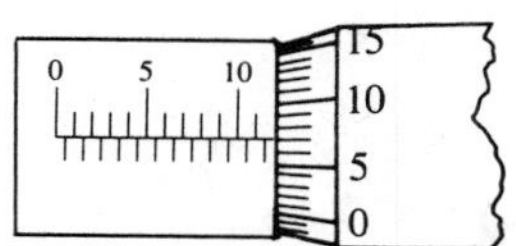


**34.** ________ cm

**35.** ________ mm

# CHAPTER 8

# The Powers and Roots of Numbers

Two mathematical operations, known as raising to a power and taking a root, frequently appear in technical formulas. In this chapter you will learn to find a power of any number, find the root of a number, write numbers in scientific notation, and multiply and divide numbers in scientific notation.

## Section 8-1 The Power of a Number

**1**

This example shows a number being multiplied by itself four times.

$$2 \times 2 \times 2 \times 2 =$$

This multiplication can more conveniently be written as

$$2 \times 2 \times 2 \times 2 = 2^4$$

The symbol $2^4$ is read as "two to the fourth power." The number 2 is called the *base* (the number to be multiplied), and the number 4 is called the *exponent* and means "multiply the base times itself four times."

In general, the number $y^x$ means that the base $y$ can be raised to the $x$th power, where $x$ is any exponent.

In the number $4^6$

**a.** the exponent is ______ **b.** the base is ______

In the number $m^k$

**c.** the exponent is ______ **d.** the base is ______

**2**

This example shows a number being multiplied by itself four times in the denominator of a fraction.

$$\frac{1}{2 \times 2 \times 2 \times 2} = \frac{1}{2^4}$$

This multiplication can be written as

$$\frac{1}{2^4} = 2^{-4}$$

**1. a.** 6
**b.** 4
**c.** $k$
**d.** $m$

Write the following fractions as negative powers.

a. $\frac{1}{10^4} =$ b. $\frac{1}{5^2} =$ c. $\frac{1}{X^k}$

Write these negative powers as fractions.

d. $10^{-7} =$ e. $9^{-3} =$ f. $H^{-m} =$

**3**

A number can be raised to the second power (or squared) by multiplying the number by itself.

**Example** $12^2 = 12 \times 12 = 144$

It is more convenient to use the [x2] key on your calculator. When using the [x2] key, the [=] key *is not* used.

**Example** Find $16^2$.

| *Enter* | *Press* | *Display* |
|---|---|---|
| 16 | [x2] | 256 |

The answer is 256.

**Example** Find $(-0.4)^2$.

| *Enter* | *Press* | *Display* |
|---|---|---|
| 0.4 | [+/-] [x2] | 0.16 |

The answer is 0.16.

When squaring a negative number, the sign will be positive.

Find the square of these numbers.

a. 14.2 b. 0.5 c. −6.4 d. −0.21

2. a. $10^{-4}$
b. $5^{-2}$
c. $X^{-k}$
d. $\frac{1}{10^7}$
e. $\frac{1}{9^3}$
f. $\frac{1}{H^m}$

**4**

A base can be raised to any power by using the [$y^x$] key. The [=] key *is* used when using the [$y^x$] key. Examples of using the calculator to find a power of any number are shown below.

**Example** Find $3^6$.

| | *Enter* | *Press* | *Display* |
|---|---|---|---|
| Base | 3 | [$y^x$] | 3 |
| Exponent | 6 | [=] | 729 |

$3^6 = 729$.

3. a. 201.64
b. 0.25
c. 40.96
d. 0.0441

**Example** Find $(0.7)^3$.

| *Enter* | *Press* | *Display* |
|---|---|---|
| 0.7 | $y^x$ | 0.7 |
| 3 | = | 0.343 |

$(0.7)^3 = 0.343$.

Find the indicated power of each number, rounding to three decimal places.

**a.** $2.2^5$ **b.** $0.3^4$ **c.** $0.1^3$ **d.** $1.002^6$

**5**

Answers to frame 4.

**a.** 51.536 **b.** 0.008 **c.** 0.001 **d.** 1.012

## Section 8-2 The Square Root of a Number

**1**

The square root of any number $x$ is the number whose square is $x$.

**Examples** The square root of 36 is 6, because $6 \times 6 = 36$.

The square root of 81 is 9, because $9 \times 9 = 81$.

Complete the following.

**a.** The square root of 49 is ________.

**b.** The square root of 100 is ________.

**c.** The square root of 400 is ________.

**2**

Positive numbers have both a negative and a positive square root. This book will only concern itself with positive square roots. The symbol $\sqrt{x}$ is read "the square root of the number $x$" and can be found on a calculator. The = key *is not* used with the $\sqrt{x}$ key.

1. **a.** 7
**b.** 10
**c.** 20

**Example** Find $\sqrt{169}$.

| *Enter* | *Press* | *Display* |
|---|---|---|
| 169 | $\sqrt{x}$ | 13 |

$\sqrt{169} = 13$.

**Example** Find $\sqrt{0.2}$.

| *Enter* | *Press* | *Display* |
|---|---|---|
| 0.2 | $\sqrt{x}$ | 0.4472136 |

$\sqrt{0.2} = 0.447$ (rounded).

Find the square root of these numbers. Round to the thousandths place.

**a.** 121 **b.** 900 **c.** 18 **d.** 0.35

**3**

---

Certain types of formulas contain squares and square roots. Here is an example of such a formula.

$$a = \sqrt{b^2 + c^2}$$

Find $a$ when $b = 3$ and $c = 4$.

$$a = \sqrt{3^2 + 4^2} \longleftarrow \text{Do the indicated operations under the square root sign first.}$$
$$a = \sqrt{9 + 16}$$
$$a = \sqrt{25}$$
$$a = 5$$

On the calculator, the problem above would look like this.

| *Enter* | *Press* | *Display* |
|---|---|---|
| 3 | [x²] [+] | 9 |
| 4 | [x²] [=] [√x] | 5 |

Using the formula, $a = \sqrt{b^2 + c^2}$, find $a$ when

**a.** $b = 4$, $c = 5$ **b.** $b = 5$, $c = 6$ **c.** $b = 2.2$, $c = 3.1$ **d.** $b = 0.5$, $c = 0.3$

**2. a.** 11
**b.** 30
**c.** 4.243
**d.** 0.592

**4**

---

Answers to frame 3.

**a.** 6.403 **b.** 7.810 **c.** 3.81 **d.** 0.583

# Exercise Set, Sections 8-1–8-2

Round all answers in this exercise set to three decimal places.

## The Power of a Number

Square each number.

**1.** 13.6 ______ **2.** 0.7 ______ **3.** −2.5 ______ **4.** −0.42 ______

Find the indicated power of each number.

**5.** $1.6^4$ ______ **6.** $0.2^3$ ______ **7.** $2.4^5$ ______ **8.** $2.01^6$ ______

## The Square Root of a Number

Find the square root of each number.

**9.** $\sqrt{131}$ ______ **10.** $\sqrt{700}$ ______ **11.** $\sqrt{1.62}$ ______ **12.** $\sqrt{0.28}$ ______

Using the formula, $a = \sqrt{b^2 + c^2}$, find $a$ when

**13.** $b = 4$, $c = 6$

**14.** $b = 2$, $c = 4.2$

**15.** $b = 0.3$, $c = 0.4$

# Supplementary Exercise Set, Sections 8-1–8-2

Round all answers in this exercise set to three decimal places.

Square each number.

**1.** 11.5 **2.** 0.4 **3.** −1.8 **4.** −0.24

Find the indicated power of each number.

**5.** $1.2^4$ **6.** $0.3^3$ **7.** $2.6^5$ **8.** $2.02^6$

Find the square root of each number.

**9.** $\sqrt{111}$ **10.** $\sqrt{450}$ **11.** $\sqrt{1.78}$ **12.** $\sqrt{0.39}$

Using the formula, $a = \sqrt{b^2 + c^2}$, find $a$ when

**13.** $b = 3$, $c = 5$

**14.** $b = 1.5$, $c = 2$

**15.** $b = 0.4$, $c = 0.2$

# Section 8-3 Powers of Ten and Scientific Notation

## 1

Numbers to any power with a base of ten are said to be written in power-of-ten form.

$10^5$ (the base is 10, the exponent is 5)

$10^{-3}$ (the base is 10, the exponent is −3)

What are the base and exponent of these numbers written in power-of-ten form?

**a.** $10^{-7}$
base ______
exponent ______

**b.** $10^4$
base ______
exponent ______

## 2

Numbers that are multiples of ten are easy to write in power-of-ten form.

$100 = 10 \times 10$ or $10^2$

$1{,}000 = 10 \times 10 \times 10$ or $10^3$

**1. a.** base = 10
exponent = −7
**b.** base = 10
exponent = 4

Note that the number of zeros in the numbers corresponds to the exponent on the base ten.

100 has two zeros and is $10^2$.

1,000 has three zeros and is $10^3$.

Write these numbers in power-of-ten form.

**a.** 10,000 ______ **b.** 1,000,000 ______

**3**

By noting the number of zeros in numbers that are multiples of ten, the following sequence is obtained.

2. **a.** $10^4$
**b.** $10^6$

$$1{,}000 = 10^3$$
$$100 = 10^2$$
$$10 = 10^1$$
$$1 = 10^0$$

Write these numbers in power-of-ten form, or write the power-of-ten form as a regular number.

**a.** $10^0$ ______ **b.** 10 ______ **c.** 10,000 ______ **d.** 1 ______

**4**

Numbers other than multiples of ten can be written in a form called *scientific notation.* Scientific notation states a number between one and ten multiplied by a power-of-ten.

3. **a.** 1
**b.** $10^1$
**c.** $10^4$
**d.** $10^0$

**Example** $3{,}100 = 3.1 \times 1{,}000 = 3.1 \times 10^3$

To arrive at the scientific-notation form of 3,100, the decimal at the end of 3,100 was moved *to the left* so that only one whole-number digit remained to the left of it. The number of places the decimal has been moved becomes the exponent on the base 10.

3,100. ⟶ 3.100

The decimal was moved three places to the left, so 3,100 becomes $3.1 \times 10^3$. The two zeros at the end of 3.100 are not needed and are dropped.

Here are more examples.

$203{,}000. = 2.03000 \times 10^5$ or $2.03 \times 10^5$

$8960. = 8.960 \times 10^3$ or $8.96 \times 10^3$

Write these numbers in scientific notation, dropping unneeded zeros.

**a.** 6,960 ______ **b.** 123,000 ______

**c.** 80,200,000 ______ **d.** 43,700 ______

**5**

Decimal numbers also can be written in scientific notation. The decimal point is moved *to the right* so that one whole-number digit is to the left of it. The number of places the decimal point is moved becomes the negative exponent on the base 10.

**4. a.** $6.96 \times 10^3$
**b.** $1.23 \times 10^5$
**c.** $8.02 \times 10^7$
**d.** $4.37 \times 10^4$

$$0.0017 = 0001.7 \times 10^{-3} \quad \text{or} \quad 1.7 \times 10^{-3}$$

The zeros before the whole-number digit are not needed and are dropped. Here are more examples.

$$0.000832 = 00008.32 \times 10^{-4} \quad \text{or} \quad 8.32 \times 10^{-4}$$

$$0.0000197 = 1.97 \times 10^{-5}$$

Remember to move the decimal point far enough so that one whole-number digit remains to the left of it.

Write these numbers in scientific notation.

**a.** 0.0046 ______ **b.** 0.000983 ______

**c.** 0.00000402 ______ **d.** 0.0000297 ______

**6**

Answers to frame 5.

**a.** $4.6 \times 10^{-3}$ **b.** $9.83 \times 10^{-4}$ **c.** $4.02 \times 10^{-6}$ **d.** $2.97 \times 10^{-5}$

# Section 8-4 Multiplying and Dividing Numbers in Scientific-Notation Form

**1**

Multiplication of numbers in scientific notation uses the following law of exponents.

$$y^a \cdot y^b = y^{a+b}$$

**Examples** $10^1 \cdot 10^3 = 10^{(+1)+(+3)}$ or $10^4$

$10^2 \cdot 10^{-5} = 10^{(+2)+(-5)}$ or $10^{-3}$

Multiply these numbers.

**a.** $10^2 \times 10^5 =$ ______ **b.** $10^{-7} \times 10^3 =$ ______

**c.** $10^{-3} \times 10^5 =$ ______ **d.** $10^{-2} \times 10^{-3} =$ ______

## 2

Using the law of exponents illustrated in frame 1, numbers in scientific notation can be multiplied as follows:

$$(2 \times 10^3)(4 \times 10^2) =$$

*Step 1.* First multiply the numbers not in power-of-ten form.

$$(2 \times 10^3)(4 \times 10^2)$$

multiply

$$2 \times 4 = 8$$

*Step 2.* Multiply the numbers with exponents by adding the exponents.

$$(2 \times 10^3)(4 \times 10^2)$$

add

$$10^{(+3)+(+2)} = 10^5$$

Combine the steps.

$$(2 \times 10^3)(4 \times 10^2) = 8 \times 10^5$$

Multiply these numbers.

**a.** $(2 \times 10^2)(3 \times 10^3) =$ ______ **b.** $(3 \times 10^{-4})(3 \times 10^2) =$ ______

**c.** $(4 \times 10^6)(1 \times 10^3) =$ ______ **d.** $(3 \times 10^{-2})(2 \times 10^{-3}) =$ ______

**1. a.** $10^7$
**b.** $10^{-4}$
**c.** $10^2$
**d.** $10^{-5}$

## 3

In this example, an extra step must be included to turn the answer into scientific-notation form. Problems worked in scientific-notation form must have answers in that form.

$$(9.2 \times 10^{-7})(3.5 \times 10^{-2}) = ?$$

$$32.2 \times 10^{(-7)+(-2)} = 32.2 \times 10^{-9}$$

This answer is not in scientific-notation form because there are two digits to the left of the decimal point. The decimal point is moved so that only one digit remains to the left of it.

*Step 3.* $32.2 \times 10^{-9} = 3.22 \times 10^{(-9)+(+1)} = 3.22 \times 10^{-8}$

A *(+1) is added* to the exponent because the decimal point was moved one place to the *left.*

Multiply these numbers, putting the answers into scientific notation.

**a.** $(2 \times 10^2)(7 \times 10^1) =$ ______ **b.** $(8 \times 10^3)(7 \times 10^4) =$ ______

**c.** $(1.5 \times 10^{-3})(8 \times 10^5) =$ ______ **d.** $(4.2 \times 10^{-1})(9.8 \times 10^{-5}) =$ ______

**2. a.** $6 \times 10^5$
**b.** $9 \times 10^{-2}$
**c.** $4 \times 10^9$
**d.** $6 \times 10^{-5}$

**4**

The calculator can be used to perform multiplication of numbers in scientific notation. The number is entered into the calculator using the [EE] key.

3. a. $1.4 \times 10^4$
b. $5.6 \times 10^8$
c. $1.2 \times 10^3$
d. $4.116 \times 10^{-5}$

Enter $3.5 \times 10^5$ into the calculator.

| *Enter* | *Press* | *Display* | |
|---|---|---|---|
| 3.5 | [EE] | 3.5 | 00 |
| 5 | | 3.5 | 05 |

exponent on base 10 (arrow to 05)

Enter $2.7 \times 10^{-3}$.

| *Enter* | *Press* | *Display* | |
|---|---|---|---|
| 2.7 | [EE] | 2.7 | 00 |
| 3 | [+/−] | 2.7 | −03 |

exponent on base 10 (arrow to −03)

Write these calculator-display numbers in scientific notation.

**a.** 4.2 03 = ________ **b.** 2.1 −05 = ________

**5**

Here are examples of multiplication of numbers in scientific notation using the calculator.

4. a. $4.2 \times 10^3$
b. $2.1 \times 10^{-5}$

**Example** $(4.1 \times 10^3)(2.3 \times 10^2) = ?$

| *Enter* | *Press* | *Display* | |
|---|---|---|---|
| 4.1 | [EE] | 4.1 | 00 |
| 3 | [×] | 4.1 | 03 |
| 2.3 | [EE] | 2.3 | 00 |
| 2 | [=] | 9.43 | 05 |

$(4.1 \times 10^3)(2.3 \times 10^2) = 9.43 \times 10^5$.

**Example** $(9.2 \times 10^{-7})(3.5 \times 10^{-2}) = ?$

| *Enter* | *Press* | *Display* | |
|---|---|---|---|
| 9.2 | [EE] | 9.2 | 00 |
| 7 | [+/−] [×] | 9.2 | −07 |
| 3.5 | [EE] | 3.5 | 00 |
| 2 | [+/−] [×] | 3.22 | −08 |

$(9.2 \times 10^{-7})(3.5 \times 10^{-2}) = 3.22 \times 10^{-8}$.

Use your calculator to multiply these numbers.

**a.** $(9 \times 10^2)(2 \times 10^1) =$ **b.** $(8 \times 10^7)(3 \times 10^{-4}) =$

**c.** $(2.2 \times 10^{-3})(2.1 \times 10^{-5}) =$ **d.** $(6.2 \times 10^{-1})(2.8 \times 10^1) =$

## 6

Dividing numbers in scientific notation uses the following law of exponents.

$$\frac{y^a}{y^b} = y^{a-b}$$

Note that the exponent in the denominator is subtracted from the exponent in the numerator. Do not reverse the order.

**Examples** $\frac{10^3}{10^2} = 10^{(+3)-(+2)} = 10^{(+3)+(-2)} = 10^1$

$$\frac{10^{-4}}{10^{-2}} = 10^{(-4)-(-2)} = 10^{(-4)+(+2)} = 10^{-2}$$

Divide each of these numbers.

**a.** $\frac{10^5}{10^2} =$ ______ **b.** $\frac{10^{-2}}{10^{-5}} =$ ______

**c.** $\frac{10^3}{10^{-2}} =$ ______ **d.** $\frac{10^{-5}}{10^3} =$ ______

5. a. $1.8 \times 10^4$
b. $2.4 \times 10^4$
c. $4.62 \times 10^{-8}$
d. $1.736 \times 10^1$

## 7

Numbers in scientific notation are divided using the following steps.

$$\frac{4 \times 10^4}{2 \times 10^2} = ?$$

*Step 1.* First divide the numbers not in power-of-ten form by dividing the number in the numerator by the number in the denominator.

$$\frac{\boxed{4} \times 10^4}{\boxed{2} \times 10^2} \quad \text{or} \quad \frac{4}{2} = 2$$

*Step 2.* Subtract the exponent in the denominator from the exponent in the numerator.

$$\frac{4 \times 10^4}{2 \times 10^2} \quad \text{or} \quad 10^{(+4)-(+2)} = 10^{(+4)+(-2)} = 10^2$$

Now combine the steps.

$$\frac{4 \times 10^4}{2 \times 10^2} = 2 \times 10^2$$

6. a. $10^3$
b. $10^3$
c. $10^5$
d. $10^{-8}$

Here is another example:

$$\frac{3.52 \times 10^5}{1.1 \times 10^2} = ?$$

*Step 1.* $\frac{3.52}{1.1} = 3.2$

*Step 2.* $\frac{10^5}{10^2} = 10^{(+5)-(+2)}$

$= 10^{(+5)+(-2)}$

$= 10^3$

Combine steps.

$$\frac{3.52 \times 10^5}{1.1 \times 10^2} = 3.2 \times 10^3$$

Divide each of these numbers.

**a.** $\frac{8 \times 10^5}{2 \times 10^3} =$

**b.** $\frac{9 \times 10^5}{3 \times 10^{-2}} =$

**c.** $\frac{6 \times 10^{-8}}{3 \times 10^2} =$

**d.** $\frac{8 \times 10^{-9}}{4 \times 10^{-2}} =$

**e.** $\frac{7.2 \times 10^7}{3 \times 10^2} =$

**f.** $\frac{6.6 \times 10^5}{2.2 \times 10^{-4}} =$

**g.** $\frac{8.2 \times 10^{-9}}{4.1 \times 10^{-5}} =$

**h.** $\frac{5.0 \times 10^{-3}}{2.5 \times 10^2} =$

**8**

When the numerator is smaller than the denominator, the answer will not be in scientific notation.

**7.** **a.** $4 \times 10^2$
**b.** $3 \times 10^7$
**c.** $2 \times 10^{-10}$
**d.** $2 \times 10^{-7}$
**e.** $2.4 \times 10^5$
**f.** $3 \times 10^9$
**g.** $2 \times 10^{-4}$
**h.** $2 \times 10^{-5}$

**Example** $\frac{3.2 \times 10^7}{8 \times 10^2} =$

*Step 1.* $\frac{3.2}{8} = 0.4$

*Step 2.* $\frac{10^7}{10^2} = 10^{(+7)-(+2)}$

$= 10^{(+7)+(-2)}$

$= 10^5$

Combine steps.

$$\frac{3.2 \times 10^7}{8 \times 10^2} = 0.4 \times 10^5$$

A third step must be added to shift the decimal point one place to the right.

*Step 3.* $0.4 \times 10^5 = 4.0 \times 10^{(+5)+(-1)} = 4 \times 10^4$

A *(−1) is added* to the exponent because the decimal point was moved one place to the *right.*

Divide each number, putting the answer in scientific notation.

**a.** $\dfrac{2 \times 10^5}{4 \times 10^2} =$ **b.** $\dfrac{2 \times 10^3}{5 \times 10^{-2}} =$

**c.** $\dfrac{3.6 \times 10^{-3}}{9 \times 10^2} =$ **d.** $\dfrac{1.2 \times 10^{-3}}{6 \times 10^{-3}} =$

## 9

Here are examples using a calculator.

**8. a.** $5 \times 10^2$
**b.** $4 \times 10^4$
**c.** $4 \times 10^{-6}$
**d.** $2 \times 10^{-1}$

**Example** $\dfrac{8.8 \times 10^3}{2.2 \times 10^5} =$

| *Enter* | *Press* | *Display* | |
|---|---|---|---|
| 8.8 | [EE] | 8.8 | 00 |
| 3 | [÷] | 8.8 | 03 |
| 2.2 | [EE] 5 [=] | 4 | −02 |

$$\frac{8.8 \times 10^3}{2.2 \times 10^5} = 4 \times 10^{-2}$$

**Example** $\dfrac{2.2 \times 10^{-6}}{8.8 \times 10^{-3}} =$

| *Enter* | *Press* | *Display* | |
|---|---|---|---|
| 2.2 | [EE] | 2.2 | 00 |
| 6 | [+/−] [÷] | 2.2 | −06 |
| 8.8 | [EE] | 8.8 | 00 |
| 3 | [+/−] [=] | 2.5 | −04 |

$$\frac{2.2 \times 10^{-6}}{8.8 \times 10^{-3}} = 2.5 \times 10^{-4}$$

Divide each of these numbers, rounding to two decimal places.

**a.** $\dfrac{4 \times 10^{-3}}{5 \times 10^{-2}} =$ ________ **b.** $\dfrac{3 \times 10^5}{6 \times 10^3} =$ ________

**c.** $\dfrac{8.7 \times 10^{-3}}{4.1 \times 10^5} =$ ________ **d.** $\dfrac{6.9 \times 10^6}{3 \times 10^{-3}} =$ ________

## 10

Answers to frame 9.

**a.** $8 \times 10^{-2}$ **b.** $5 \times 10^1$ **c.** $2.12 \times 10^{-8}$ **d.** $2.3 \times 10^9$

# Exercise Set, Sections 8-3–8-4

## Powers of Ten and Scientific Notation

Write each of these numbers in scientific notation.

**1.** 1,000 ______ **2.** 10 ______ **3.** 1 ______ **4.** 5,870 ______ **5.** 183,000 ______

**6.** 402,000 ______ **7.** 0.02 ______ **8.** 0.0043 ______ **9.** 0.000097 ______

## Multiplying and Dividing Numbers in Scientific-Notation Form

Multiply these numbers.

**10.** $10^3 \times 10^4 =$ ______

**11.** $10^{-5} \times 10^3 =$ ______

**12.** $10^4 \times 10^{-3} =$ ______

**13.** $10^{-2} \times 10^{-3} =$ ______

**14.** $(2 \times 10^1)(3 \times 10^{-3}) =$ ______

**15.** $(4 \times 10^{-2})(2 \times 10^5) =$ ______

**16.** $(2 \times 10^3)(6 \times 10^1) =$ ______

**17.** $(8 \times 10^{-3})(8 \times 10^{-2}) =$ ______

**18.** $(7.2 \times 10^4)(1.8 \times 10^{-1}) =$ ______

**19.** $(6.5 \times 10^{-3})(3 \times 10^{-2}) =$ ______

**20.** $(8.5 \times 10^2)(4.1 \times 10^3) =$ ______

Divide each of these numbers.

**21.** $\frac{10^5}{10^2} =$ ______ **22.** $\frac{7 \times 10^{-4}}{2 \times 10^2} =$ ______ **23.** $\frac{5 \times 10^{-3}}{1.25 \times 10^{-1}} =$ ______ **24.** $\frac{1.8 \times 10^5}{6 \times 10^2} =$ ______

# Supplementary Exercise Set, Sections 8-3–8-4

Write each of these numbers in power-of-ten or scientific-notation form.

**1.** 10,000 ______ **2.** 1 ______ **3.** 10 ______ **4.** 3,680 ______ **5.** 119,000 ______

**6.** 789,000 ______ **7.** 0.07 ______ **8.** 0.00038 ______ **9.** 0.000027 ______

Multiply each of these numbers.

**10.** $10^2 \times 10^4 =$ ______

**11.** $10^{-4} \times 10^3 =$ ______

**12.** $10^3 \times 10^{-5} =$ ______

**13.** $10^{-2} \times 10^{-1} =$ ______

**14.** $(2 \times 10^2)(4 \times 10^{-4}) =$ ______

**15.** $(3 \times 10^{-6})(2 \times 10^2) =$ ______

**16.** $(2 \times 10^3)(7 \times 10^2) =$ ______

**17.** $(6 \times 10^{-3})(6 \times 10^{-4}) =$ ______

**18.** $(6.2 \times 10^4)(1.8 \times 10^{-1}) =$ ______

**19.** $(3.5 \times 10^{-4})(3 \times 10^{-3}) =$ ______

**20.** $(9.5 \times 10^2)(3.1 \times 10^5) =$ ______

Divide each of these numbers.

**21.** $\dfrac{10^7}{10^5} =$ ______

**22.** $\dfrac{5 \times 10^{-6}}{2 \times 10^2} =$ ______

**23.** $\dfrac{3.75 \times 10^{-4}}{1.25 \times 10^{-1}} =$ ______

**24.** $\dfrac{1.5 \times 10^6}{5 \times 10^2} =$ ______

**25.** $\dfrac{4 \times 10^{-5}}{5 \times 10^2} =$ ______

# Summary

1. In the number $y^x$, $y$ is the base and can be raised to the exponent, $x$.
2. A number is squared by multiplying the number by itself.
3. The square root of a number $x$ is the number whose square is $x$.
4. Writing numbers in scientific notation.

   *Step 1.* Move the decimal point until there is only one whole-number digit to the left of it.
   *Step 2.* Count the number of spaces the decimal point has been moved; this number becomes the exponent on the base 10.
   *Step 3.* If the original number is a whole number, the exponent is positive. If the original number is a decimal number, the exponent is negative.

5. Multiplying numbers in scientific notation.

   *Step 1.* Multiply the two initial digits.
   *Step 2.* Add the exponents on the base 10.
   *Step 3.* Adjust the answer, if necessary, to put it into scientific notation.

6. Dividing numbers in scientific notation.

   *Step 1.* Divide the two initial digits.
   *Step 2.* Subtract the exponent on the base 10 in the denominator from the exponent on the base 10 in the numerator.
   *Step 3.* Adjust the answer, if necessary, to put it into scientific notation.

# Chapter 8 Self-Test

## The Powers and Roots of Numbers

Square each number.

**1.** 14.2 ______

**2.** −0.19 ______

**3.** 0.82 ______

Find the indicated power of each number.

**4.** $2^4$ ______ **5.** $0.16^3$ ______ **6.** $1.7^5$ ______

Find the square root of each of these numbers.

**7.** 387 ______ **8.** 0.984 ______ **9.** 1.87 ______

Using the formula, $a = \sqrt{b^2 + c^2}$, find $a$ when

**10.** $a =$ ______

$b = 5$

$c = 7$

**11.** $a =$ ______

$b = 0.11$

$c = 0.23$

Write each number in power-of-ten or scientific-notation form.

**12.** 10 ______ **13.** 4,260 ______ **14.** 0.00081 ______

Multiply each of these numbers.

**15.** $10^2 \times 10^6 =$ ______

**16.** $10^{-2} \times 10^{-5} =$ ______

**17.** $(2 \times 10^{-2})(3 \times 10^{-5}) =$ ______

**18.** $(3.1 \times 10^5)(1.2 \times 10^{-2}) =$ ______

**19.** $(8 \times 10^4)(7 \times 10^1) =$ ______

**20.** $(8.7 \times 10^2)(2.1 \times 10^{-4}) =$ ______

Divide each of these numbers.

**21.** $\dfrac{10^8}{10^4} =$ ______ **22.** $\dfrac{6 \times 10^{-8}}{2 \times 10^2} =$ ______ **23.** $\dfrac{5.6 \times 10^{-4}}{1.4 \times 10^{-2}} =$ ______

**24.** $\dfrac{1.2 \times 10^5}{4 \times 10^2} =$ ______ **25.** $\dfrac{1 \times 10^{-6}}{5 \times 10^2} =$ ______

# CHAPTER 9

# Geometry

Modern technology requires technicians to have a basic understanding of geometry, for they must measure, weigh, form, and manipulate geometric objects. This chapter covers angles, areas, perimeters, volume, density, circles, spheres, and other composite geometric shapes.

## Section 9-1 Points, Lines, and Angles

**1**

The most basic element in geometry is a point. A point has no dimensions and is represented by a dot.

Which of the following represents a point? (Circle your selection.)

**a.** * **b.** · **c.** #

**2**

A *line* is a set of points and is represented as shown below. **1. b**

Notice that the line has arrows on both ends. These arrows indicate that the line extends infinitely in both directions.

**3**

The part of a line between two points is called a *line segment.*

*A* *B*

The line above is denoted by $\overleftrightarrow{AB}$.
The line segment above is denoted by $\overline{AB}$.

Referring to Figure 9-1, write the correct symbols for the following.

**a.** the line ______________

**b.** the line segment ______________

*C* *D*

**Figure 9-1**

**4**

Point *A* below—and all of the points to one side of *A*—is called a *ray*. **3. a.** $\overleftrightarrow{CD}$ **b.** $\overline{CD}$

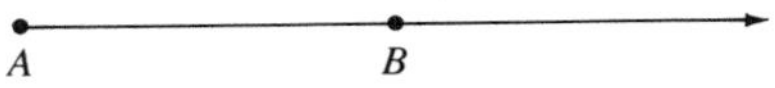

*A* *B*

This ray is denoted by the symbol $\overrightarrow{AB}$.

Below, point $D$, and all of the points to the left of $D$, is a ray denoted by $\overrightarrow{DC}$. Point $C$, and all of the points to the right of $C$, is a ray denoted by $\overrightarrow{CD}$.

Referring to Figure 9-2, write the correct symbols for the following.

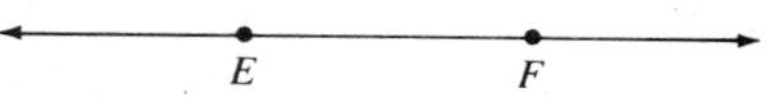


**Figure 9-2**

**a.** point $E$ and all of the points to the right of $E$ ____________

**b.** point $F$ and all of the points to the left of $F$ ____________

## 5

Two rays with a common end point are called an *angle*. The figure below shows the angle formed by the rays $\overrightarrow{OA}$ and $\overrightarrow{OB}$.

4. a. $\overrightarrow{EF}$
b. $\overrightarrow{FE}$

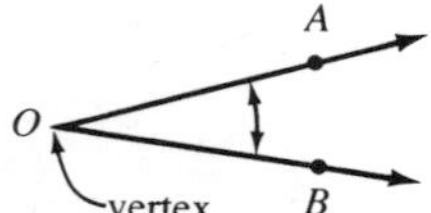


The common end point of an angle is called a *vertex*. In the figure above, $O$ is the vertex, and $\overrightarrow{OA}$ and $\overrightarrow{OB}$ are the sides of the angle.

Referring to the figure below:

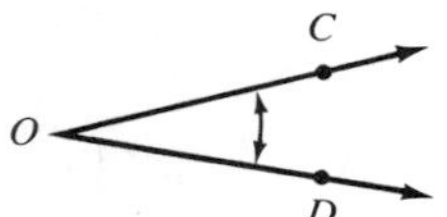


**a.** write the vertex of the angle. ____________

**b.** write the sides of the angle. ____________

## 6

The symbol $\angle$ is used to denote an angle. The angle in Figure 9-3 can be named in several different ways:

5. a. $O$
b. $\overrightarrow{OC}$ and $\overrightarrow{OD}$

**1.** by the $\angle$ symbol and the letter denoting the vertex:

$\angle A$

**2.** by the $\angle$ symbol and the three points listed, with the vertex as the center point:

$\angle BAC$ or $\angle CAB$

**3.** by the Greek letter that represents the angle:

$\theta$ (theta)

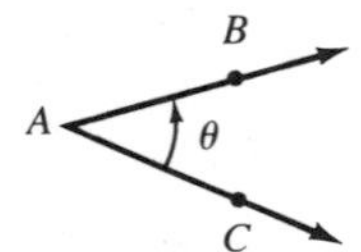


**Figure 9-3**

Write the symbol for the angle in Figure 9-4 in three different ways.

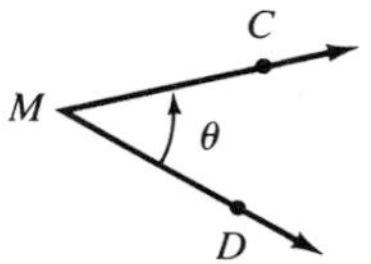


**Figure 9-4**

____________ ____________ ____________

6. $\angle M$
$\angle CMD$
$\theta$

## 7

Angles are most often measured by degrees. Some common degrees of angles are shown in Figure 9-5.

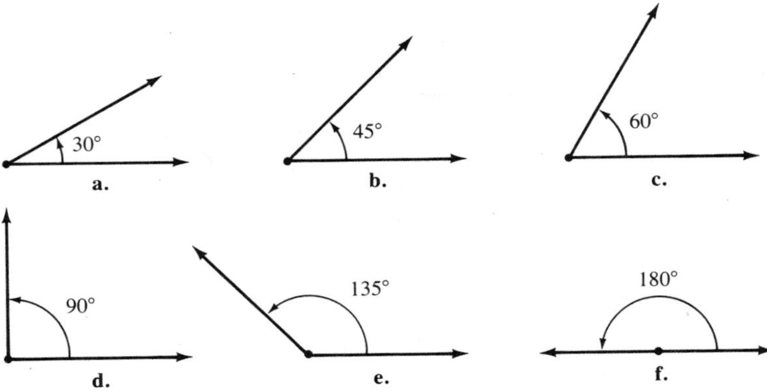


**Figure 9-5** Frequently-used angles

## 8

The protractor is used to measure angles. (A sketch of a protractor is shown in Figure 9-6.)
The following steps are used to measure an angle with a protractor.

*Step 1.* The vertex of the angle is placed at the center mark of the protractor.
*Step 2.* One of the rays lies along the base line of the protractor and passes through 0.
*Step 3.* The other ray passes through the protractor and gives the measure of the angle.

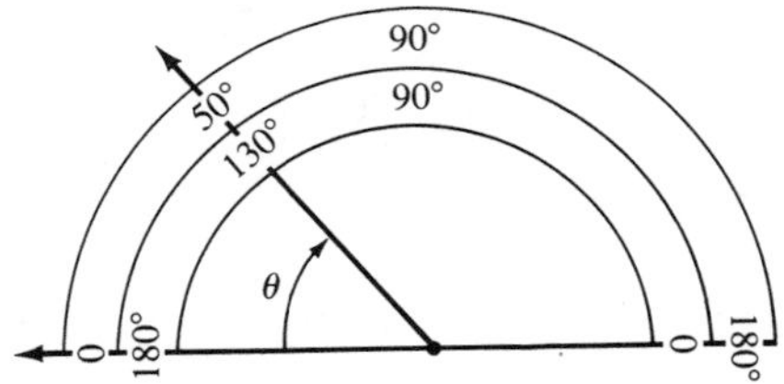


**Figure 9-6**

Angle $\theta$ is less than 90°; therefore, the 50° measure is read from the protractor.

Use a protractor to measure the angles below.

**a.** ______________ **b.** ______________

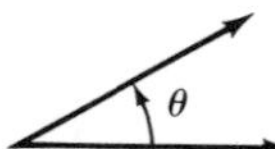


θ

## 9

Use a protractor to measure angle $\theta$. ______________

8. a. 30°
b. 140°

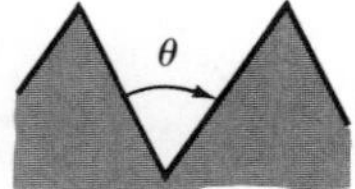


**Figure 9-7** External thread

## 10

Use a protractor to measure the angles in Figure 9-8.

9. 60°

**a.** $\angle AOB$ ______ **b.** $\angle AOC$ ______ **c.** $\angle BOC$ ______

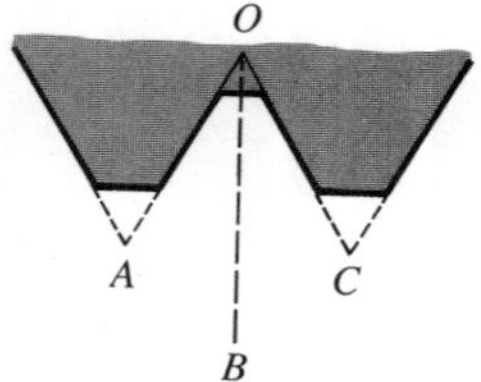


**Figure 9-8** Internal thread

## 11

Use a protractor to measure the angles in Figure 9-9.

10. a. 30°
b. 60°
c. 30°

**a.** $\angle\theta$ ______ **b.** $\angle AOB$ ______ **c.** $\angle BOC$ ______

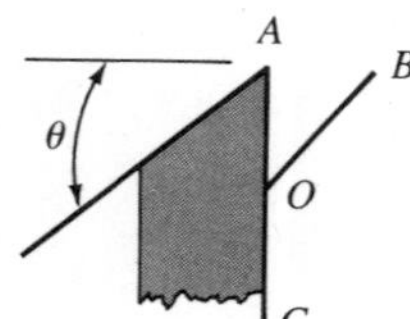


**Figure 9-9**

## 12

Answers to frame 11.

**a.** 40° **b.** 40° **c.** 140°

# Section 9-2 Polygons, Areas, and Perimeters

**1**

A closed-plane figure formed by three or more segments (sides) is called a *polygon.*

Each polygon is given a specific name according to the number of sides it has. The most common examples are shown below.

| *Polygon* | *Number of sides* | *Name* |
|---|---|---|
|  | Three | Triangle |
|  | Four | Quadrilateral |
|  | Five | Pentagon |
|  | Six | Hexagon |
|  | Eight | Octagon |

**2**

There are special quadrilaterals (four-sided polygons) that occur in industry, science, and technology. These are shown below.

| *Special Quadrilateral* | *Description* | *Special Name* |
|---|---|---|
| 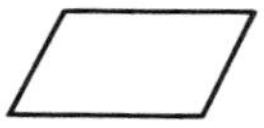 | A quadrilateral with opposite sides equal and parallel | Parallelogram |
| 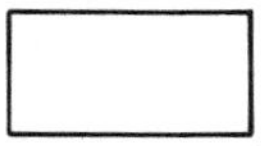 | A quadrilateral with opposite sides equal and four right angles | Rectangle |
| 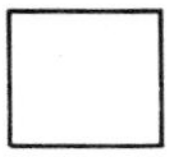 | A quadrilateral with four sides equal and four right angles | Square |
| 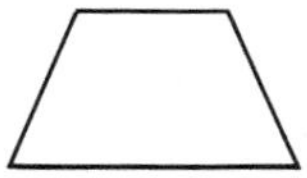 | A quadrilateral with only one pair of opposite sides parallel | Trapezoid |

## 3

This section deals with computing areas and perimeters of quadrilaterals.

Figure 9-10 is a rectangle. It has four 90° angles and two pairs of equal sides. The shorter sides of a rectangle are usually labeled the *width.* The larger sides of the rectangle are labeled the *length.* Thus, the length of the rectangle in Figure 9-10 is 5 m.

What is the width of the rectangle in Figure 9-10? __________

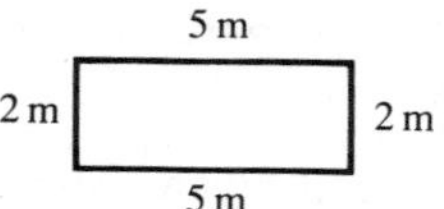


**Figure 9-10**

## 4

The distance around a rectangle is called the *perimeter.* The perimeter is found by adding the four sides (two lengths and two widths) of the rectangle. Therefore, the formula for the perimeter of a rectangle is

$$P = 2l + 2w$$

where $P =$ perimeter
$l =$ length
$w =$ width

Using the formula above, find the perimeter of the rectangle in Figure 9-10.

**3.** 2 m

## 5

The area of a rectangle is found by multiplying the length times the width. The formula for the area of a rectangle is

$$A = lw$$

where $A =$ area
$l =$ length
$w =$ width

Use the formula above to find the area of the rectangle in Figure 9-10.

**4.** $P = 2(5 \text{ m}) + 2(2 \text{ m})$
$P = 10 \text{ m} + 4 \text{ m}$
$P = 14 \text{ m}$

## 6

Find the perimeter and the area of the rectangle in Figure 9-11 (a calculator may be helpful).

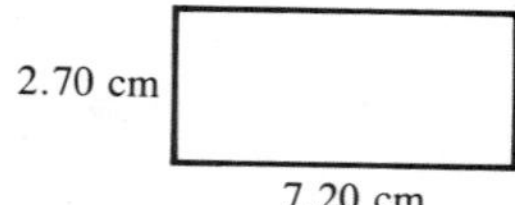


**Figure 9-11**

**a.** $P =$ **b.** $A =$

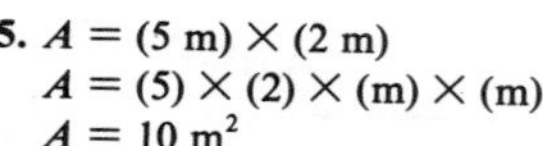

**5.** $A = (5 \text{ m}) \times (2 \text{ m})$
$A = (5) \times (2) \times (\text{m}) \times (\text{m})$
$A = 10 \text{ m}^2$

## 7

A square is a special rectangle that has four equal sides. The side of a square is usually denoted as *s.*
The perimeter of a square is found by the formula $P = 4s$.
The area of a square is found by the formula $A = s^2$.

**6. a.** $P = 19.80$ cm
**b.** $A = 19.4 \text{ cm}^2$

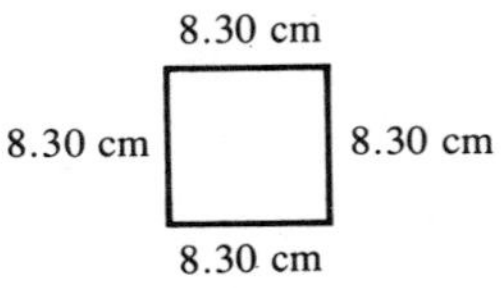

8.30 cm

8.30 cm 8.30 cm

8.30 cm

**Figure 9-12**

**a.** Find the perimeter of the square in Figure 9-12.

**b.** Find the area of the square in Figure 9-12.

## 8

A parallelogram is shown in Figure 9-13; it has two pairs of equal sides opposite each other. *Notice that the height is different from the side.*
The area of a parallelogram is found by the formula $A = bh$, where $b$ = base and $h$ = height.

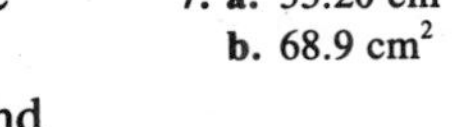
**7. a.** 33.20 cm
**b.** 68.9 $cm^2$

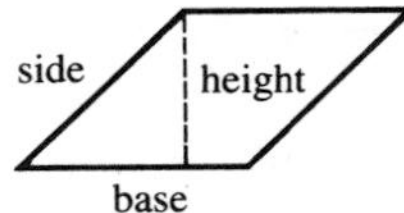

**Figure 9-13** Parallelogram

Referring to Figure 9-13, find the area when the base is 10.2 cm and the height is 4.70 cm.

## 9

**a.** Find the area of a parallelogram with a base of 7.20 in. and a height of 4.80 in.

**8.** $A = 47.9$ $cm^2$

**b.** Find the area of a parallelogram with a height of 14.2 m and a base of 22.5 m.

## 10

The base or height of a parallelogram can be found by rearranging the formula for the area of a parallelogram.

**9. a.** 34.6 $in.^2$
**b.** 320 $m^2$

**Example** Find the height of the parallelogram in Figure 9-14 when the base is 10.2 cm and the area is 66.3 $cm^2$.

*Step 1.* First rearrange the formula.

$$A = bh$$

$$\left(\frac{1}{b}\right)A = \left(\frac{1}{b}\right)bh$$

$$\frac{A}{b} = h$$

*Step 2.* Substitute the data into the rearranged formula.

$$h = \frac{66.3 \text{ cm}^2}{10.2 \text{ cm}} = 6.50 \text{ cm}$$

**a.** Rearrange the formula $A = bh$ to solve for $b$.

**b.** Find $b$ when $h = 11.0$ in. and $A = 286$ in.

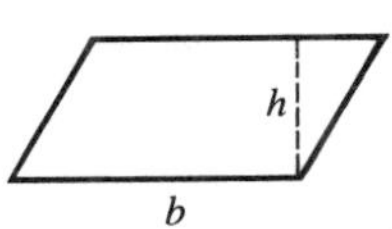

**Figure 9-14**

## 11

The formula for finding the perimeter of a rectangle is also used to find the perimeter of a parallelogram. (A rectangle is just a special case of a parallelogram.)

**10. a.** $\frac{A}{h} = b$

**b.** 26.0 in.

$$P = 2a + 2b$$

**Example** Find the perimeter of the parallelogram in Figure 9-15 when $b_1 = 42$ cm and $a_1 = 36$ cm. (Remember, in a parallelogram the opposite sides will always be equal. That is, $a_1$ will always equal $a_2$, and $b_1$ will always equal $b_2$.)

$$P = 2a + 2b$$

$$P = 2(36 \text{ cm}) + 2(42 \text{ cm})$$

$$P = 72 \text{ cm} + 84 \text{ cm}$$

$$P = 156 \text{ cm}$$

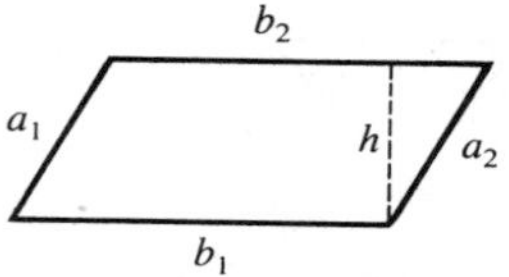


**Figure 9-15**

**a.** Find the perimeter of the parallelogram when $a_1 = 14.2$ in. and $b_2 = 12.5$ in.

**b.** Find the perimeter of the parallelogram when $a_1 = 12$ in. and $b_2 = 10$ in.

## 12

The geometric shape in Figure 9-16 is a trapezoid. The area is found using the formula

**11. a.** 53.4 in.
**b.** 44 in.

$$A = \frac{(a + b)h}{2}$$

**Example** Find the area of the trapezoid in Figure 9-16 when $b = 4.2$ in., $a = 3.8$ in., and $h = 1.7$ in.

$$A = \frac{(a + b)h}{2}$$

$$A = \frac{(4.2 \text{ in.} + 3.8 \text{ in.})(1.7 \text{ in.})}{2}$$

$$A = \frac{(8.0 \text{ in.})(1.7 \text{ in.})}{2}$$

$$A = (4.0 \text{ in.})(1.7 \text{ in.})$$

$$A = 6.8 \text{ in.}^2$$

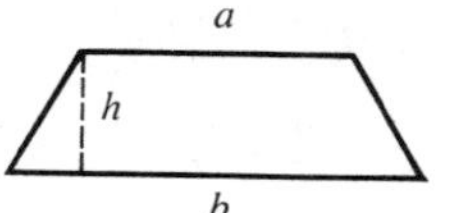


**Figure 9-16** Trapezoid

**a.** Find the area of the trapezoid in Figure 9-16 when $b = 12$ in., $a = 8$ in., and $h = 7$ in.

**b.** Find the area of the trapezoid in Figure 9-16 when $b = 9.10$ m, $a = 7.90$ m, and $h = 7.50$ m.

## 13

Answers to frame 12.

**a.** 70 in.$^2$

**b.** 63.8 m$^2$

# Exercise Set, Sections 9-1–9-2

## Points, Lines, and Angles

Use a protractor to measure the following angles in Figure 9-17.

1. $\angle AOB$ ______________
2. $\angle BAO$ ______________
3. $\angle ABO$ ______________

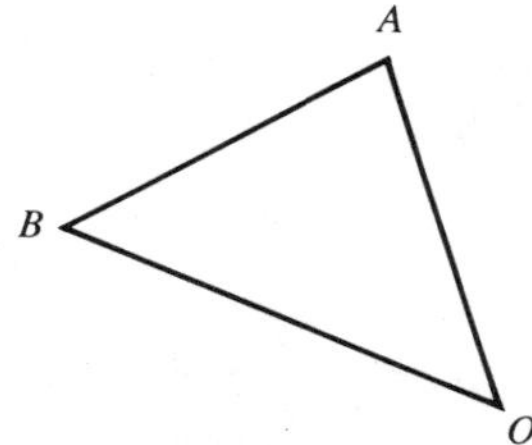


**Figure 9-17**

Use a protractor to measure the following angles in Figure 9-18.

4. $\angle AOB$ ______________
5. $\angle AOC$ ______________
6. $\angle AOD$ ______________
7. $\angle BOC$ ______________
8. $\angle BOD$ ______________
9. $\angle COD$ ______________

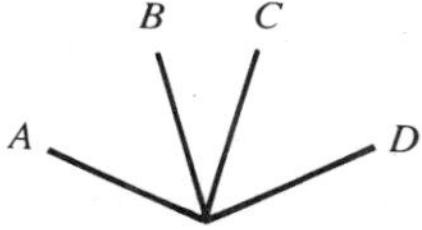


**Figure 9-18**

## Polygons, Areas, and Perimeters

Given Figure 9-19, find the following:

10. perimeter ______________
11. area ______________

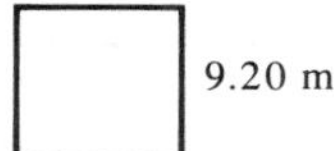


**Figure 9-19** Rectangle

Given Figure 9-20, find the following:

12. perimeter ______________
13. area ______________

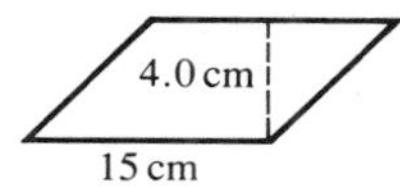


**Figure 9-20** Square

Given Figure 9-21, find the following:

14. area ______________



**Figure 9-21** Parallelogram

Given Figure 9-22, find the following:

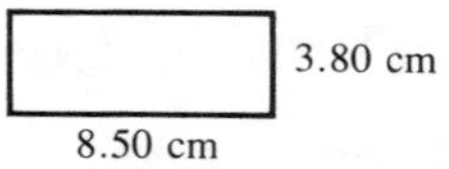

**15.** perimeter ______________

**16.** area ______________

3.80 cm

8.50 cm

**Figure 9-22** Rectangle

Given Figure 9-23, find the following:

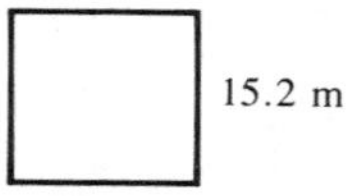

**17.** perimeter ______________

**18.** area ______________

15.2 m

**Figure 9-23** Square

Given Figure 9-24, find the following:

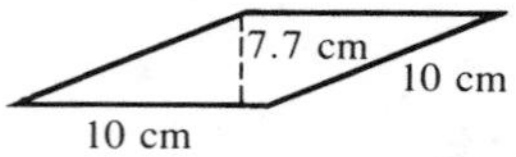

**19.** area ______________

**20.** perimeter ______________

7.7 cm

10 cm

10 cm

**Figure 9-24** Parallelogram

Given Figure 9-25, find the following:

**21.** area ______________

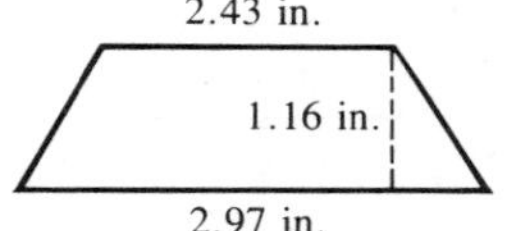


**Figure 9-25** Trapezoid

# Supplementary Exercise Set, Sections 9-1–9-2

Use a protractor to measure the following angles in Figure 9-26.

**1.** $\angle AOB$ ______________

**2.** $\angle AOC$ ______________

**3.** $\angle AOD$ ______________

**4.** $\angle BOC$ ______________

**5.** $\angle BOD$ ______________

**6.** $\angle COD$ ______________

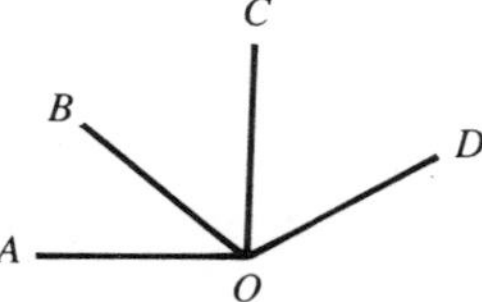


**Figure 9-26**

Given the rectangle in Figure 9-27, find the following:

**7.** perimeter ______________

**8.** area ______________

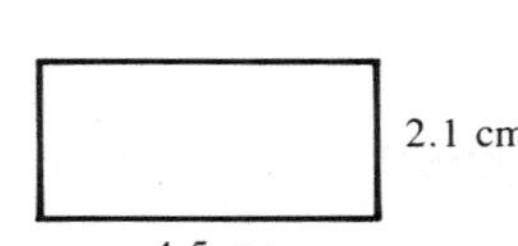


**Figure 9-27**

Given the square in Figure 9-28, find the following:

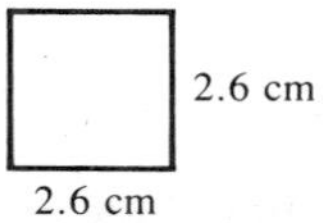

**9.** perimeter ____________

**10.** area ____________

2.6 cm

2.6 cm

**Figure 9-28**

Given the parallelogram in Figure 9-29, find the following:

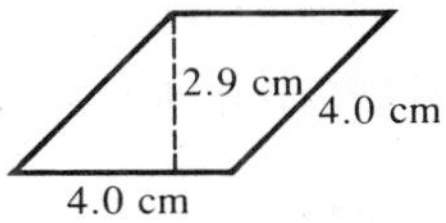

**11.** perimeter ____________

**12.** area ____________

2.9 cm

4.0 cm

4.0 cm

**Figure 9-29**

Given Figure 9-30, find the following:

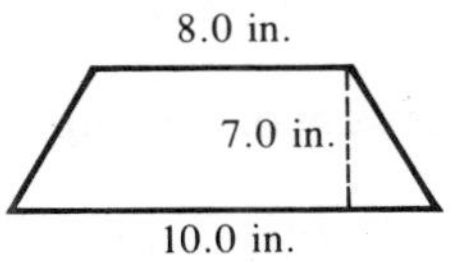

**13.** area ____________

8.0 in.

7.0 in.

10.0 in.

**Figure 9-30**

# Section 9-3 Regular Polygons

**1**

Polygons with equal sides and equal angles occur regularly in technical work. These *regular polygons* are shown below.

Equilateral triangle

Square

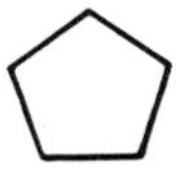

Regular pentagon

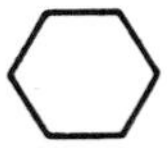

Regular hexagon

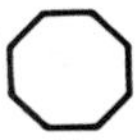

Regular octagon

The square was discussed in the previous section, and the equilateral triangle will be discussed in a later section. This section will describe methods for finding the sum of the angles and the perimeters of all regular polygons.

**a.** A polygon with equal sides and equal angles is called a ____________ polygon.

## 2

The sum of the angles of any polygon can be determined by using a formula.

1. a. regular

$$T° \text{ (polygon)} = (m - 2)(180°)$$

where $T°$ is the sum of the angles and $m$ is the number of sides of the polygon.

The formula is used to find the sum of the angles in a triangle.

$$T° \text{ (triangle)} = (3 - 2)(180°)$$
$$T° \text{ (triangle)} = (1)(180°)$$
$$T° \text{ (triangle)} = 180°$$

The sum of the angles of a rectangle can be found as follows:

$$T° \text{ (rectangle)} = (4 - 2)(180°)$$
$$T° \text{ (rectangle)} = (2)(180°)$$
$$T° \text{ (rectangle)} = 360°$$

**a.** Find the sum of the angles of a pentagon.

$T°$ (pentagon) = ___________

**b.** Find the sum of the angles of a hexagon.

$T°$ (hexagon) = ___________

**c.** Find the sum of the angles of an octagon.

$T°$ (octagon) = ___________

**d.** Find the sum of the angles of a quadrilateral.

$T°$ (quadrilateral) = ___________

## 3

Now that we know how to find the sum of the angles in a polygon, it is possible to find the size of *each* angle in a regular polygon.

2. a. 540°
b. 720°
c. 1,080°
d. 360°

For example, a regular pentagon has five equal angles and an angle sum of 540°. The size of each angle can be found as follows:

$$\text{Size of each angle} = \frac{\text{sum of the angles}}{\text{number of angles}}$$

$$\text{Size of each angle} = \frac{540°}{5}$$

$$\text{Size of each angle} = 108°$$

Therefore, each angle in a *regular* pentagon is always 108°.

**a.** Find the size of each angle in a *regular* hexagon. ___________

**b.** Find the size of each angle in a *regular* octagon. ___________

**c.** Find the size of each angle in an equilateral triangle. ___________

**d.** Find the size of each angle in a square. ___________

**4**

The perimeter of any polygon is the sum of the sides of that polygon.

For example, the perimeter of a five-sided polygon (pentagon) with sides of 1 in., 3 in., 4 in., 2 in., and 6 in. is

perimeter = 1 in. + 3 in. + 4 in. + 2 in. + 6 in.

perimeter = 16 in.

The perimeter of a regular pentagon with each side measuring 7 m is

perimeter = 7 m + 7 m + 7 m + 7 m + 7 m

perimeter = 35 m

The perimeter of any *regular* polygon can be found by multiplying the length of a side of that polygon times the number of sides. Therefore, the perimeter of the *regular* pentagon with each side measuring 7 m can also be found as follows:

perimeter (regular polygon) = (number of sides) (length of side)

perimeter (regular pentagon) = (5) (7 m)

perimeter (regular pentagon) = 35 m

**a.** Find the perimeter of a regular hexagon with a side of 25 ft. ____________

**b.** Find the perimeter of a regular octagon with a side of 41 mm. ____________

**c.** Find the perimeter of an equilateral triangle with a side of 25 yd. ____________

**d.** Find the perimeter of a regular seven-sided polygon with a side of 15 cm.

____________

**3. a.** 120°
**b.** 135°
**c.** 60°
**d.** 90°

**5**

Answers to frame 4.

**a.** 150 ft **b.** 328 mm
**c.** 75 yd **d.** 105 cm

# Section 9-4 Volume and Density

**1**

Volume is measured in cubic units. A common unit for measuring volume in technologies is the cubic centimeter, which is abbreviated $cm^3$. Figure 9-31 illustrates one cubic centimeter.

The volume of a cube is found by the formula $V = s^3$, where $V$ = volume and $s$ = side of a cube.

The volume of a cube with sides of 4.0 cm is

$$V = (4.0\text{ cm})^3$$

$$V = (4.0\text{ cm})(4.0\text{ cm})(4.0\text{ cm})$$

$$V = 64\text{ cm}^3$$

Find the volume of a cube with sides of 3.0 cm.

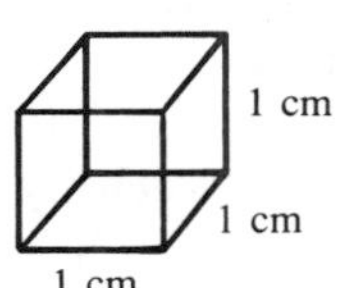


**Figure 9-31** One cubic centimeter (1 $cm^3$)

## 2

The volume of a rectangular solid is found by the formula $V = lwh$, where $l =$ length, $w =$ width, and $h =$ height.

1. $V = 27\ \text{cm}^3$

The volume of the rectangular solid in Figure 9-32 is found below.

$$V = lwh$$

$$V = (8.0\ \text{in.})(5.0\ \text{in.})(2.0\ \text{in.})$$

$$V = 80\ \text{in.}^3$$

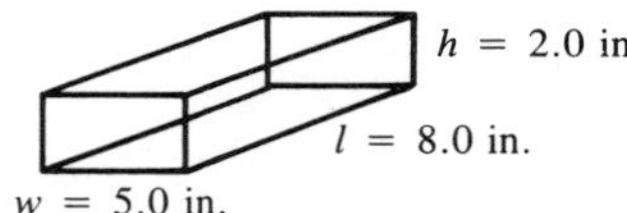


**Figure 9-32** Rectangular solid

**a.** Find the volume of a rectangular solid with a length of 3.0 cm, a width of 2.5 cm, and a height of 4.5 cm.

**b.** Find the volume of a room with a length of 15 ft, a width of 10 ft, and a height of 8.0 ft.

## 3

The volume of a cylinder is found by multiplying the area of the end (also called the *base*) by the length (also called the *height*).

2. **a.** $34\ \text{cm}^3$ **b.** $1{,}200\ \text{ft}^3$

The area of the end $= \pi r^2$.

$$V = \pi r^2 h$$

**Example** Find the volume of the cylinder in Figure 9-33, when $r = 3.00$ cm and $h = 12.0$ cm.

$$V = \pi r^2 h$$

$$V = (3.14)(3.00\ \text{cm})^2(12.0\ \text{cm})$$

$$V = (3.14)(9.00\ \text{cm}^2)(12.0\ \text{cm})$$

$$V = 339\ \text{cm}^3$$

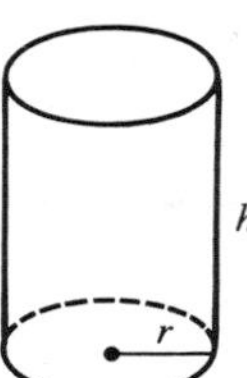


**Figure 9-33** Cylinder

**a.** Find the volume of the cylinder in Figure 9-33, when $r = 1.7$ in. and $h = 2.0$ in.

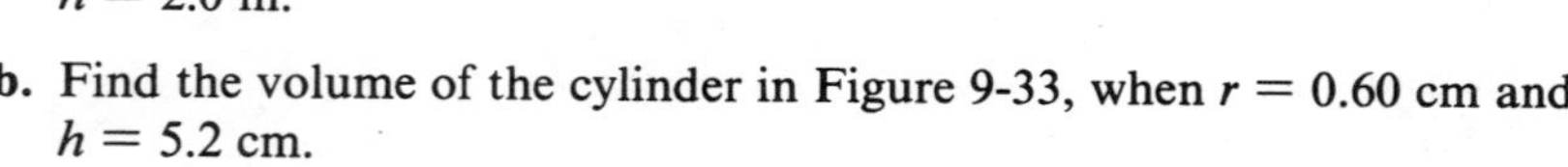

**b.** Find the volume of the cylinder in Figure 9-33, when $r = 0.60$ cm and $h = 5.2$ cm.

## 4

The density of a liquid is its weight per unit volume. In the metric system, density is usually stated in grams per cubic centimeter.

3. **a.** $18\ \text{in.}^3$ **b.** $5.9\ \text{cm}^3$

**Examples** The density of water is $1\ \text{g/cm}^3$. (Its weight is 1 g; its volume is $1\ \text{cm}^3$.) The density of mercury is $13.6\ \text{g/cm}^3$.

How much does $1\ \text{cm}^3$ of mercury weigh? ______________

## 5

Density can be used to find the weight of a given volume of liquid, using the following formula.

4. 13.6 g

$$\text{weight} = \text{density} \times \text{volume}$$

This formula, along with dimensional analysis, can be used to solve the following applied problem.

A lab technician is asked to find the weight of 32.0 $cm^3$ of mercury.

The five-step procedure for problem solving can be used.

*Step 1.* Determine what is being asked for.

weight of mercury = $W$

*Step 2.* Determine what information is already known.

density of mercury = 13.6 $g/cm^3$
volume = 32.0 $cm^3$

*Step 3.* Find the mathematical model that describes the relationship.

weight = density × volume

*Step 4.* Substitute the data into the model.

$W = 13.6\ g/cm^3 \times 32.0\ cm^3$

*Step 5.* Do the calculations.

$W = (13.6\ g/\cancel{cm^3})(32.0\ \cancel{cm^3})$
$W = 435\ g$

Use the five-step procedure to solve the following problem.

Find the weight of 2.5 $cm^3$ of mercury.

**a.** ______________________

**b.** ______________________

______________________

**c.** ______________________

**d.** ______________________

**e.** ______________________

## 6

Answers to frame 5.

**a.** weight of mercury = $W$
**b.** density of mercury = 13.6 $g/cm^3$
volume of mercury = 2.5 $cm^3$
**c.** weight of mercury = density × volume
**d.** weight of mercury = 13.6 $g/cm^3$ × 2.5 $cm^3$
**e.** weight of mercury = 34 g

# Exercise Set, Sections 9-3–9-4

## Regular Polygons

**1.** Find the size of each angle in the regular polygons below.

**a.** Regular hexagon ____________

**b.** Regular octagon ____________

**c.** Regular pentagon ____________

2. Find the perimeter of a pentagon with sides of 8 ft, 7 ft, 4 ft, 5 ft, and 3 ft.

3. Find the perimeter of a regular hexagon with a side of 17.5 mm.

4. Find the perimeter of a regular octagon with a side of 42.8 cm.

5. Find the perimeter of a regular pentagon with a side of 78 ft.

## Volume and Density

6. Find the volume of a cube 3.20 cm on each side.

7. Find the volume of a container with the following dimensions.

$l = 15.2$ cm
$w = 10.1$ cm
$h = 5.40$ cm

8. Find the volume of a rectangular solid with the following dimensions.

$l = 1.0$ mm
$w = 0.90$ mm
$h = 0.40$ mm

9. Find the volume of a room with a length of 25 ft, a width of 20 ft, and a height of 8 ft.

10. Find the weight of 25 $cm^3$ of water. (Density of water is 1 $g/cm^3$.)

11. Find the weight of 25.0 $cm^3$ of mercury. (Density of mercury is 13.6 $g/cm^3$.)

12. Find the weight of 10.8 $cm^3$ of water.

13. Find the weight of 2.10 $cm^3$ of mercury.

## Density

Use the five-step procedure to solve the following problems.

14. A lab technician has to find the weight of the water in a container 4 cm by 5 cm by 8 cm.

15. A medical lab technician is required to find the weight of mercury in a container with the following dimensions.

$l = 3.10$ cm
$w = 2.40$ cm
$h = 1.20$ cm

# Supplementary Exercise Set, Sections 9-3–9-4

1. Find the volume of a cube 4.6 cm on each side.

2. Find the volume of a container with a length of 9.60 cm, a width of 8.50 cm, and a height of 6.20 cm.

3. Find the volume of a solid with a length of 1.7 mm, a width of 0.8 mm, and a height of 0.3 mm.

4. Find the volume of a room with a length of 20 m, a width of 15 m, and a height of 3 m.

5. Find the weight of 15 $cm^3$ of water.

6. Find the weight of 30.0 $cm^3$ of mercury. (The density of mercury is 13.6 $g/cm^3$.)

7. Find the weight of 10 $cm^3$ of alcohol. (The density of alcohol is 0.8 $g/cm^3$.)

8. Find the weight of 0.29 $cm^3$ of mercury. (The density of mercury is 13.6 $g/cm^3$.)

9. Find the weight of mercury in a container with a length of 2.30 cm, a width of 3.20 cm, and a height of 1.30 cm.

# Section 9-5 Triangles

**1**

A triangle is a three-sided closed geometric figure. The sum of the three angles in a triangle will always be 180°.

Figure 9-34a shows a triangle with angles of 60°, 70°, and 50°. Notice that $60° + 70° + 50° = 180°$.

An unknown angle in a triangle can be found if the other two angles are known. Figure 9-34b shows a triangle with known angles of 64° and 73° and with angle $C$ unknown. The sum of these three angles is 180°. Therefore, this equation can be written: $64° + 73° + \angle C = 180°$. Angle $C$ can be found by rearranging the equation.

$$\angle C = 180° - 64° - 73°$$

$$\angle C = 43°$$

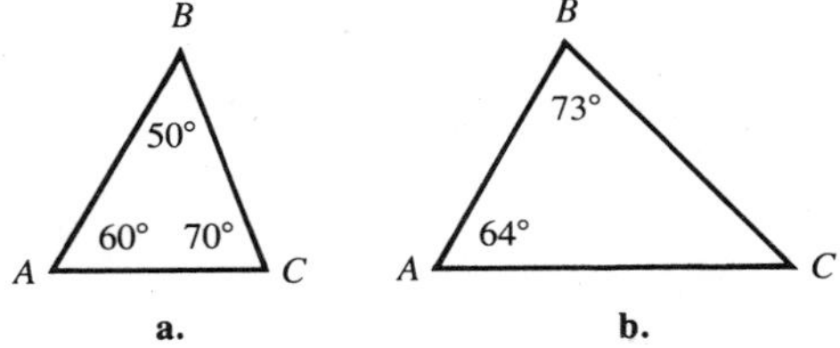


**Figure 9-34** Triangles

**a.** Find $\angle C$ when $\angle A = 14°$ and $\angle B = 39°$.

$\angle C =$ ______

**b.** Find $\angle B$ when $\angle A = 90°$ and $\angle C = 45°$.

$\angle B =$ ______

**c.** Find $\angle A$ when $\angle B = 58°$ and $\angle C = 79°$.

$\angle A =$ ______

**d.** Find $\angle A$ when $\angle B = 31.2°$ and $\angle C = 49.7°$.

$\angle A =$ ______

**1. a.** 127°
**b.** 45°
**c.** 43°
**d.** 99.1°

## 2

The area of a triangle is given by the formula

$$A = \frac{1}{2}(\text{altitude})(\text{base})$$

Figure 9-35 shows a triangle with a base of 5 meters and an altitude of 8 meters. The area can be found as follows.

$$A = \frac{1}{2}(8 \text{ m})(5 \text{ m})$$

$$A = \frac{1}{2}(40 \text{ m}^2)$$

$$A = 20 \text{ m}^2$$

**a.** Find the area of a triangle with a base of 18 in. and an altitude of 7.0 in.

**b.** Find the area of a triangle with a base of 7.50 mm and an altitude of 21.0 mm.

**c.** Find the area of a triangle with a base of 37 ft and an altitude of 14 ft.

**d.** Find the area of a triangle with a base of 9.10 m and an altitude of 4.40 m.

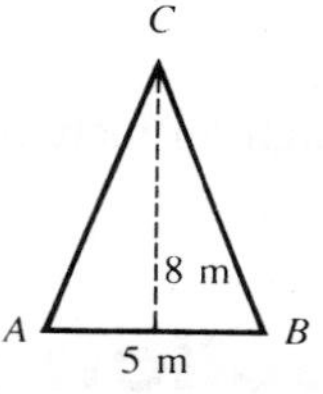


**Figure 9-35**

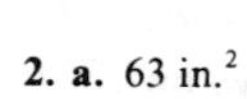

**2. a.** 63 in.$^2$
**b.** 78.8 mm$^2$
**c.** 259 ft$^2$
**d.** 20.0 m$^2$

## 3

A triangle that contains a right (90°) angle is called a *right triangle.* The right angle is indicated by the symbol ⌉. The *altitude* and *base* of a right triangle are also called the *legs of the triangle.* The side opposite the right angle is the *hypotenuse.* In Figure 9-36, side *BA* is the hypotenuse because it is opposite the right angle *C.*

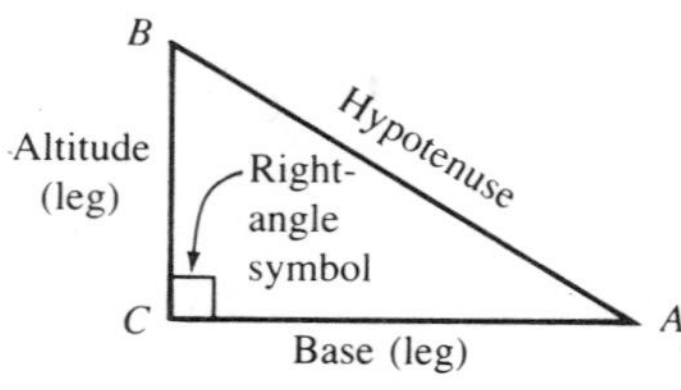


**Figure 9-36**

## 4

Capital letters are used to indicate the angles in a triangle. Lowercase letters are used to indicate the sides. *The lowercase letter corresponding to the capital letter of the angle is always used to indicate the side opposite that angle.* In Figure 9-37, side *a* is opposite angle *A*, side *b* is opposite angle *B*, and side *c* is opposite angle *C*.

Write the lowercase letter that indicates the hypotenuse in Figure 9-37.

Hypotenuse = ______

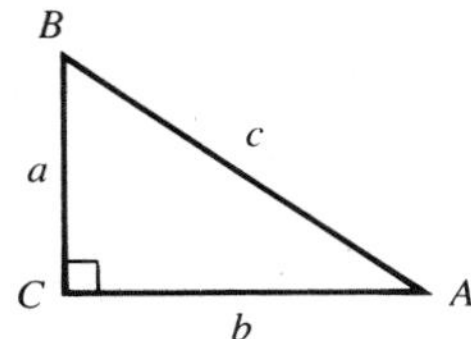


**Figure 9-37**

## 5

The area of a right triangle is found using the formula

$$A = \frac{1}{2}ab$$

The area of a right triangle is easy to find because one leg of the triangle is the base and the other leg is the altitude.

**a.** Find the area of the right triangle in Figure 9–37, when $a = 9.0$ ft and $b = 12$ ft.

$A =$ ______

**b.** Find the area of the right triangle in Figure 9–37, when $a = 6.0$ m and $b = 8.0$ m.

$A =$ ______

**4.** *c*

## 6

Right triangles have a special property, which is described by the Pythagorean theorem. This theorem states that, in a right triangle, the sum of the squares of the two legs is equal to the square of the hypotenuse. This fact is illustrated in Figure 9-38.

The Pythagorean theorem can be written as $c^2 = a^2 + b^2$ for a right triangle with a hypotenuse of $c$.

If the legs of a right triangle are 3 m and 4 m, the hypotenuse can be found as follows:

$$c^2 = a^2 + b^2$$
$$c = \sqrt{a^2 + b^2}$$
$$c = \sqrt{(3\text{ m})^2 + (4\text{ m})^2}$$
$$c = \sqrt{9\text{ m}^2 + 16\text{ m}^2}$$
$$c = \sqrt{25\text{ m}^2} \quad \text{or} \quad c = 5\text{ m}$$

**a.** Find the hypotenuse of a right triangle when the legs are 6.0 ft and 11 ft.

**b.** Find the hypotenuse of a right triangle when the legs are 7.0 m and 9.0 m.

**5. a.** $A = 54\text{ ft}^2$
**b.** $A = 24\text{ m}^2$

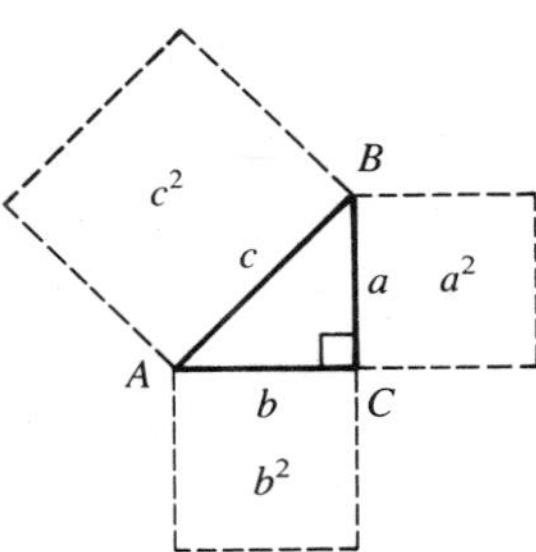


**Figure 9-38**
The Pythagorean theorem

**7**

The formula for the Pythagorean theorem can be rearranged to solve for any missing side in a right triangle.

6. **a.** 13 ft
**b.** 11 m

$$a^2 + b^2 = c^2$$
$$a^2 + b^2 + (-a^2) = c^2 + (-a^2)$$
$$b^2 = c^2 - a^2$$
$$b = \sqrt{c^2 - a^2}$$

Solve $a^2 + b^2 = c^2$ for $a$.

**8**

Solve the following problems.

7. $a = \sqrt{c^2 - b^2}$

**a.** Given a right triangle with a hypotenuse of 15 in. and one leg of 9.0 in., find the other leg.

**b.** Given a right triangle with $c = 14$ m and $b = 10$ m, find side $a$.

**9**

Answers to frame 8.

**a.** $b = \sqrt{(15 \text{ in.})^2 - (9.0 \text{ in.})^2}$

$b = \sqrt{225 \text{ in.}^2 - 81 \text{ in.}^2}$

$b = \sqrt{144 \text{ in.}^2}$ or 12 in.

**b.** $a = \sqrt{(14 \text{ m})^2 - (10 \text{ m})^2}$

$a = \sqrt{196 \text{ m}^2 - 100 \text{ m}^2}$

$a = \sqrt{96 \text{ m}^2}$ or 9.8 m

**10**

The calculator can be used to solve problems involving right triangles and the Pythagorean theorem.

Given a right triangle with legs of 25.2 m and 15.4 m, the hypotenuse can be found as follows:

Solve the Pythagorean theorem for $c$.

$c = \sqrt{a^2 + b^2}$

$a = 25.2$ m

$b = 15.4$ m

| *Enter* | *Press* | *Display* |
|---|---|---|
| 25.2 | [$x^2$] [+] | 635.04 |
| 15.4 | [$x^2$] | 237.16 |
| | [=] | 872.2 |
| | [$\sqrt{\ }$] | 29.533032 |

$c = 29.5$ m

**a.** Use the calculator to find the hypotenuse when the legs are 6.28 ft and 5.10 ft.

**b.** Use the calculator to find one leg of a right triangle when the other leg is 15.0 cm and the hypotenuse is 18.0 cm. (Use $a = \sqrt{c^2 - b^2}$.)

**11**

Answers to frame 10.

**a.** 8.90 ft **b.** 9.95 cm

# Exercise Set, Section 9-5

## Triangles

1. Find $\angle B$ when $\angle A = 120°$ and $\angle C = 30°$.

2. Find $\angle A$ when $\angle B = 48.2°$ and $\angle C = 62.4°$.

3. Find the area of a triangle with a base of 1.86 in. and an altitude of 1.17 in.

4. Find the area of a triangle with a base of 3.42 m and an altitude of 2.40 m.

5. Find the area of a right triangle with a base of 8.61 ft and an altitude of 12.1 ft.

6. Find the area of a right triangle with a base of 6.94 cm and an altitude of 5.70 cm.

## Pythagorean Theorem

7. Solve $a^2 + b^2 = c^2$ for $c$.

8. Solve $a^2 + b^2 = c^2$ for $a$.

Use Figure 9-39 to solve the following problems.

9. Find the hypotenuse when $a = 5.40$ ft and $b = 7.55$ ft.

10. Find the hypotenuse when $a = 11.15$ mm and $b = 15.70$ mm.

11. Find the other leg when the hypotenuse equals 8.76 in. and $b = 5.42$ in.

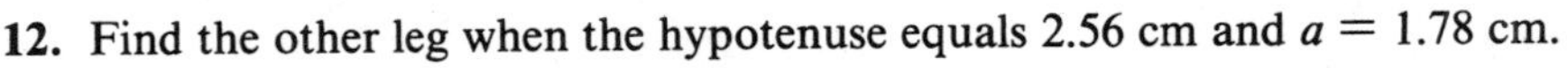

12. Find the other leg when the hypotenuse equals 2.56 cm and $a = 1.78$ cm.

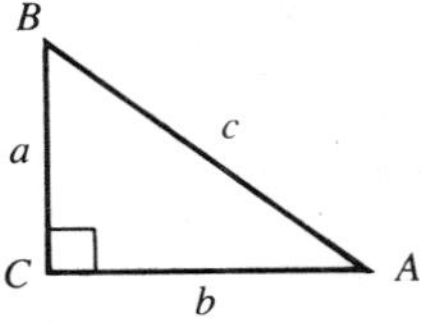


**Figure 9-39**

# Supplementary Exercise Set, Section 9-5

1. Find $\angle A$ when $\angle B = 110°$ and $\angle C = 45°$.

2. Find $\angle C$ when $\angle A = 52.6°$ and $\angle B = 72.9°$.

3. Find the area of a triangle with a base of 1.84 ft and an altitude of 2.74 ft.

4. Find the area of a right triangle with a base of 8.62 m and an altitude of 11.3 m.

5. Solve $a^2 + b^2 = c^2$ for $b$.

Use Figure 9-40 to solve the following problems.

6. Find the hypotenuse when $a = 6.87$ m and $b = 9.02$ m.

7. Find the other leg when the hypotenuse equals 7.15 in. and $b = 6.30$ in.

8. Find the other leg when the hypotenuse equals 15.49 cm and $a = 12.02$ cm.

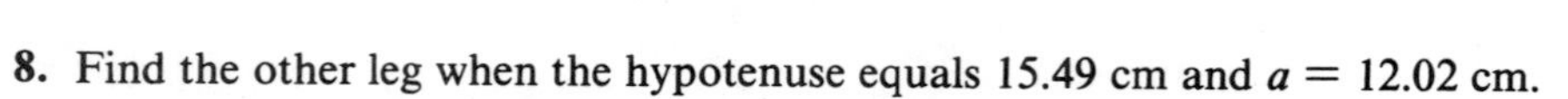

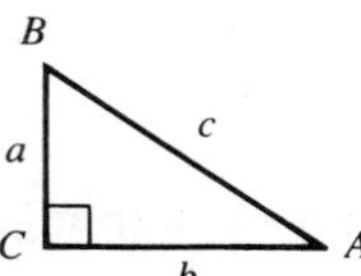

**Figure 9-40**

## Section 9-6 Circles and Spheres

### 1

Point "$0$" is the center of the circle in Figure 9-41. The line $d$ is called the *diameter,* and the line $r$ is the *radius.* In every circle, the diameter is twice the length of the radius. Therefore, the formula for the diameter of a circle is $d = 2r$.

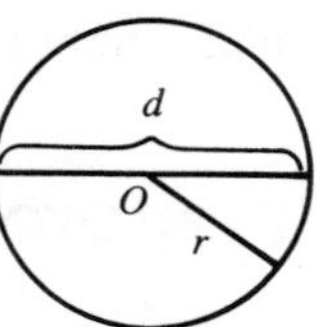

**Figure 9-41**

**a.** Using the formula above, find the diameter of a circle with a radius of 13.2 ft.

**b.** Using the formula above, find the radius of a circle with a diameter of 18.4 cm. (Note: Solve for $r$.)

**1. a.** 26.4 ft
**b.** 9.2 cm

### 2

The perimeter of a circle is called its *circumference.* The ratio of the circumference of a circle to its diameter is always a constant (3.1415925 . . . ). The Greek letter $\pi$ (pronounced "pie") is used to represent the constant. In this book, $\pi$ will always be rounded to 3.14.

The ratio of circumference to diameter can be shown as

$$\frac{c}{d} = \pi$$

This can be solved for $c$ to get the formula for the circumference of a circle:

$$c = \pi d$$

where

$c$ = circumference

$\pi$ = constant, pi (3.14)

**a.** Using the formula above, find the circumference of a circle with a diameter of 7.10 in.

**b.** Using the formula above, find the circumference of a circle with a diameter of 8.20 cm.

**3**

Most calculators have a special $\boxed{\pi}$ button. This is more accurate than the 3.14-rounded pi used in this book. It is therefore suggested that the $\boxed{\pi}$ button be used when possible. When the $\boxed{\pi}$ button is used, your answer will vary slightly from the answer given in this book.

2. **a.** $c = 22.3$ in.
   **b.** $c = 25.7$ cm

**4**

Because $d = 2r$, the formula for the circumference of a circle can be written not only as

$$c = \pi d$$

but also as

$$c = 2\pi r$$

**a.** Using the $c = 2\pi r$ formula, find the circumference of a circle with a radius of 9.21 in.

**b.** Using the same formula, find the circumference of a circle with a radius of 4.25 cm.

**5**

The area of a circle can be found using the formula

4. **a.** $c = 57.8$ in.
   **b.** $c = 26.7$ cm

$$A = \pi r^2$$

**a.** Use this formula to find the area of a circle with a radius of 3.27 in.

**b.** Use this formula to find the area of a circle with a radius of 15.6 cm.

**6**

The shaded area in Figure 9-42 is called a *sector* of a circle. Angle $\theta$ is called the *central angle* of the sector.

5. **a.** 33.6 $\text{in.}^2$
   **b.** 764 $\text{cm}^2$

The area of a sector of a circle can be found by using the following proportion:

$$\frac{\text{central angle}}{360°} = \frac{\text{area of sector}}{\text{area of circle}}$$

The proportion above can be used to find the area of the sector with a central angle of 60° when the circle has an area of 48 $\text{cm}^2$. The five-step procedure for problem solving can be used to find the area of the sector.

*Step 1.* Determine what is being asked for.

area of the sector $= A$

*Step 2.* Determine what information is already known.

central angle of sector $= 60°$
area of circle $= 48\ \text{cm}^2$ (This is found by using $A = \pi r^2$.)

*Step 3.* Find the mathematical model that describes the relationship.

$$\frac{\text{central angle}}{360°} = \frac{\text{area of sector}}{\text{area of circle}}$$

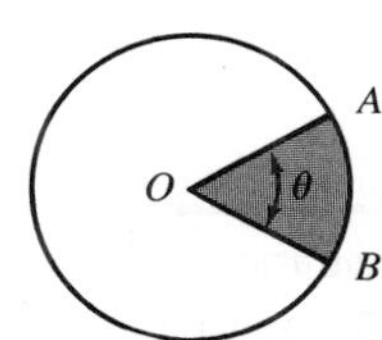


**Figure 9-42**

*Step 4.* Substitute the data into the model.

$$\frac{60°}{360°} = \frac{A}{48 \text{ cm}^2}$$

*Step 5.* Do the calculations.

$$A = \frac{(48 \text{ cm}^2)(60°)}{(360°)}$$

$$A = 8.0 \text{ cm}^2$$

## 7

Use the five-step procedure to solve the following problem.

Find the area of the sector with a central angle of 45° when the circle has an area of 89.6 $\text{cm}^2$.

**a.**

**b.**

**c.**

**d.**

**e.**

## 8

In most cases, the diameter or radius of a circle will be given instead of the area. Then the area of the circle will be calculated using the formula $A = \pi r^2$.

**Example** Find the area of the sector with a central angle of 27° when the circle has a radius of 5.00 cm.

*Step 1.* area of sector $= A$

*Step 2.* central angle of sector $= 27°$
area of circle $A = \pi r^2$

$$A = (3.14)(5.00 \text{ cm})^2$$
$$A = (3.14)(25.0 \text{ cm}^2)$$
$$A = 78.5 \text{ cm}^2$$

*Step 3.* $\dfrac{\text{central angle}}{360°} = \dfrac{\text{area of sector}}{\text{area of circle}}$

*Step 4.* $\dfrac{27°}{360°} = \dfrac{A}{78.5 \text{ cm}^2}$

*Step 5.* $A = \dfrac{(27°)(78.5 \text{ cm}^2)}{(360°)}$

$$A = 5.9 \text{ cm}^2$$

**7. a.** area of sector $= A$
**b.** central angle $= 45°$
area of circle $= 89.6 \text{ cm}^2$
**c.** $\dfrac{\text{central angle}}{360°} = \dfrac{\text{area of sector}}{\text{area of circle}}$
**d.** $\dfrac{45°}{360°} = \dfrac{A}{89.6 \text{ cm}^2}$
**e.** $A = 11 \text{ cm}^2$

## 9

The geometric shape in Figure 9-43 is called a *sphere.* It is defined as a geometric solid formed by a closed curved surface with all the points on the surface the same distance from the center. In short, it is a ball or globe.

The formula for the volume of a sphere is

$$V = \frac{4}{3}\pi r^3$$

**Example** Find the volume of the sphere in Figure 9-43 when $r = 3.00$ cm.

$$V = \frac{4(3.14)(3.00\text{ cm})^3}{3}$$

$$V = 113\text{ cm}^3$$

**a.** Find the volume of the sphere in Figure 9-43 when $r = 2.50$ in.

**b.** Find the volume of the sphere in Figure 9-43 when $r = 4.20$ cm.

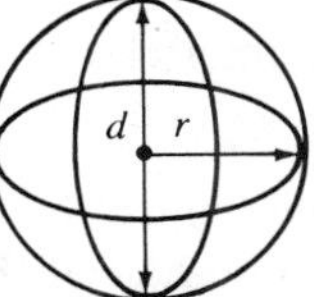

**Figure 9-43** Sphere

**9. a.** 65.4 in.$^3$
**b.** 310 cm$^3$

**10**

The surface area of a sphere is found using the following formula.

$$A = 4\pi r^2$$

**Example** Find the surface area of the sphere in Figure 9-43 when $r = 2.10$ in.

$$A = 4\pi r^2$$

$$A = 4(3.14)(2.10\text{ in.})^2$$

$$A = 55.4\text{ in.}^2$$

**a.** Find the surface area of the sphere in Figure 9-43 when $r = 1.50$ cm.

**b.** Find the surface area of the sphere in Figure 9-43 when $r = 6.50$ in.

**11**

Answers to frame 10.

**a.** 28.3 cm$^2$

**b.** 531 in.$^2$

## Section 9-7 Composite Geometric Shapes

**1**

Many shapes are more involved than the simple geometric figures in the previous sections of this chapter. This section will include composite geometric shapes and will illustrate some strategies for finding the areas of these shapes.

Consider the shape in Figure 9-44. To find the area of the figure, it is convenient to think of it as two separate geometric shapes, a rectangle and a semicircle. This can be illustrated using the shapes below.

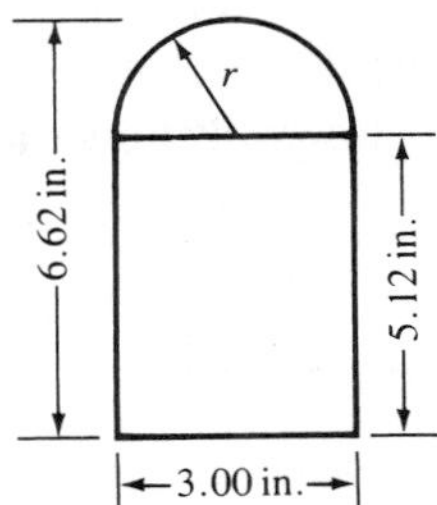

**Figure 9-44**

It is helpful to reduce composite geometric shapes to combinations of simple shapes.

The area of a rectangle is $A$ (rectangle) $= l \times w$

$$l = 5.12 \text{ in.}$$
$$w = 3.00 \text{ in.}$$
$$A \text{ (rectangle)} = (5.12 \text{ in.})(3.00 \text{ in.}) = 15.4 \text{ in.}^2$$

The area of a semicircle is one-half the area of a circle or

$$A \text{ (semicircle)} = \frac{\pi r^2}{2}$$

The radius of the semicircle in Figure 9-44 is 1.50 in. (This measurement is found either by taking one-half of the diameter 3.00 in. or by subtracting 5.12 in. from 6.62 in.)

$$\text{Therefore, } A \text{ (semicircle)} = \frac{\pi(1.50 \text{ in.})^2}{2} = \frac{(3.14)(2.25 \text{ in.}^2)}{2} = 3.53 \text{ in.}^2$$

The area of the complete figure is found by adding the areas of the two parts.

$$A \text{ (total)} = A \text{ (rectangle)} + A \text{ (semicircle)}$$
$$A \text{ (total)} = 15.4 \text{ in.}^2 + 3.53 \text{ in.}^2$$
$$A \text{ (total)} = 18.93 \text{ in.}^2$$

Use the strategy above to find the area of Figure 9-45.

Area = __________

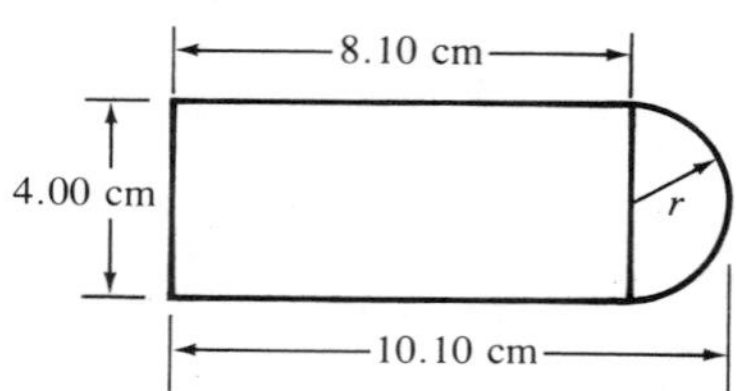

**Figure 9-45**

**2**

Answer to frame 1.

Area (rectangle) = 32.4 cm$^2$; Area (semicircle) = 6.28 cm$^2$; Area (total) = 38.7 cm$^2$

**3**

A semicircle can be on either side of a rectangle, as in Figure 9-46. It is convenient to think of the shape as the sum of the areas of a rectangle and a circle. The illustration below will show this.

+ =

The radius of each semicircle is 2.01 cm; therefore the radius of the circle formed by joining the two semicircles is also 2.01 cm.

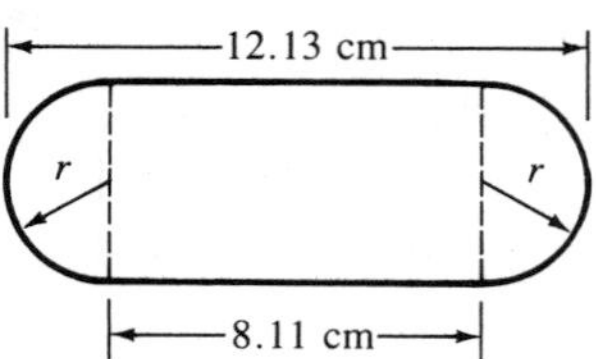

**Figure 9-46**

The area of the circle (two semicircles) is $A = \pi r^2$.

$$A_c = (3.14)(2.01 \text{ cm})^2 = 12.7 \text{ cm}^2$$

The area of the rectangle is $A = l \times w$. The length of the rectangle is 8.11 cm. The width of the rectangle is 4.02 cm. (The width of the rectangle is the same as the diameter of the circle.)

$$A_r = (8.11 \text{ cm})(4.02 \text{ cm}) = 32.6 \text{ cm}^2$$

The area of the complete figure is found by adding the area of the parts.

$$A_t = A_c + A_r$$

$$A_t = 12.7 \text{ cm}^2 + 32.6 \text{ cm}^2$$

$$A_t = 45.3 \text{ cm}^2$$

Using the strategy above, find the area of Figure 9-47.

Area = ________

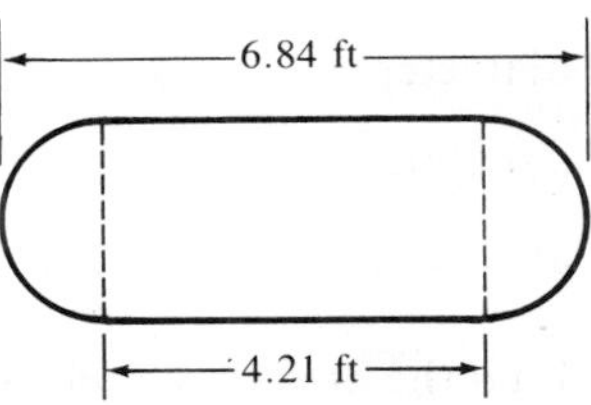

**Figure 9-47**

**4**

Answer to frame 3.

Area (circle) = $(3.14)(1.32 \text{ ft})^2 = 5.47 \text{ ft}^2$; Area (rectangle) = $(4.21 \text{ ft})(2.64 \text{ ft}) = 11.1 \text{ ft}^2$; Area (total) = $16.6 \text{ ft}^2$

**5**

The area of a geometric shape such as the one in Figure 9-48 is found by dividing the shape into smaller rectangles.

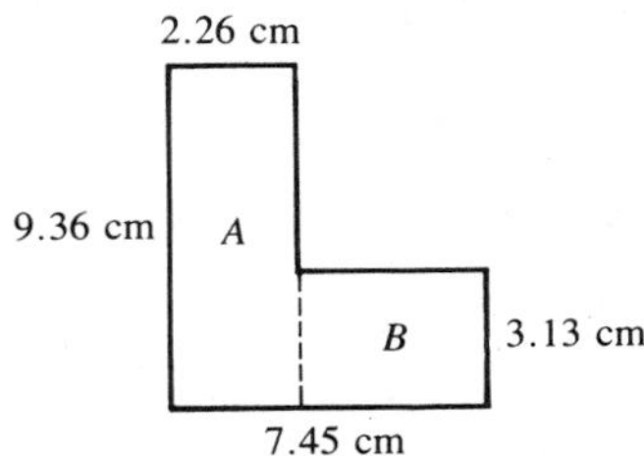

**Figure 9-48**

$$\text{Area of } A = (2.26 \text{ cm})(9.36 \text{ cm}) = 21.2 \text{ cm}^2$$

$$\text{Area of } B = (3.13 \text{ cm})(7.45 \text{ cm} - 2.26 \text{ cm}) = (3.13 \text{ cm})(5.19 \text{ cm}) = 16.2 \text{ cm}^2$$

$$\text{Total area} = 21.2 \text{ cm}^2 + 16.2 \text{ cm}^2 = 37.4 \text{ cm}^2$$

Use a method similar to the method above to find the area of Figure 9-49.

Area = ________

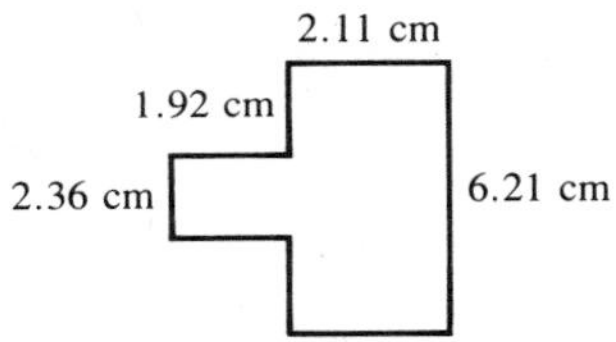

**Figure 9-49**

**6**

Answers to frame 5.

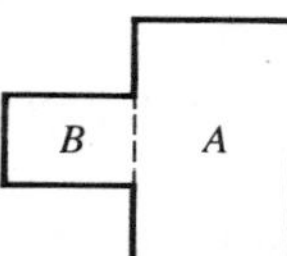


Area of $A = (6.21 \text{ cm})(2.11 \text{ cm}) = 13.1 \text{ cm}^2$; Area of $B = (2.36 \text{ cm})(1.92 \text{ cm}) = 4.53 \text{ cm}^2$;
Area (total) $= 13.1 \text{ cm}^2 + 4.53 \text{ cm}^2 = 17.6 \text{ cm}^2$

# Exercise Set, Sections 9-6–9-7

## Circles and Spheres

Calculate the designated dimensions of the circle in Figure 9-50.

**1.** diameter ______________

**2.** circumference ______________

**3.** area ______________

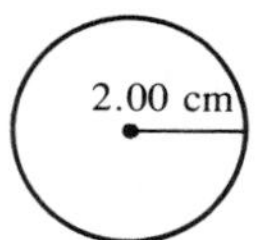


**Figure 9-50**

**4.** Find the area of a circle with a radius of 4.23 in.

**5.** Find the circumference of a circle with a diameter of 17.2 cm.

Use the five-step procedure to solve the following problem.

**6.** What is the area of a sector with a central angle of 31° when the circle has a diameter of 8.9 in.?

Find the volume and the surface area for the sphere in Figure 9–51.

**7.** volume = ______________

**8.** surface area = ______________

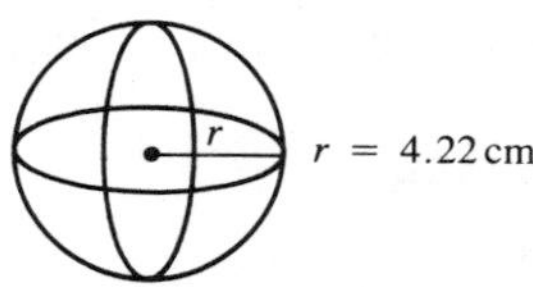


**Figure 9-51**

## Composite Geometric Shapes

**9.** Find the area of the shape in Figure 9–52.

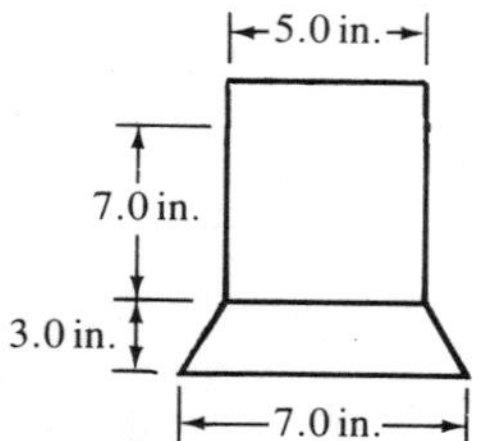


**Figure 9-52**

**10.** Find the area of the shape in Figure 9–53.

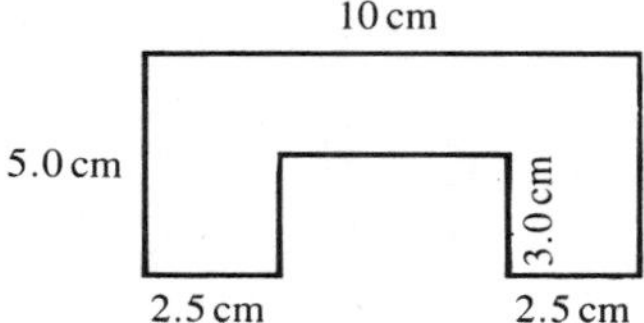


**Figure 9-53**

# Supplementary Exercise Set, Sections 9-6–9-7

Calculate the following for a circle with a radius of 1.50 cm.

**1.** diameter ____________ **2.** circumference ____________ **3.** area ____________

**4.** Find the area of a circle with a radius of 8.60 mm.

**5.** Find the circumference of a circle with a diameter of 9.11 in.

**6.** What is the area of a sector with a central angle of 30° when the circle has a diameter of 1.6 cm?

**7.** What is the volume of a sphere with a radius of 5.25 in.?

**8.** Find the surface area of the sphere in Exercise 7.

**9.** Find the area of the shape in Figure 9–54.

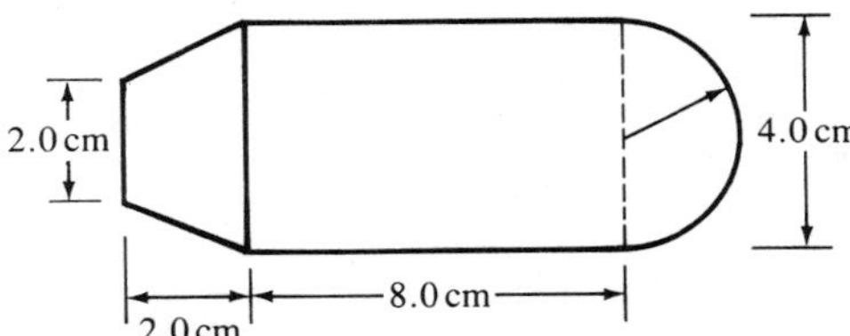


**Figure 9-54**

**10.** Find the area of the shape in Figure 9–55.

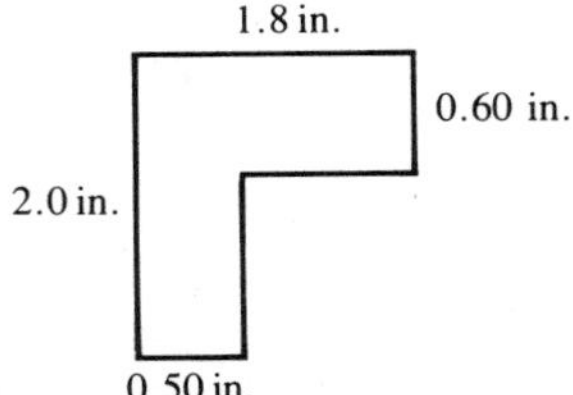


**Figure 9-55**

# Summary

1. A line is a set of points.
2. An angle is two rays with a common end point.
3. Angles are represented by the following:
   a. the symbol $\angle$
   b. the symbol $\angle$ and the three points of the angle (for example, $\angle AOB$)
   c. the lowercase Greek theta, $\theta$
4. The perimeter of a rectangle $= 2l + 2w$.
5. The perimeter of a square $= 4s$.
   The area of a square $= s^2$.

6. The area of a parallelogram $= bh$.
7. The area of a trapezoid $= \dfrac{(a + b)h}{2}$.
8. The volume of a cube $= s^3$.
   The volume of a rectangular solid $= lwh$.
9. The weight of a given amount of liquid is its volume times its density.
10. The diameter of a circle $= 2r$.
11. The volume of a sphere $= \dfrac{4}{3}\pi r^3$.
    The surface area of a sphere $= 4\pi r^2$.

## Summary of Geometric Formulas

Rectangle $A = lw$
$P = 2l + 2w$

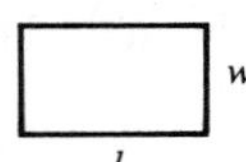


Parallelogram $A = bh$
$P = 2a + 2b$

Trapezoid $A = \dfrac{(a + b)h}{2}$

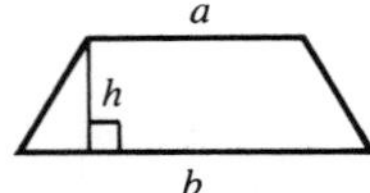


Square $A = s^2$
$P = 4s$

Cube $V = s^3$

Rectangular solid $V = lwh$

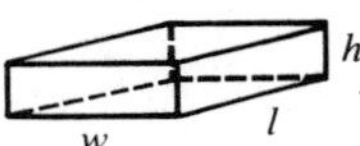


Cylinder $V = \pi r^2 h$

Circle $A = \pi r^2$
$c = \pi d$

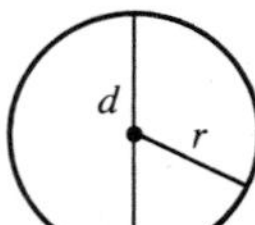


Sphere $V = \dfrac{4}{3}\pi r^3$
$S = 4\pi r^2$

# Chapter 9 Self-Test

## Geometry

Use a protractor to measure the angles in Figure 9-56.

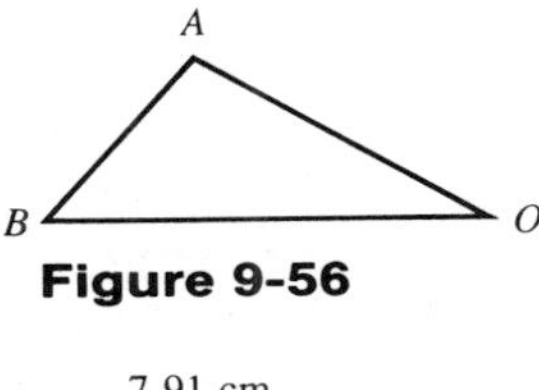


Figure 9-56

**1.** $\angle AOB =$ __________

**2.** $\angle ABO =$ __________

**3.** $\angle BAQ =$ __________

Given the rectangle in Figure 9-57, find the following.

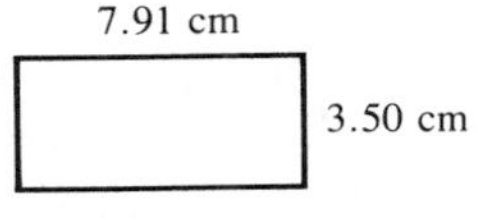

Figure 9-57

**4.** perimeter __________

**5.** area __________

Given the square in Figure 9-58, find the following.

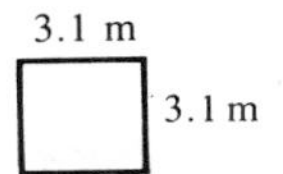


Figure 9-58

**6.** perimeter __________

**7.** area __________

Find the volume of the following.

**8.** A cube 2.71 cm on each side.

**9.** A rectangle 18.9 cm long, 8.10 cm wide, and 3.20 cm high.

**10.** A sphere with a radius of 0.722 in.

Compute the following.

**11.** The weight of water in a container 3 cm × 5 cm × 7 cm.

**12.** The area of a circle with a radius of 4.81 cm.

**13.** The circumference of a circle with a diameter of 9.28 in.

**14.** The area of a sector with a central angle of 40°, when the circle has a diameter of 20 in.

**15.** The area of the shape in Figure 9-59.

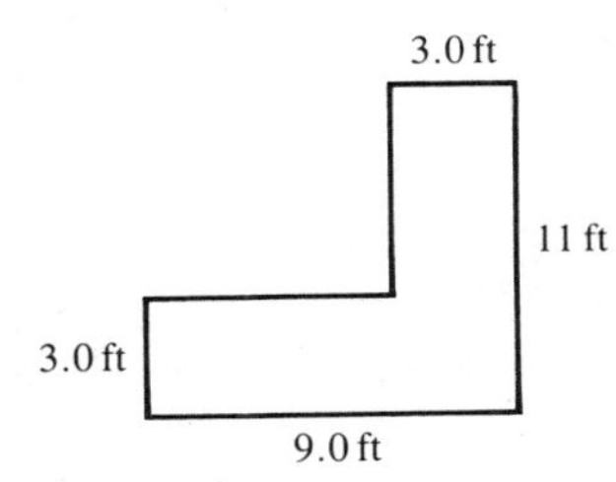


Figure 9-59

# CHAPTER 10

# Graphing

Graphing is a skill that technically trained people must master. The graphing of data and equations allows us to obtain a visual image of the information. In this chapter, you will learn how to make bar and circle graphs, how to graph tables of values using the rectangular coordinate system, how to prepare tables from functions of equations, and how to graph functions of equations.

## Section 10-1 Bar, Line, and Circle Graphs

### 1

Numerical data are often displayed in tables, which help to organize the information for the reader. The bar graph goes one step beyond the table in that it uses a picture to display the numerical data. The information in Table 10-1 will be used to construct the bar graph in Figure 10-1.

**Table 10-1** The amount of sales per year by a manufacturing company

| *Year* | *Sales in millions of dollars* |
|---|---|
| 1978 | 110 |
| 1979 | 140 |
| 1980 | 80 |
| 1981 | 30 |
| 1982 | 40 |

To construct a bar graph from Table 10-1, we need to follow these steps.

*Step 1.* Determine whether you want the bars positioned horizontally or vertically on the graph. To construct a graph with vertical bars, we will put the data dealing with the amount of sales on the vertical axis of the graph, and the bars will then be vertical.

*Step 2.* Clearly label the data represented on each axis.

*Step 3.* Determine the scale to be used. The data on the amount of sales range from 30 million to 140 million. A good scale to use would be 0 to 150 million.

If we make one mark for every 10 million dollars, we will need 15 marks ($150 \div 10 = 15$). If we make a mark every $\frac{1}{8}$ inch, the graph will be almost 2 inches high ($15 \times \frac{1}{8}'' = 1\frac{7}{8}''$ or almost 2″).

Figure 10-1 illustrates these three steps.

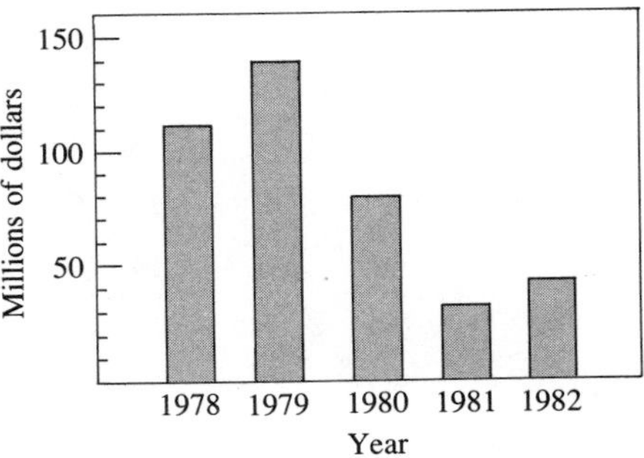


**Figure 10-1**

**2**

We could have made the bar graph with the bars in a horizontal position. Use the information in Table 10-1 to complete Figure 10-2.

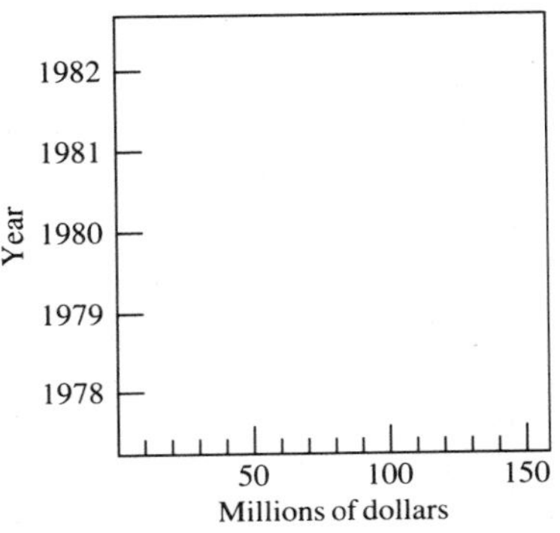


**Figure 10-2**

**3**

Answer to frame 2.

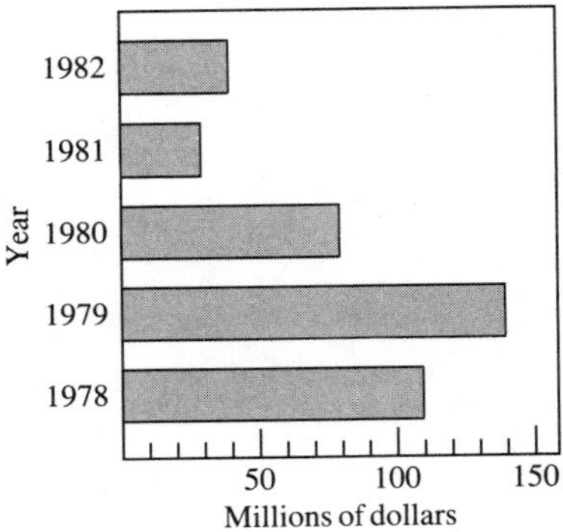

**4**

Use the information from Table 10-2 to complete Figure 10-3.

**Table 10-2** Number of machine parts produced by each employee per month

| *Employee number* | *Parts produced per month* |
|---|---|
| 27 | 1,800 |
| 35 | 1,650 |
| 42 | 1,470 |
| 53 | 1,580 |

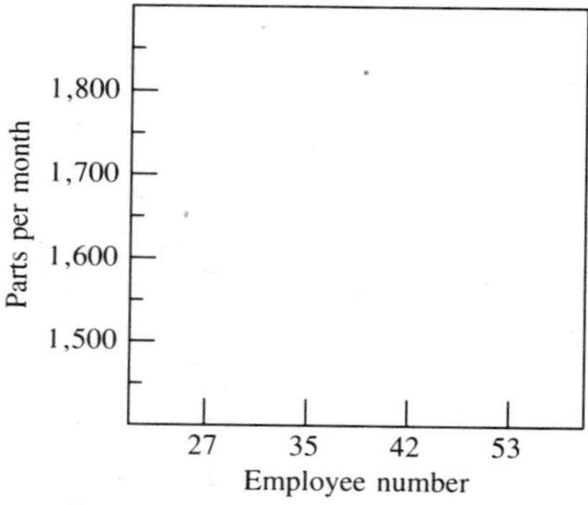


**Figure 10-3**

**5**

Answer to frame 4.

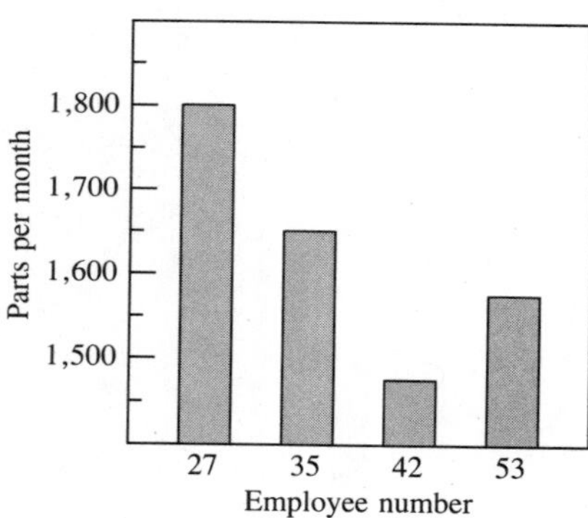


**6**

The use of a line graph is another method to illustrate data. Line graphs are useful for showing trends. Table 10-3 shows the number of housing permits issued in a community over a six-month period.

**Table 10-3** The number of housing permits issued per month

| *Month* | *Permits issued* |
|---|---|
| Jan. | 10 |
| Feb. | 15 |
| Mar. | 30 |
| Apr. | 90 |
| May | 85 |
| June | 60 |

The information in Table 10-3 is plotted as a line graph in Figure 10-4 using the following steps.

*Step 1.* Determine which data will be placed on each axis. In Figure 10-4, the month is placed on the horizontal axis and the number of permits is placed on the vertical axis.
*Step 2.* Place a dot on the graph corresponding to the number of permits issued each month.
*Step 3.* Connect the dots with straight lines.

These steps are shown in Figure 10-4.

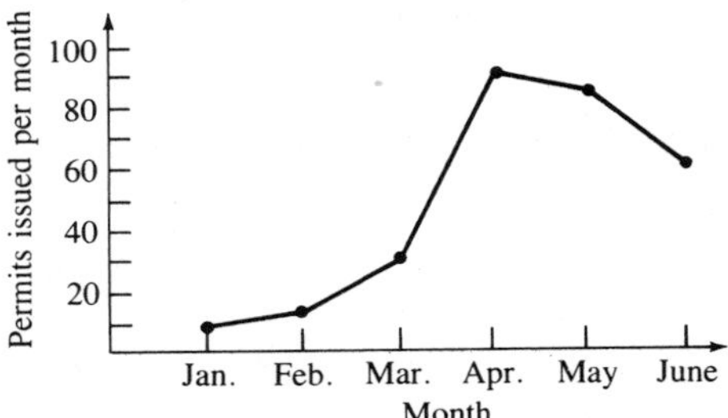


**Figure 10-4**

By looking at the line graph, you can see that most building permits in this locality are issued in April and May. Use the data in Table 10-4 to construct a line graph in Figure 10-5.

**Table 10-4** Average mean temperature per month in the city of Maplewood

| *Month* | *Mean temperature (°F)* |
|---|---|
| Apr. | 62 |
| May | 68 |
| June | 76 |
| July | 88 |
| Aug. | 82 |
| Sept. | 72 |

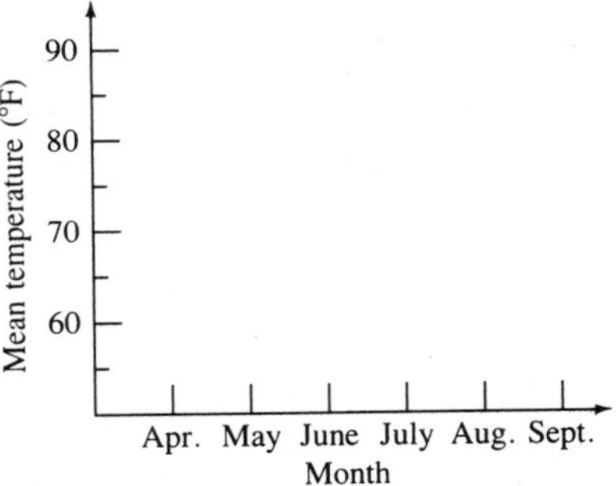


**Figure 10-5**

## 7

Answer to frame 6.

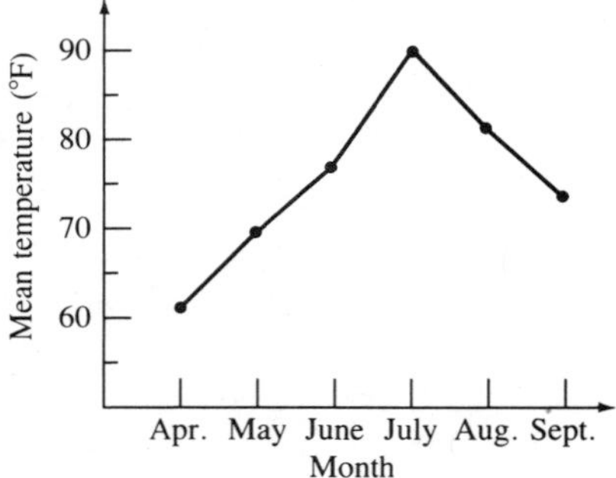


## 8

The circle graph helps us visualize what percentage of the whole is represented by each part. Table 10-5 shows the percentage of engines shipped by a company on each type of transportation.

**Table 10-5** Percentage of engines shipped by each type of transportation

| *Type* | *Percent* |
|---|---|
| Truck | 60 |
| Rail | 30 |
| Air | 10 |

The steps needed to make a circle graph from this information are shown below.

*Step 1.* The percentage of engines shipped on each type of transportation is proportional to a sector of a circle area. The size of this sector in degrees can be found by multiplying the decimal equivalent of each type of transportation times the 360° in a circle.

| *Type* | *Percent* | *Decimal equivalent* | *Size of sector in degrees* |
|---|---|---|---|
| Truck | 60 | $0.6 \times 360° =$ | 216° |
| Rail | 30 | $0.3 \times 360° =$ | 108° |
| Air | 10 | $0.1 \times 360° =$ | 36° |

*Step 2.* The size of the sector in degrees for each type of transportation can now be used to draw a circle graph.

**a.** Use a compass to draw a circle.

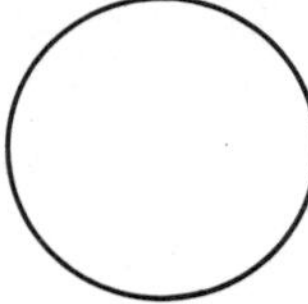

**b.** Draw a radius.

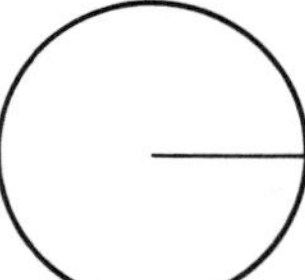

c. Use a protractor to measure 216° from the first radius. This area represents the percentage of engines shipped by truck.

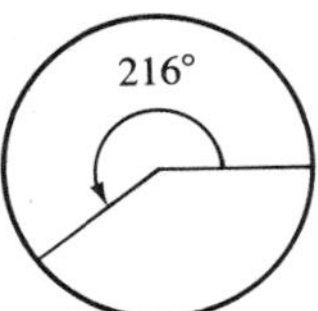


d. Next measure 108° from the second radius. This area represents the percentage of engines shipped by rail.

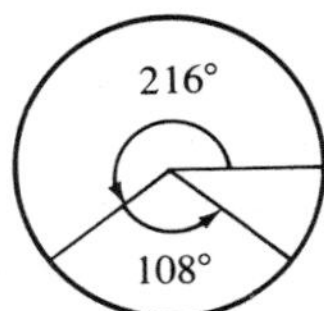


e. The remaining sector should measure 36° and represents the percentage of engines shipped by air. Label each sector.

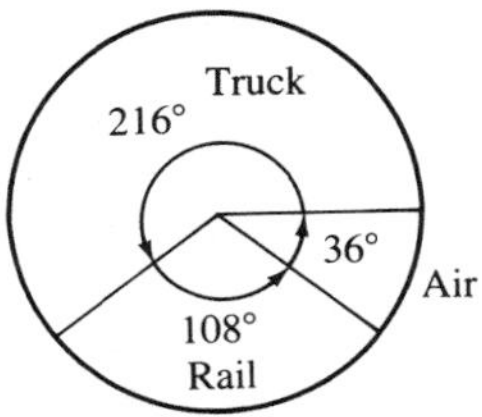


Use the information in Table 10-6 to construct a circle graph.

**Table 10-6** The cost of operating an automobile, by percent

| *Type of expense* | *Percent of total cost* |
|---|---|
| Fuel | 45 |
| Insurance | 30 |
| Maintenance | 20 |
| Registration | 5 |

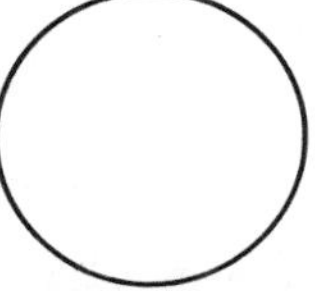

**9**

Answers to frame 8.

| | |
|---|---|
| Fuel | 0.45 × 360° = 162° |
| Insurance | 0.30 × 360° = 108° |
| Maintenance | 0.20 × 360° = 72° |
| Registration | 0.05 × 360° = 18° |

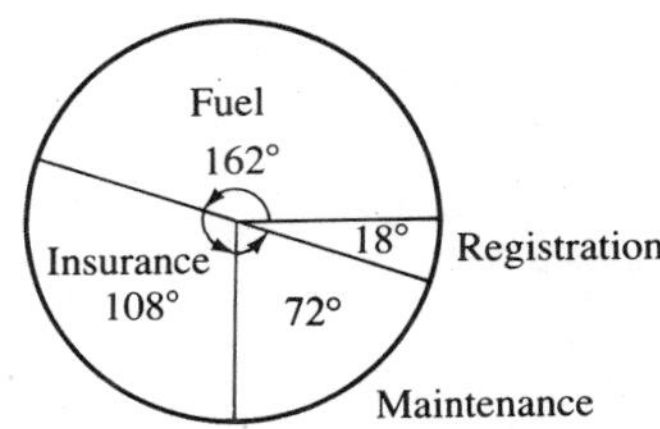


# Exercise Set, Section 10-1

## Bar Graphs

1. Construct a vertical bar graph displaying the data from the table below.

| *Month* | *Number of machines sold* |
|---|---|
| Jan. | 240 |
| Feb. | 200 |
| Mar. | 180 |
| Apr. | 90 |

2. Construct a horizontal bar graph displaying the data from the table below.

| *Year* | *Total sales* |
|---|---|
| 1978 | $280,000 |
| 1979 | 300,000 |
| 1980 | 300,000 |
| 1981 | 400,000 |
| 1982 | 420,000 |

## Line Graphs

3. Construct a line graph showing the monthly high and low temperatures in River City over a six-month period. There will be two graphs, one for the high temperatures and one for the low temperatures. Put the months on the horizontal axis and the temperatures on the vertical axis.

| *Month* | *High (° F)* | *Low (° F)* |
|---|---|---|
| Jan. | 55 | 30 |
| Feb. | 58 | 38 |
| Mar. | 62 | 45 |
| Apr. | 72 | 57 |
| May | 80 | 62 |
| June | 88 | 65 |

## Circle Graphs

4. A manufacturing company has four factories. Make a circle graph showing the percentage of total sales from each plant.

| *Plant* | *Percent of total sales* |
|---|---|
| *A* | 20 |
| *B* | 40 |
| *C* | 30 |
| *D* | 10 |

5. A typical family spent the following on monthly expenses. Construct a circle graph illustrating these data.

| *Item* | *Percent of income* |
|---|---|
| Housing | 35 |
| Food | 30 |
| Clothing | 25 |
| Savings | 5 |
| Entertainment | 5 |

# Supplementary Exercise Set, Section 10-1

1. Construct a vertical bar graph displaying the data from the table below. The data show how Americans spend their money.

| *Item* | *Billions of dollars* |
|---|---|
| Food | 210 |
| Housing | 110 |
| Transportation | 105 |
| Medical care | 60 |
| Recreation | 55 |

2. Construct a horizontal bar graph displaying the data from the table below. The data show the number of people employed by a factory over a five-year period.

| *Year* | *Number of people employed* |
|---|---|
| 1979 | 15,200 |
| 1980 | 16,000 |
| 1981 | 16,400 |
| 1982 | 18,600 |
| 1983 | 17,000 |

3. Construct a circle graph displaying the percentage of families with income in each category.

| *Annual income* | *Percent of families* |
|---|---|
| $25,000 and over | 15 |
| $15,000–$24,999 | 30 |
| $10,000–$14,999 | 25 |
| $5,000–$9,999 | 20 |
| Under $5,000 | 10 |

4. Construct a circle graph showing the percentage of each type of employee working in the United States.

| *Type* | *Percent of work force* |
|---|---|
| White collar | 53 |
| Service worker | 13 |
| Blue collar | 32 |
| Farm worker | 2 |

# Section 10-2 The Rectangular Coordinate System

## 1

Figure 10-6 shows a rectangular coordinate system. It is made up of a horizontal number line and a vertical number line. The horizontal number line is called a *horizontal axis;* the vertical number line is called a *vertical axis.* These two axes intersect at the *origin* (the "0" point), as shown in Figure 10-6. The two axes divide the space into four *quadrants.* The quadrants are labeled counterclockwise, as in 10-6. Roman numerals I, II, III, and IV are used to label the quadrants.

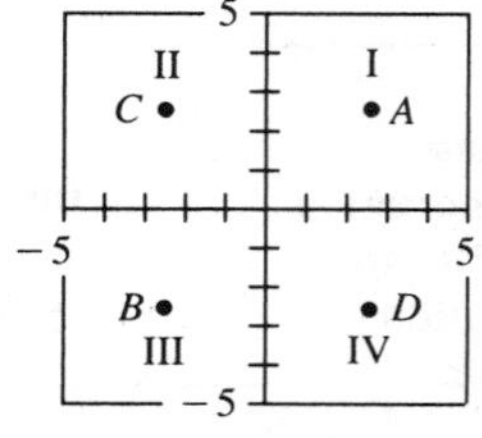

**Figure 10-6** A rectangular coordinate system

Point $A$ is in quadrant I.
Point $B$ is in quadrant III.

**a.** What quadrant is point $C$ in? ____________

**b.** What quadrant is point $D$ in? ____________

**1. a.** II
**b.** IV

## 2

The horizontal axis of the rectangular coordinate system is usually labeled as $x$ and called the *x-axis.* The vertical axis is usually labeled $y$ and called the *y-axis.* These axes also may be labeled with other letters, depending upon the nature of the information being graphed.

A point may be located on a rectangular coordinate system by using an $x$-value for a horizontal distance from the origin and a $y$-value for a vertical distance from the origin.

> **Example** Point $A$ in Figure 10-7 has a horizontal distance ($x$) of two units and a vertical distance ($y$) of three units.

The horizontal distance also may be called the $x$-coordinate; the vertical distance the $y$-coordinate.

It can therefore be said that point $A$ has an $x$-coordinate of 2 and a $y$-coordinate of 3.

Name the indicated coordinate of each of the following points in Figure 10-7.

**a.** $x$-coordinate of $B$ ____________

**b.** $y$-coordinate of $B$ ____________

**c.** $x$-coordinate of $C$ ____________

**d.** $y$-coordinate of $C$ ____________

**e.** $x$-coordinate of $D$ ____________

**f.** $y$-coordinate of $D$ ____________

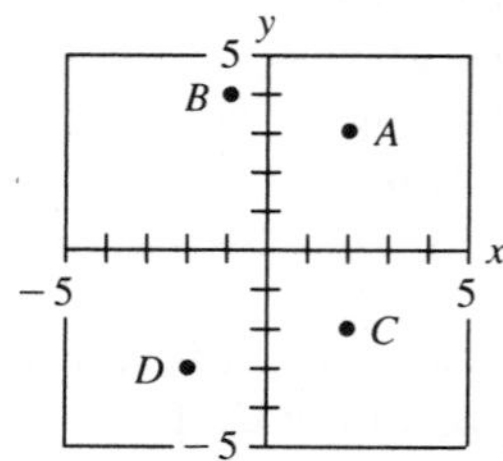

**Figure 10-7**

## 3

Locate and label each of the following points in Figure 10-8.

2. a. $-1$
b. $+4$
c. $+2$
d. $-2$
e. $-2$
f. $-3$

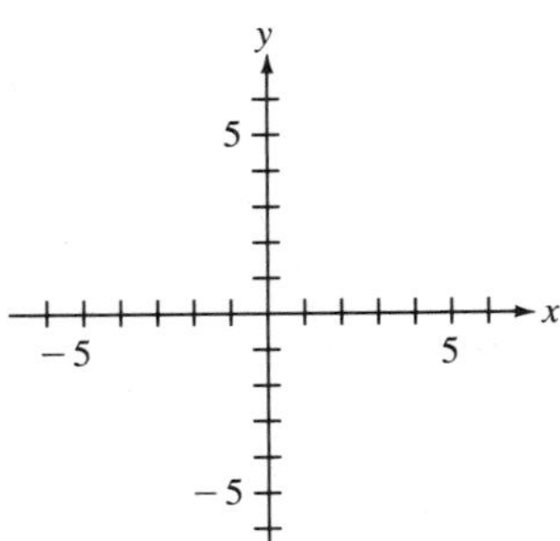


**Figure 10-8**

**a.** A point with an $x$-coordinate of 1 and a $y$-coordinate of 6.
**b.** A point with an $x$-coordinate of $-2$ and a $y$-coordinate of 4.
**c.** A point with an $x$-coordinate of $-3$ and a $y$-coordinate of $-3$.
**d.** A point with an $x$-coordinate of 4 and a $y$-coordinate of $-2$.
**e.** A point with an $x$-coordinate of 5 and a $y$-coordinate of 5.

## 4

Answers to frame 3.

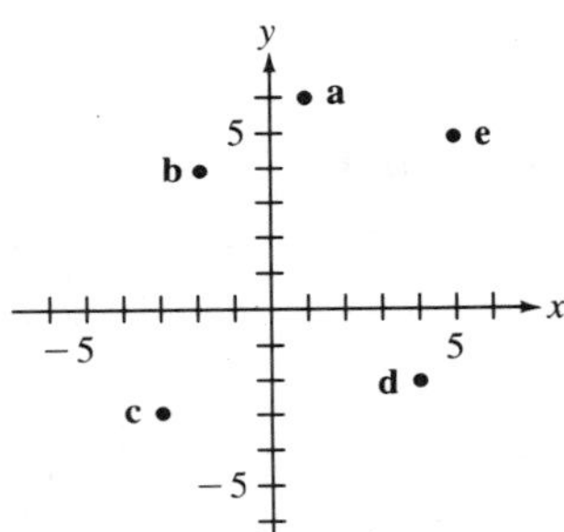


## 5

The $x$- and $y$-coordinates of a point can be expressed as a set of ordered pairs. A point with an $x$-coordinate of $-3$ and a $y$-coordinate of $+5$ can be expressed as the following ordered pair: $(-3, 5)$. An ordered pair will be enclosed by parentheses and will contain two (a pair of) numbers. The first number will always be the $x$-coordinate, and the second number will always be the $y$-coordinate. In general, an ordered pair can be expressed as $(x, y)$.

**a.** Write the ordered pair that has an $x$-coordinate of 17 and a $y$-coordinate of $-3$.

**b.** Write the ordered pair that has an $x$-coordinate of $-1$ and a $y$-coordinate of 35.2.

**c.** Write the ordered pair that has an $x$-coordinate of 21 and a $y$-coordinate of 0.

**d.** Write the ordered pair that has an $x$-coordinate of 0 and a $y$-coordinate of $\frac{1}{4}$.

## 6

Answers to frame 5.

**a.** $(17, -3)$ **b.** $(-1, 35.2)$ **c.** $(21, 0)$ **d.** $\left(0, \frac{1}{4}\right)$

# Section 10-3 Graphing a Table of Values

**1**

Data placed in table form can be graphed by using each pair of numbers to find the points and then by connecting the points with a smooth line. The first entry in the table is the $x$-coordinate and the second entry is the $y$-coordinate.

**Example** Graph the following table of values.

| $x$ | $y$ |
|---|---|
| −2 | −4 |
| −1 | −2 |
| 0 | 0 |
| 1 | 2 |
| 2 | 4 |

Each point is located using the related $x$- and $y$-coordinates, and a smooth line is drawn through the points, as shown in Figure 10-9.

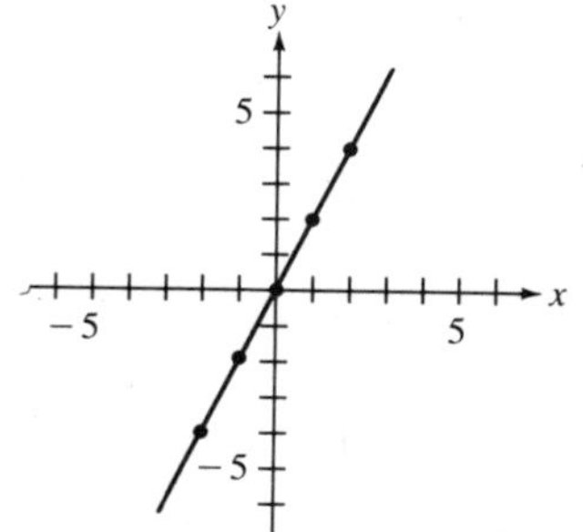

Figure 10-9

Graph this table of values using Figure 10-10.

| $x$ | $y$ |
|---|---|
| −2 | −3 |
| −1 | −1 |
| 0 | 1 |
| 1 | 3 |
| 2 | 5 |

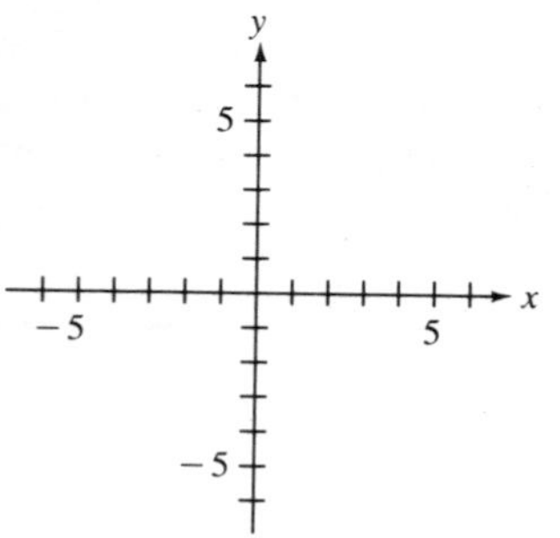

Figure 10-10

**2**

Answer to frame 1.

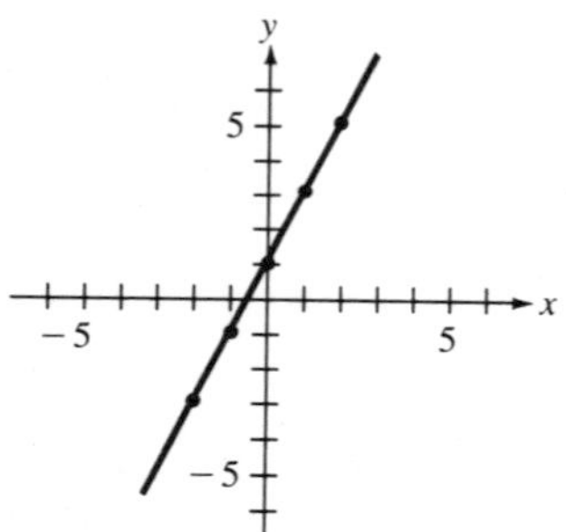


**3**

The graph resulting from a table of values can be a curved line as well as a straight line.

**Example** Graph the table of values on Figure 10-11.

| $x$ | $y$ |
|---|---|
| −3 | 9 |
| −2 | 4 |
| −1 | 1 |
| 0 | 0 |
| 1 | 1 |
| 2 | 4 |
| 3 | 9 |

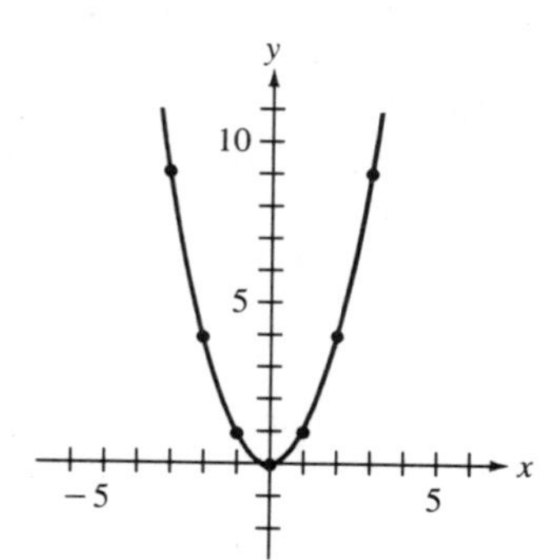

Figure 10-11

Graph the table of values on Figure 10-12.

| $a$ | $b$ |
|---|---|
| −3 | 4.5 |
| −2 | 2 |
| −1 | 0.5 |
| 0 | 0 |
| 1 | 0.5 |
| 2 | 2 |
| 3 | 4.5 |

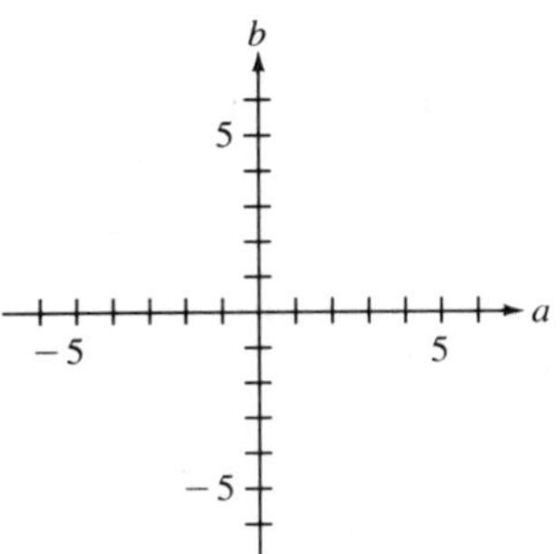


**Figure 10-12**

**4**

Answer to frame 3.

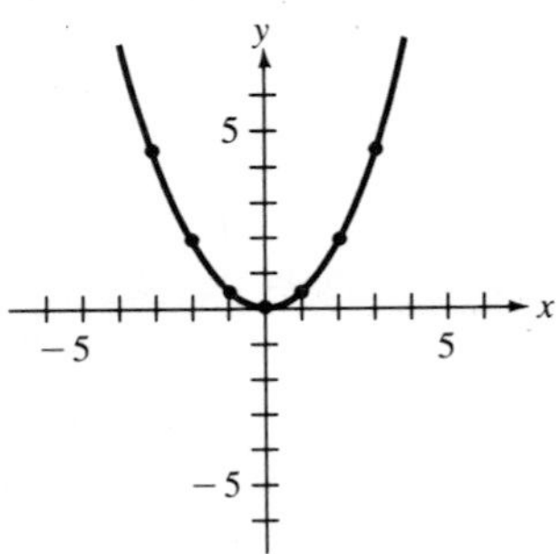


**5**

Graph the following tables of values.

**a.**

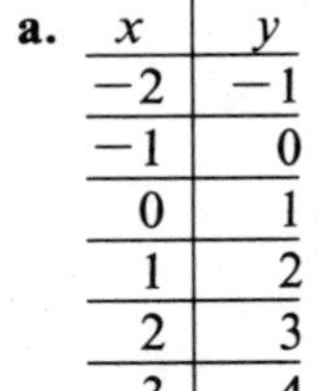

| $x$ | $y$ |
|---|---|
| −2 | −1 |
| −1 | 0 |
| 0 | 1 |
| 1 | 2 |
| 2 | 3 |
| 3 | 4 |

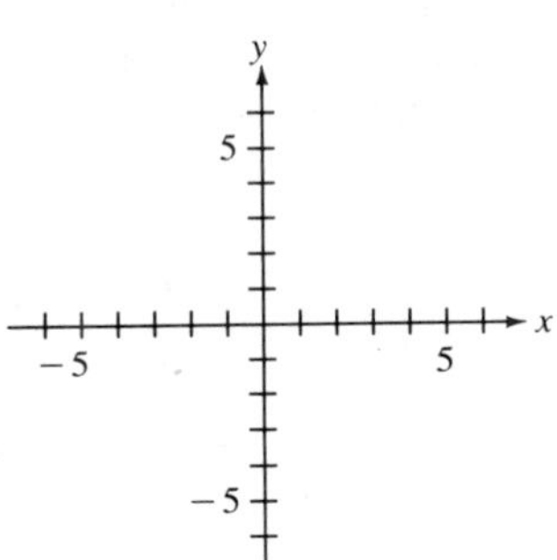


**b.**

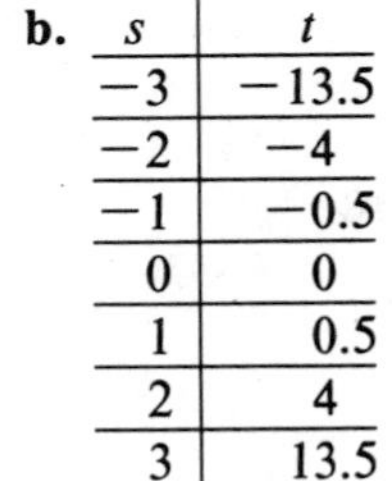

| $s$ | $t$ |
|---|---|
| −3 | −13.5 |
| −2 | −4 |
| −1 | −0.5 |
| 0 | 0 |
| 1 | 0.5 |
| 2 | 4 |
| 3 | 13.5 |

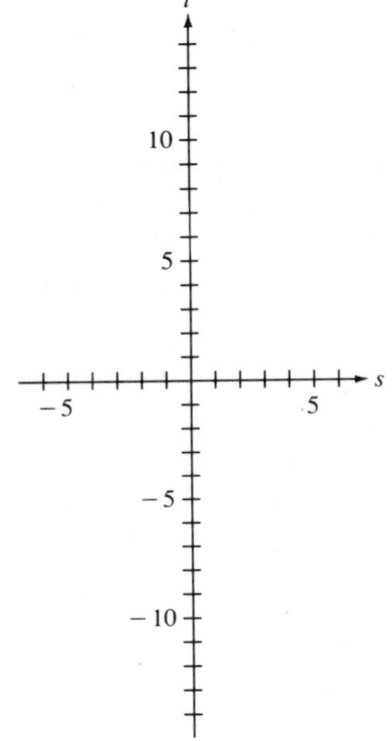

## 6

Answers to frame 5.

a.

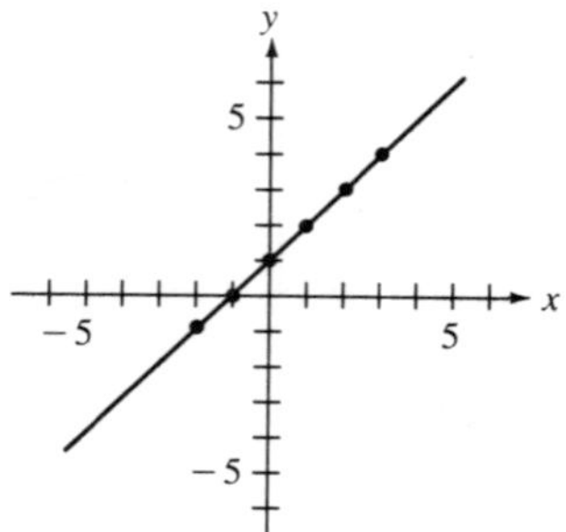


b.

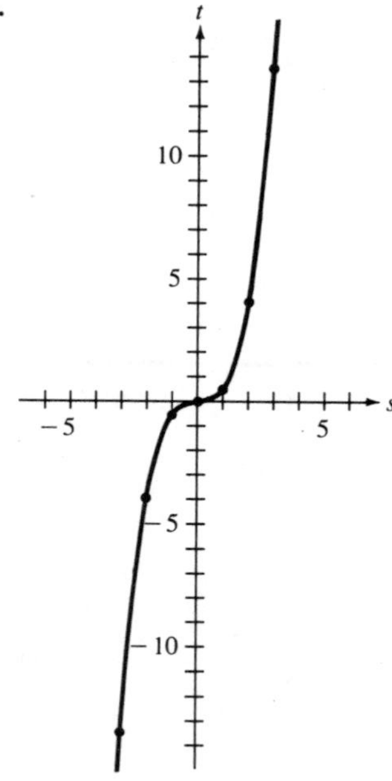


## 7

The graphs drawn previously each had one space equal to one unit on each axis. The number of units per space must often be different to keep the graph to a reasonable size. In many instances, the scale of one axis is different from the scale of the other axis.

**Example** Graph the table of values on Figure 10-13.

| $x$ | $y$ |
|---|---|
| 0 | −30 |
| 1 | −28 |
| 2 | −22 |
| 3 | −12 |
| 4 | 2 |
| 5 | 20 |
| 6 | 42 |

Each space on the $x$-axis equals one unit; on the $y$-axis, each space equals five units.

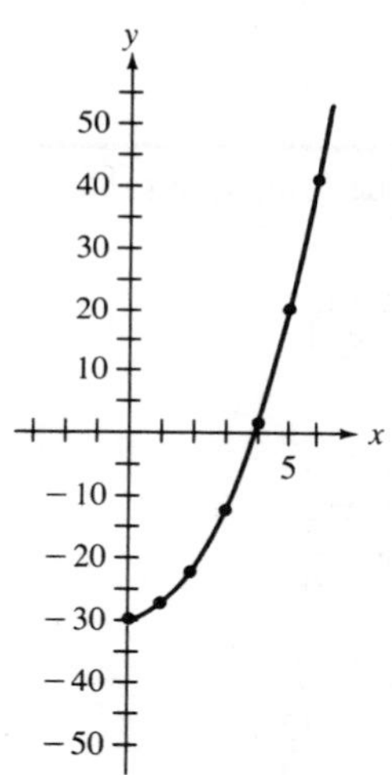


**Figure 10-13**

Graph the table of values on Figure 10-14.

| $d$ | $s$ |
|---|---|
| −3 | −54 |
| −2 | −16 |
| −1 | −2 |
| 0 | 0 |
| 1 | 2 |
| 2 | 16 |
| 3 | 54 |

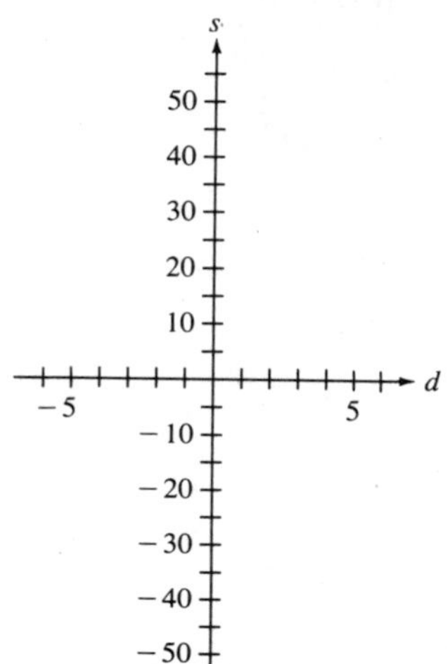


**Figure 10-14**

**8**

Answer to frame 7.

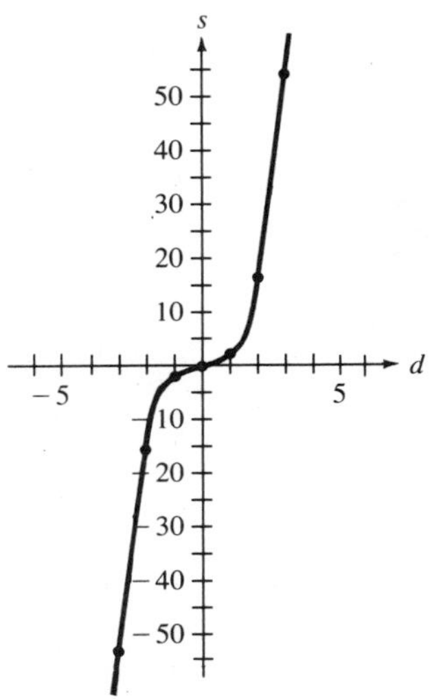


# Exercise Set, Sections 10-2–10-3

## The Rectangular Coordinate System

Name the indicated coordinate of each point in Figure 10-15.

1. $x$-coordinate of $A$ ______
2. $y$-coordinate of $A$ ______
3. $x$-coordinate of $B$ ______
4. $y$-coordinate of $B$ ______
5. $x$-coordinate of $C$ ______
6. $y$-coordinate of $C$ ______
7. $x$-coordinate of $D$ ______
8. $y$-coordinate of $D$ ______

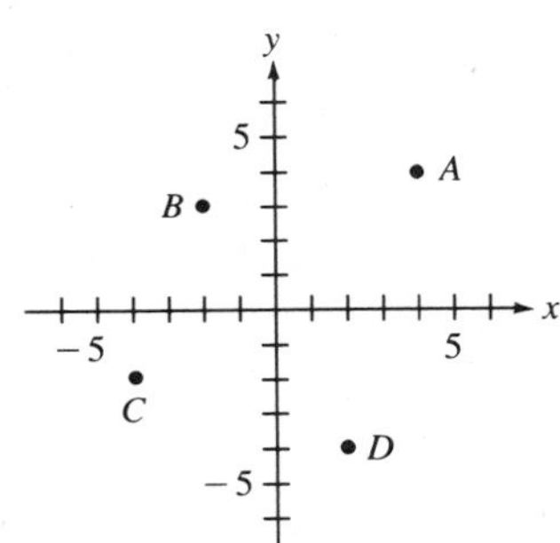


**Figure 10-15**

Locate and label each of the following points in Figure 10-16.

9. Point $A$ has an $x$-coordinate of 2 and a $y$-coordinate of 5.
10. Point $B$ has an $x$-coordinate of $-3$ and a $y$-coordinate of 4.
11. Point $C$ has an $x$-coordinate of $-2$ and a $y$-coordinate of $-2$.
12. Point $D$ has an $x$-coordinate of 3 and a $y$-coordinate of $-4$.

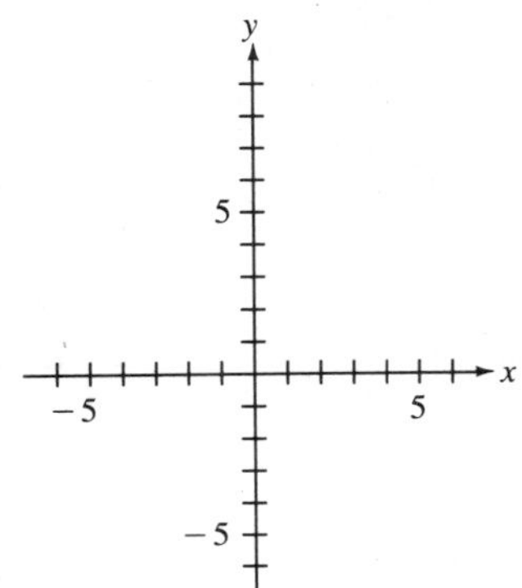


**Figure 10-16**

# Graphing a Table of Values

Graph each of the following tables of values.

**13.**

| $x$ | $y$ |
|---|---|
| 2 | 5 |
| 1 | 2 |
| 0 | −1 |
| −1 | −4 |
| −2 | −7 |

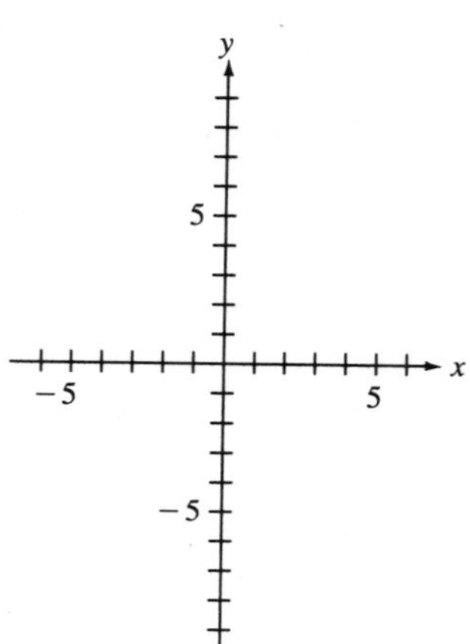

**14.**

| $x$ | $y$ |
|---|---|
| 2 | 0 |
| 1 | 2 |
| 0 | 4 |
| −1 | 6 |
| −2 | 8 |

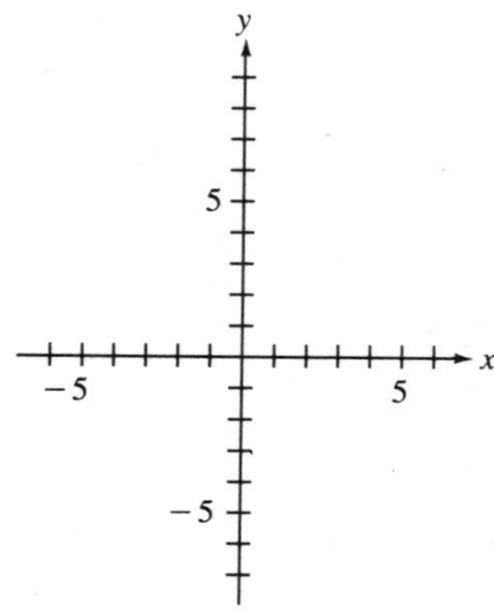

**15.**

| $x$ | $y$ |
|---|---|
| 0 | 0 |
| 1 | −0.5 |
| 2 | −2 |
| 3 | −4.5 |
| 4 | −8 |
| 5 | −12.5 |

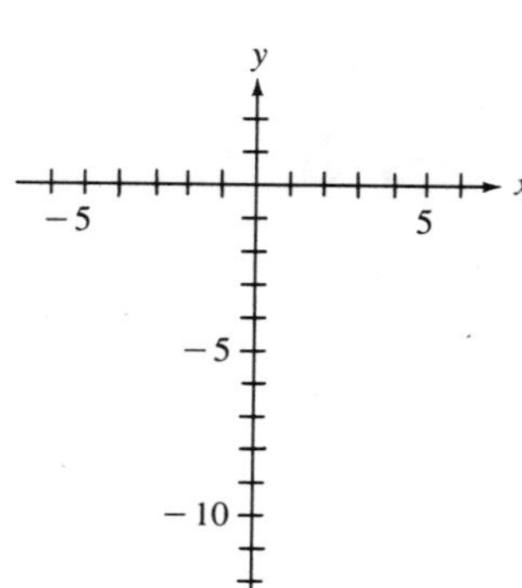

**16.**

| $x$ | $y$ |
|---|---|
| −3 | 12 |
| −2 | 7 |
| −1 | 4 |
| 0 | 3 |
| 1 | 4 |
| 2 | 7 |
| 3 | 12 |

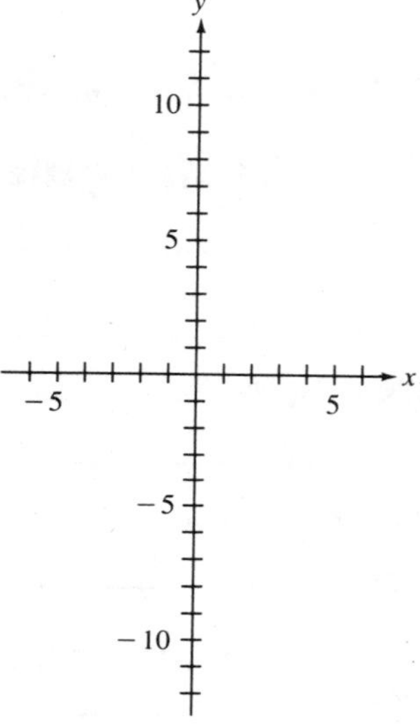

**17.**

| $t$ | $v$ |
|---|---|
| 0 | 0 |
| 1 | 22 |
| 2 | 38 |
| 3 | 45 |
| 4 | 42 |

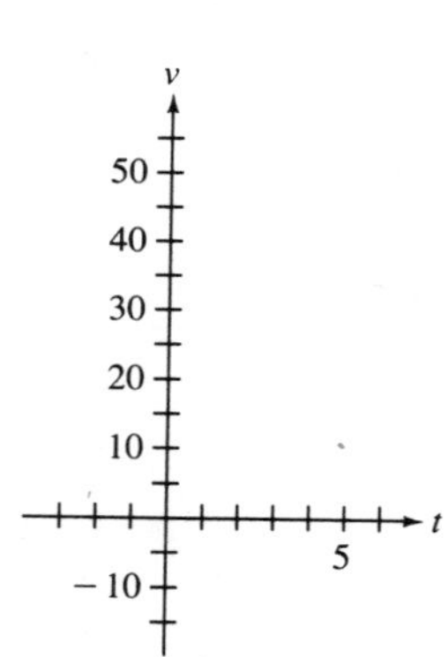

**18.**

| Volts E | milliamps ma |
|---|---|
| 1 | 40 |
| 2 | 80 |
| 3 | 120 |
| 4 | 160 |
| 5 | 200 |
| 6 | 240 |

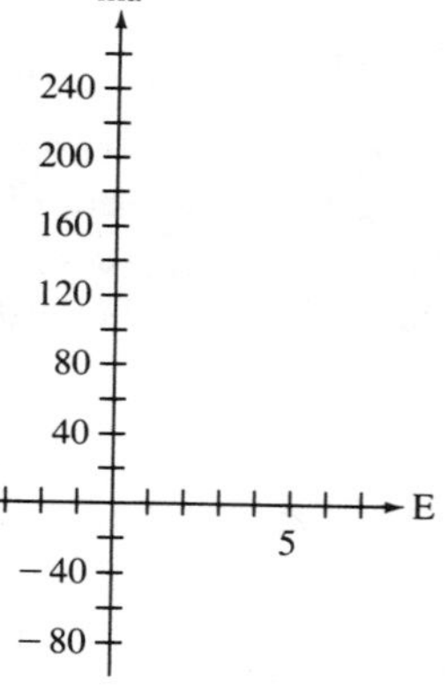

# Supplementary Exercise Set, Sections 10-2–10-3

Name the indicated coordinate of each point in Figure 10-17.

1. $x$-coordinate of $A$ ________
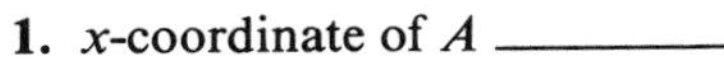
2. $y$-coordinate of $B$ ________
3. $x$-coordinate of $C$ ________
4. $y$-coordinate of $D$ ________

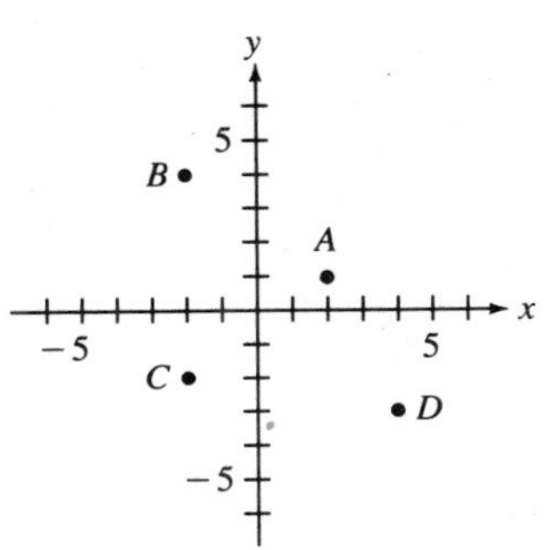


Figure 10-17

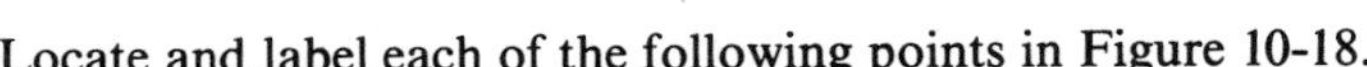

Locate and label each of the following points in Figure 10-18.

5. Point $A$ has an $x$-coordinate of 3 and a $y$-coordinate of 4.
6. Point $B$ has an $x$-coordinate of $-4$ and a $y$-coordinate of 5.
7. Point $C$ has an $x$-coordinate of $-3$ and a $y$-coordinate of $-1$.
8. Point $D$ has an $x$-coordinate of 4 and a $y$-coordinate of $-3$.

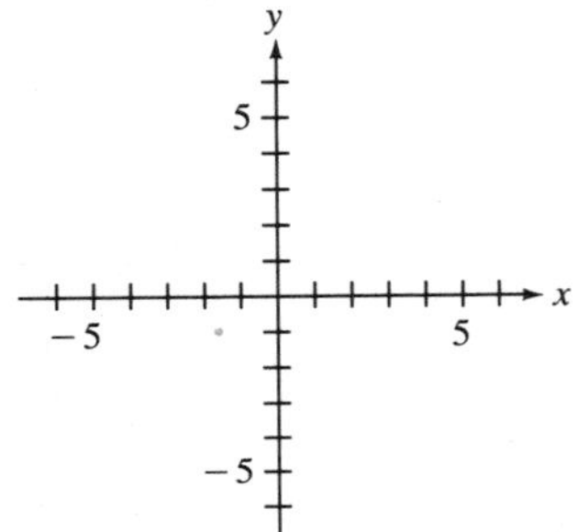


Figure 10-18

Graph each table of values.

9.

| $x$ | $y$ |
|---|---|
| $-2$ | $-8$ |
| $-1$ | $-5$ |
| 0 | $-2$ |
| 1 | 1 |
| 2 | 4 |
| 3 | 7 |

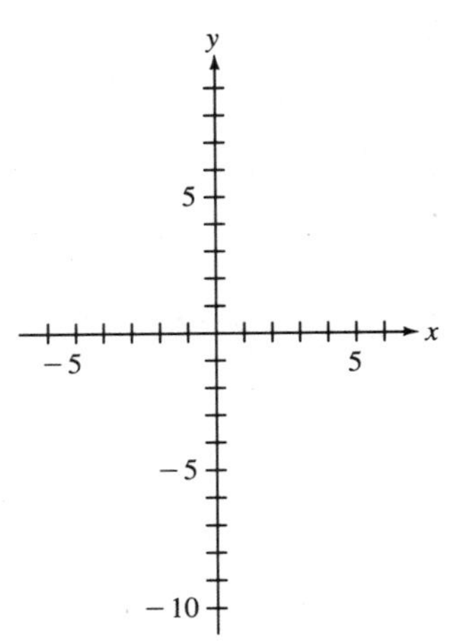


10.

| $x$ | $y$ |
|---|---|
| $-2$ | 7 |
| $-1$ | 5 |
| 0 | 3 |
| 1 | 1 |
| 2 | $-1$ |

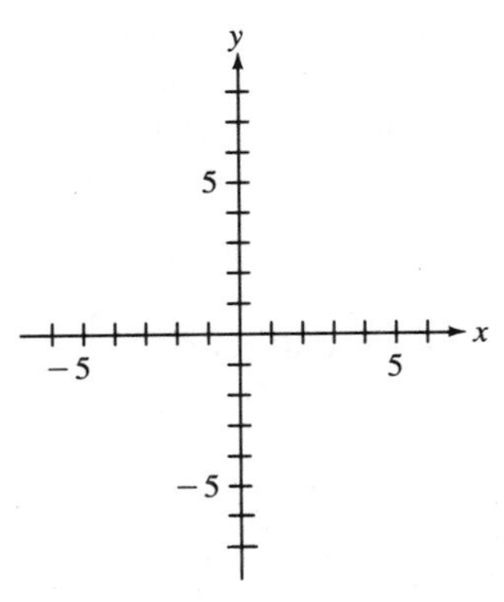

11.

| $s$ | $t$ |
|---|---|
| 0 | 0 |
| 1 | 0.4 |
| 2 | 1.6 |
| 3 | 3.6 |
| 4 | 6.4 |
| 5 | 10 |

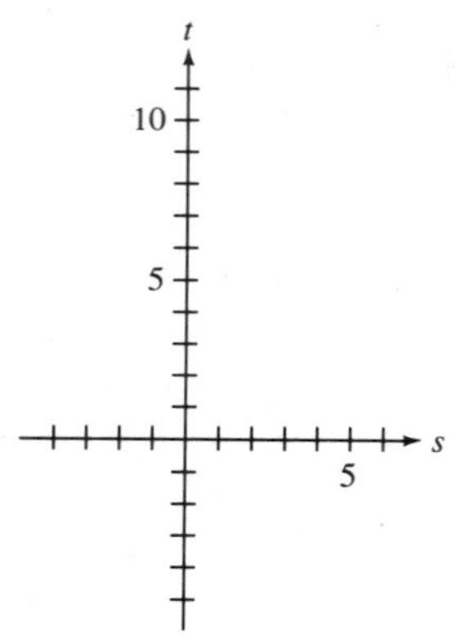


12.

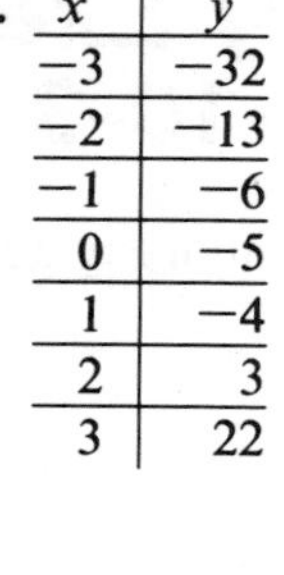

| $x$ | $y$ |
|---|---|
| −3 | −32 |
| −2 | −13 |
| −1 | −6 |
| 0 | −5 |
| 1 | −4 |
| 2 | 3 |
| 3 | 22 |

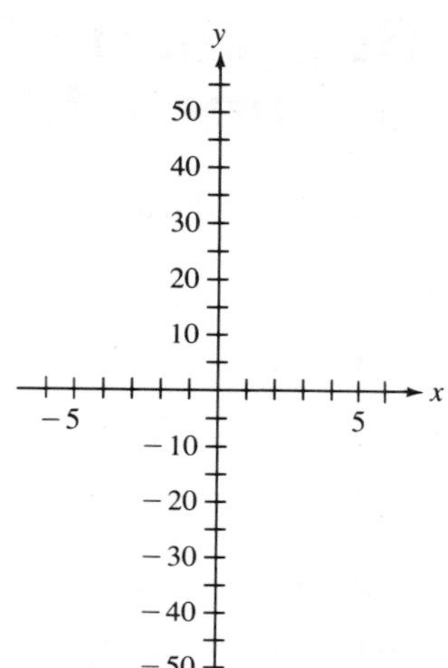


# Section 10-4 Preparing a Table of Values from a Function

**1**

When a consumer buys gasoline for a car, the price of the gasoline per gallon, the number of gallons wanted, and the amount of money needed to pay for the gasoline must be known. The price of the gasoline per gallon will not change while the tank is being filled and is regarded as constant. Using \$1.30/gal as the constant price of gasoline, the relationship between the amount of money that must be paid and the number of gallons of gasoline wanted can be expressed with an equation.

$$\text{Amount of money } (A) = \text{price per gallon} \times \text{number of gallons } (n)$$

$$A = (\$1.30/\text{gal})n \quad \text{or} \quad A = 1.30n$$

Both $A$ and $n$ are variables in the equation above. The variable $n$ (number of gallons) is called the *independent variable*, since the number of gallons can be chosen. The variable $A$ (amount of money) is called the *dependent variable*, since once a definite number of gallons are put in the gas tank, they must be paid for. Simply stated, once the value of $n$ is chosen, the value of $A$ is determined.

List the dependent and independent variables in the equations below.

**a.** $a = 3m^2$ independent = ________

dependent = ________

**b.** $y = 2x + 3$ independent = ________

dependent = ________

**c.** $b = \frac{2H}{5}$ independent = ________

dependent = ________

**2**

In frame 1, it was stated that the value of $A$ was dependent on the value of $n$. Another way to say this is that $A$ is a *function* of $n$. In equation form, this statement becomes

$$A = f(n)$$

**1. a.** independent = $m$
dependent = $a$
**b.** independent = $x$
dependent = $y$
**c.** independent = $H$
dependent = $b$

The symbol $f(n)$ does not mean $f$ times $n$. $f(n)$ means "the function of $n$" or that set of instructions that tells what to do with the independent variable $n$. In the example in frame 1,

$$A = f(n)$$

$$f(n) = 1.30n$$

The set of instructions for the $f(n)$ says to take the number of gallons of gas wanted ($n$) and multiply this by \$1.30/gal to get the amount of money ($A$) needed to pay for the gas.

Here is another example:

$$y = 2x^2 + 3$$

$$f(x) = 2x^2 + 3$$

The set of instructions for $f(x)$ is

1. Take the value of the independent variable $x$ and square it.
2. Take the squared value of $x$ and multiply by 2.
3. Add 3 to two times the squared value of $x$.

When this set of instructions is performed with a stated value of $x$, the value of the dependent variable $y$ will be obtained. Listing the set of instructions allows for easy use of a calculator or computer to solve the equation.

Here is another example:

$$y = 3x - 1$$

**a.** List the $f(x)$ — $f(x) = 3x - 1$

**b.** List the instructions of $f(x)$ — Multiply $x$ times 3.
Subtract 1 from 3 times $x$.

Try this problem.

$$y = \frac{x}{2} + 1$$

**a.** $f(x) =$ ________________

**b.** Instructions of $f(x)$ ____________________________

____________________________

## 3

Here are more problems.

$$s = 6t - 4$$

**a.** $f(t) =$ ________________

**b.** List the instructions of $f(t)$ ____________________________

____________________________

**2. a.** $\frac{x}{2} + 1$

**b.** Divide $x$ by 2;
add 1 to one-half of $x$.

$$y = 2q^2 - q$$

**c.** $f(q) =$ ______________

**d.** List the instructions of $f(q)$ ______________________

______________________

## 4

The concept of a function can be used to generate a table of values.

**Example** Compute a table of values for the following equation. Use the values shown in the table for the independent variable $x$.

| $x$ | $y$ |
|---|---|
| −3 | |
| −2 | |
| 0 | |
| 2 | |
| 3 | |

$y = 2x$

$f(x) = 2x$ (Multiply the value of $x$ times 2 to obtain the value of the dependent variable $y$.)

When $x$ is −3 $f(x) = 2(-3)$ or −6 therefore $y = -6$
When $x$ is −2 $f(x) = 2(-2)$ or −4 therefore $y = -4$
When $x$ is 0 $f(x) = 2(0)$ or 0 therefore $y = 0$
When $x$ is 2 $f(x) = 2(2)$ or 4 therefore $y = 4$
When $x$ is 3 $f(x) = 2(3)$ or 6 therefore $y = 6$

The computed values for $y$ can now be placed in the table for the corresponding values of $x$.

| $x$ | $y$ |
|---|---|
| −3 | −6 |
| −2 | −4 |
| 0 | 0 |
| 2 | 4 |
| 3 | 6 |

Compute the values for the dependent variable $B$ using the equation and the given values of the independent variable $A$.

| $A$ | $B$ |
|---|---|
| −5 | |
| 0 | |
| 5 | |
| 10 | |

$B = 2A + 3$

3. a. $6t - 4$
b. Multiply $t$ times 6. Subtract 4 from $t$ times 6.
c. $2q^2 - q$
d. Square $q$. Multiply $q^2$ times 2. Subtract $q$ from $q^2$ times 2.

## 5

Answers to frame 4.

| $A$ | $B$ |
|---|---|
| −5 | −7 |
| 0 | 3 |
| 5 | 13 |
| 10 | 23 |

$f(A) = 2A + 3$
When $A$ is −5 $B = 2(-5) + 3$ or −7
When $A$ is 0 $B = 2(0) + 3$ or +3
When $A$ is 5 $B = 2(5) + 3$ or 13
When $A$ is 10 $B = 2(10) + 3$ or 23

**6**

a. Use the equation $P = 3R + 2$ to complete the table.

| $R$ | $P$ |
|---|---|
| −4 | |
| 0 | |
| 4 | |
| 8 | |

b. Use the equation $y = 2x^2 - 4$ to complete the table.

| $x$ | $y$ |
|---|---|
| −3 | |
| −2 | |
| −1 | |
| 0 | |
| 1 | |
| 2 | |
| 3 | |

**7**

Answers to frame 6.

a.

| $R$ | $P$ |
|---|---|
| −4 | −10 |
| 0 | 2 |
| 4 | 14 |
| 8 | 26 |

b.

| $x$ | $y$ |
|---|---|
| −3 | 14 |
| −2 | 4 |
| −1 | −2 |
| 0 | −4 |
| 1 | −2 |
| 2 | 4 |
| 3 | 14 |

# Section 10-5 Graphing an Equation

**1**

An equation containing a dependent and independent variable can be graphed on the rectangular coordinate system by using the following steps.

*Step 1.* Construct a table of values using the given values of the dependent variable.
*Step 2.* Graph the corresponding points on the rectangular coordinate system.
*Step 3.* Connect the points with a smooth line.

**Example** Graph the following equation on Figure 10-19.

$$y = x - 5$$

*Step 1.* Construct a table of values.
Let the independent variable $x$ have the values −10, −5, 0, 5, and 10.

When $x = -10$ $y = -10 - 5$ or $-15$
When $x = -5$ $y = -5 - 5$ or $-10$
When $x = 0$ $y = 0 - 5$ or $-5$
When $x = 5$ $y = 5 - 5$ or $0$
When $x = 10$ $y = 10 - 5$ or $5$

Place the corresponding $x$ and $y$ values in a table.

| $x$ | $y$ |
|---|---|
| −10 | −15 |
| −5 | −10 |
| 0 | −5 |
| 5 | 0 |
| 10 | 5 |

*Step 2.* Graph the corresponding points on Figure 10-19.
*Step 3.* Connect the points with a smooth line.

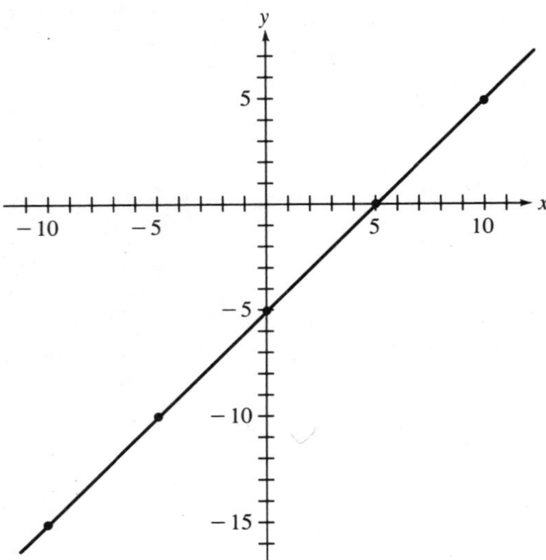


**Figure 10-19**

Construct a table of values, and graph each equation.

**a.** $y = 1.5x + 10$
Let $x = -4, -2, 0, 2, 4$, and 6.

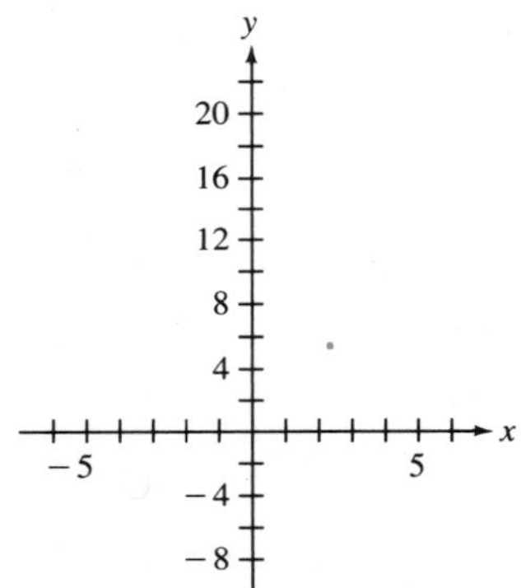


**b.** $y = -x + 5$
Let $x = -4, -2, 0, 2$, and 4.

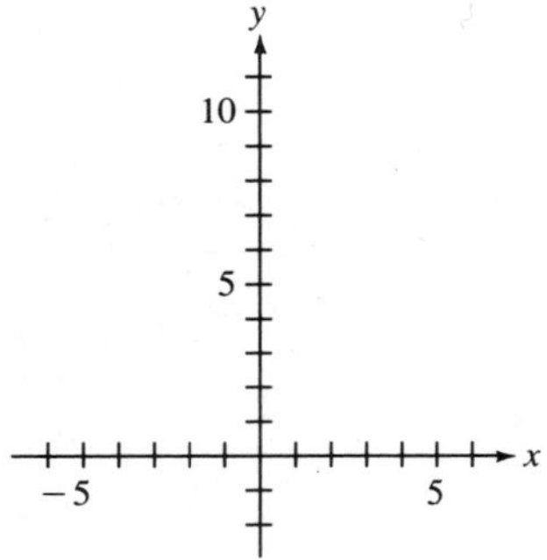

## 2

Answers to frame 1.

**a.**

| $x$ | $y$ |
|---|---|
| −4 | 4 |
| −2 | 7 |
| 0 | 10 |
| 2 | 13 |
| 4 | 16 |
| 6 | 19 |

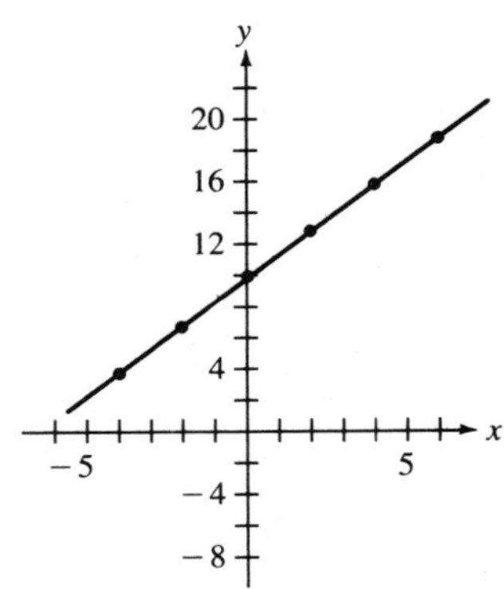

**b.**

| $x$ | $y$ |
|---|---|
| 4 | 1 |
| 2 | 3 |
| 0 | 5 |
| −2 | 7 |
| −4 | 9 |

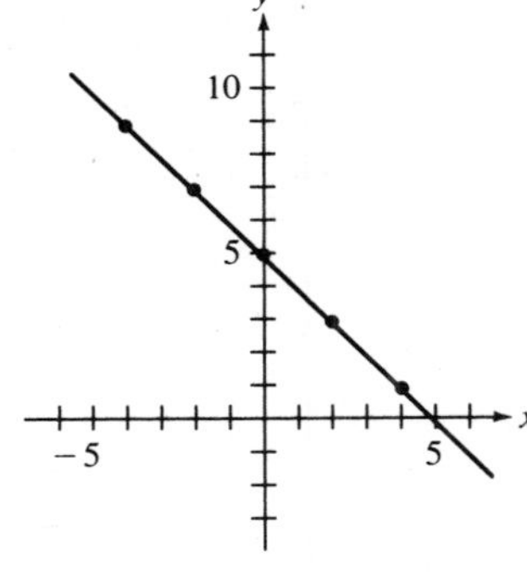

## 3

Construct a table of values, and graph each equation.

**a.** $y = 3x^2 - 2$

Let $x = 0, 1, 2, 3,$ and 4.

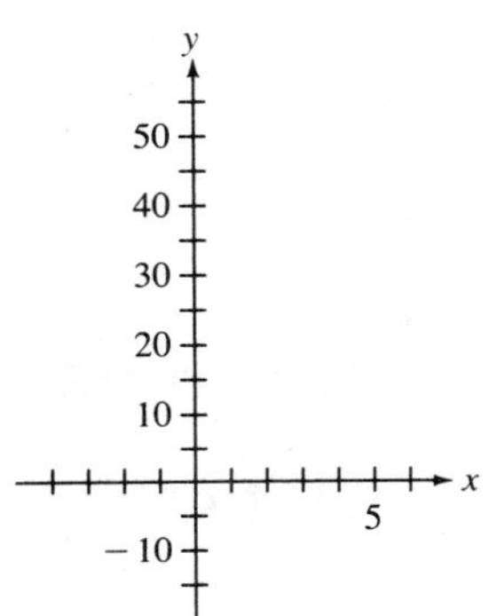

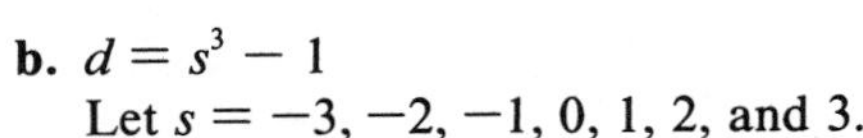

**b.** $d = s^3 - 1$

Let $s = -3, -2, -1, 0, 1, 2,$ and 3.

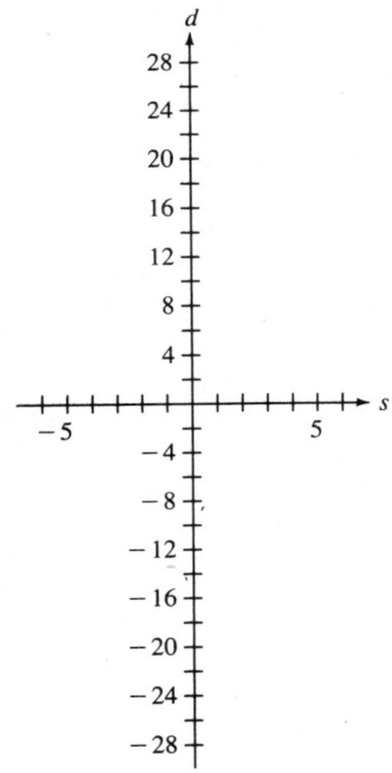

**4**

Answers to frame 3.

a.

| $x$ | $y$ |
|---|---|
| 0 | −2 |
| 1 | 1 |
| 2 | 10 |
| 3 | 25 |
| 4 | 46 |

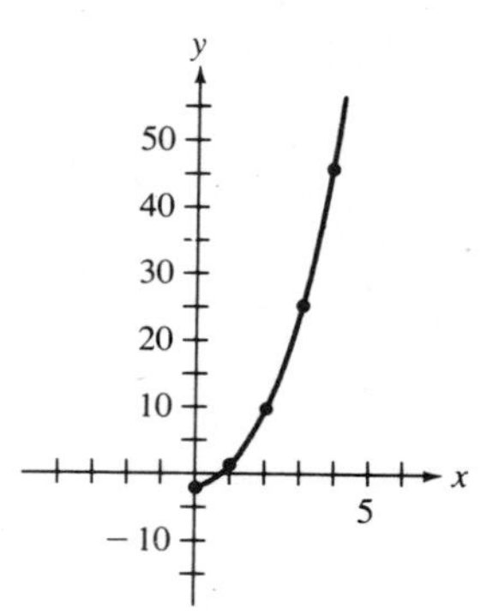

b.

| $s$ | $d$ |
|---|---|
| −3 | −28 |
| −2 | −9 |
| −1 | −2 |
| 0 | −1 |
| 1 | 0 |
| 2 | 7 |
| 3 | 26 |

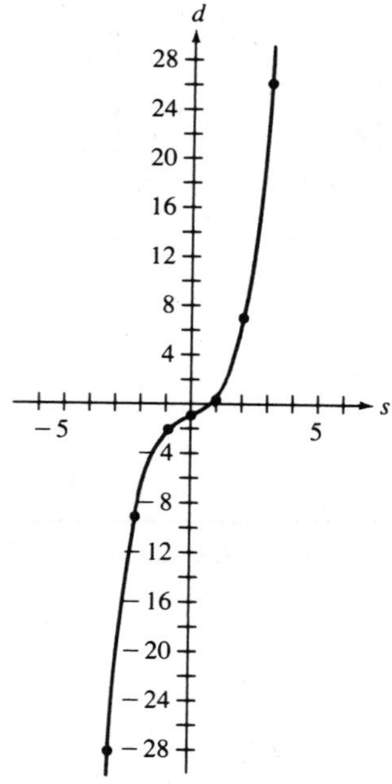

# Exercise Set, Sections 10-4–10-5

## Preparing a Table of Values

List each independent and dependent variable in the equations below.

$$V = \frac{t^2}{2} \qquad\qquad y = 0.3x + 2$$

**1.** independent = ______ **3.** independent = ______

**2.** dependent = ______ **4.** dependent = ______

List the $f(x)$ in the following equations.

$$y = 2x^3 + 4 \qquad\qquad a = x^2 - 2x$$

**5.** $f(x) =$ ______ **6.** $f(x) =$ ______

Generate a table of values for each of the given equations. Use the values shown in the table for the independent variable.

**7.** $y = 0.5x + 6$

| $x$ | $y$ |
|---|---|
| 0 | |
| 1 | |
| 2 | |
| 3 | |
| 4 | |
| 5 | |

**8.** $a = b^2 + 2$

| $b$ | $a$ |
|---|---|
| −3 | |
| −2 | |
| −1 | |
| 0 | |
| 1 | |
| 2 | |
| 3 | |

**9.** $y = -x + 4$

| $x$ | $y$ |
|---|---|
| −3 | |
| −2 | |
| −1 | |
| 0 | |
| 1 | |
| 2 | |
| 3 | |

**10.** $k = m^3 - 2$

| $m$ | $k$ |
|---|---|
| −2 | |
| −1 | |
| 0 | |
| 1 | |
| 2 | |

## Graphing an Equation

Construct a table of values, and graph each equation.

**11.** $y = 2x + 6$

| $x$ | $y$ |
|---|---|
| −3 | |
| −2 | |
| 0 | |
| 1 | |
| 2 | |
| 3 | |

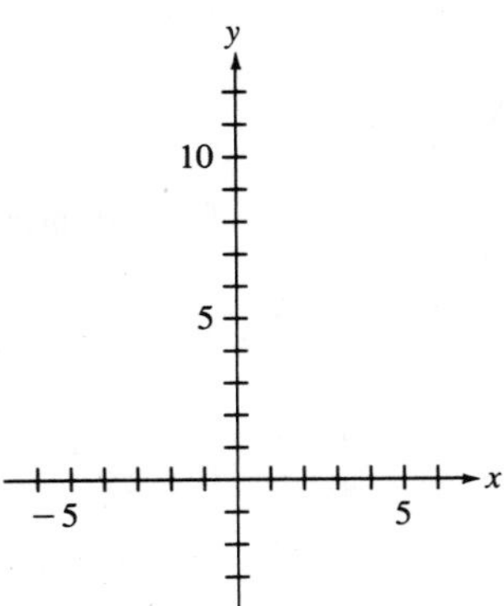


**12.** $y = -2x + 3$

| $x$ | $y$ |
|---|---|
| −4 | |
| −2 | |
| −1 | |
| 0 | |
| 1 | |
| 2 | |

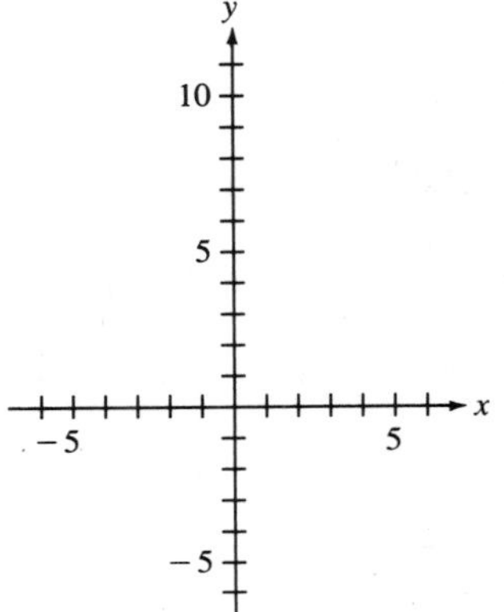


**13.** $y = 2x^2 - 1$

| $x$ | $y$ |
|---|---|
| 0 | |
| 1 | |
| 2 | |
| 3 | |

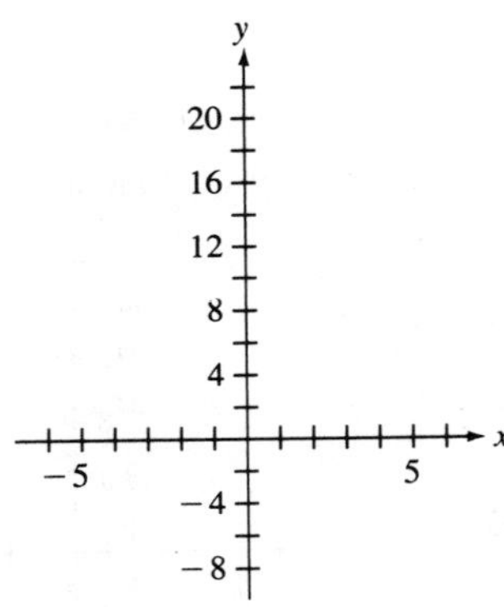


**14.** $a = b^3 + 1$

| $b$ | $a$ |
|---|---|
| −3 | |
| −2 | |
| −1 | |
| 0 | |
| 1 | |
| 2 | |
| 3 | |

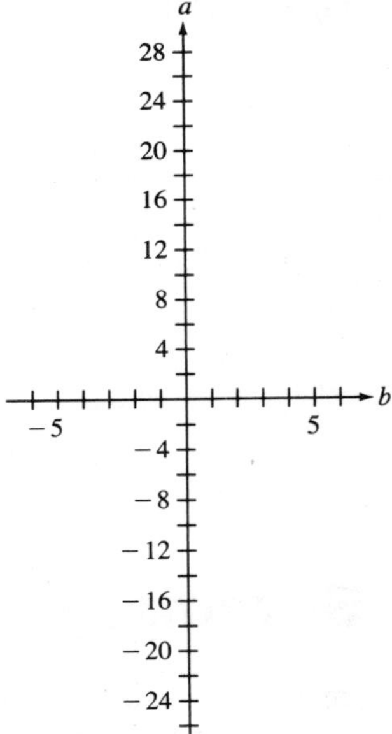

# Supplementary Exercise Set, Sections 10-4–10-5

List the dependent and independent variables in the equations below.

$V = 5m^2$ $\qquad\qquad$ $y = 3.5x + 0.3$

1. dependent = ______
2. independent = ______
3. dependent = ______
4. independent = ______

List the $f(x)$ in the following equations.

$y = \sqrt{x} + 3$ $\qquad\qquad$ $b = x^3 - x^2 + 1$

5. $f(x) =$ ______
6. $f(x) =$ ______

Generate a table of values for the given equations. Use the values shown in the table for the independent variable.

7. $y = -x + 7$

| $x$ | $y$ |
|---|---|
| −2 | |
| −1 | |
| 0 | |
| 1 | |
| 2 | |

8. $a = b^3 - 0.5$

| $b$ | $a$ |
|---|---|
| −2 | |
| −1 | |
| 0 | |
| 1 | |
| 2 | |

Construct a table of values, and graph each equation.

9. $y = -x - 1$

| $x$ | $y$ |
|---|---|
| 0 | |
| 1 | |
| 3 | |
| 4 | |

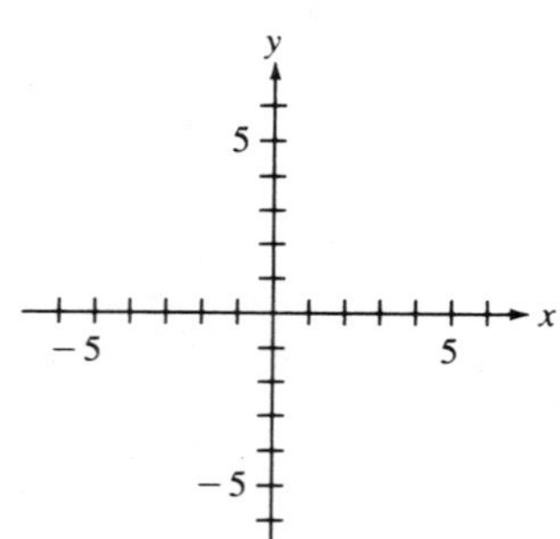


10. $y = 2x^2 - 2$

| $x$ | $y$ |
|---|---|
| −3 | |
| −2 | |
| −1 | |
| 0 | |
| 1 | |
| 2 | |
| 3 | |

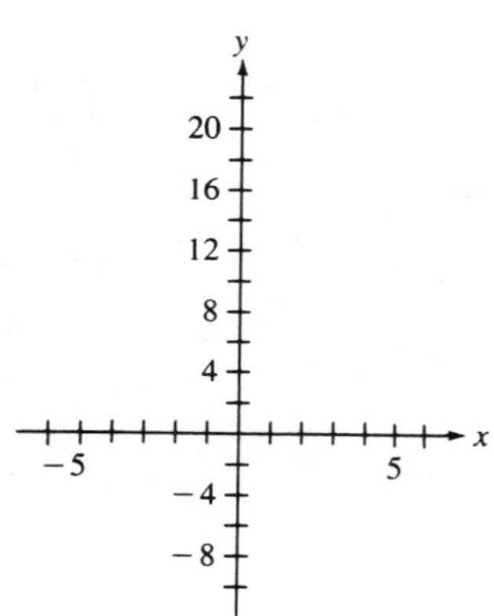


# Summary

1. The rectangular coordinate system contains a vertical and a horizontal number line. The system is used to plot equations containing two variables.
2. An ordered pair of numbers is enclosed by parentheses. The first number is always the $x$-coordinate, and the second number is always the $y$-coordinate.
3. Data placed in table form can be graphed by finding each pair of numbers on the rectangular coordinate system and then by connecting the points with a smooth line.
4. The function of $n$, written $f(n)$, means the set of instructions that tells what to do with the independent variable $n$.
5. To graph an equation, use the following steps.
   a. Construct a table of values.
   b. Graph the corresponding points on the rectangular coordinate system.
   c. Connect the points with a smooth line.

6. Use the following steps to create a bar graph.
   a. Determine whether you want the bars positioned horizontally or vertically.
   b. Label each axis.
   c. Determine the scale to be used.
7. Use the following steps to draw a circle graph.
   a. Multiply the decimal equivalent of each quantity by 360° to get the size of each sector in degrees.
   b. Draw a radius on a circle and measure the angles for each sector.
   c. Label each sector.

# Chapter 10 Self-Test

## Graphing

Name the indicated coordinate of each point in Figure 10-20.

1. $x$-coordinate of $A$ ______
2. $y$-coordinate of $B$ ______
3. $x$-coordinate of $C$ ______
4. $y$-coordinate of $D$ ______

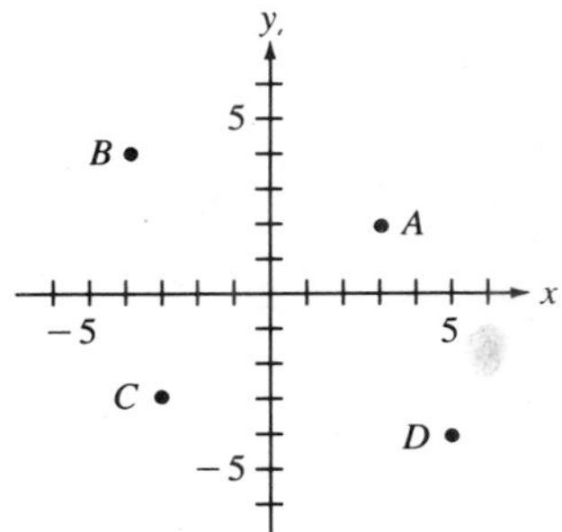


Figure 10-20

Locate and label each of the following points in Figure 10-21.

5. Point $A$ has an $x$-coordinate of 4 and a $y$-coordinate of $-3$.
6. Point $B$ has an $x$-coordinate of $-3$ and a $y$-coordinate of 4.

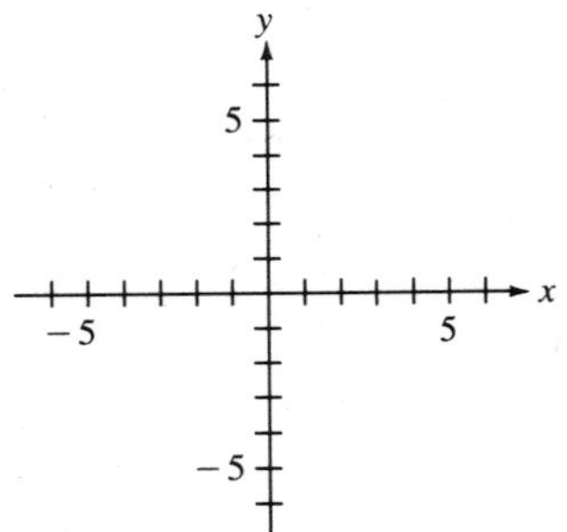


Figure 10-21

Graph each table of values.

7.

| $x$ | $y$ |
|---|---|
| −4 | 8 |
| −2 | 6 |
| 0 | 4 |
| 2 | 2 |
| 4 | 0 |

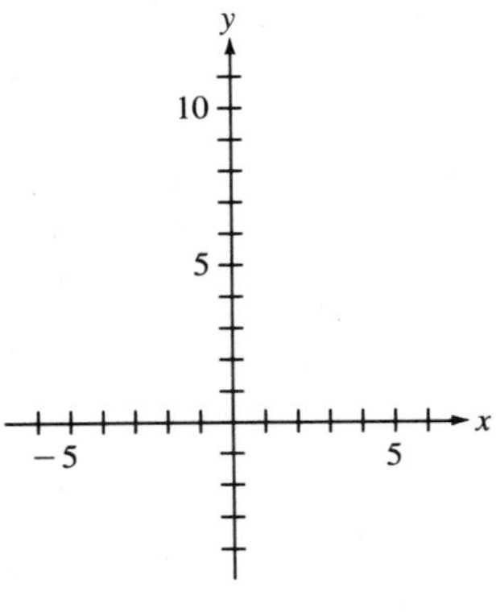


8.

| $x$ | $y$ |
|---|---|
| 0 | −4 |
| 1 | −3 |
| 2 | 0 |
| 3 | 5 |
| 4 | 12 |

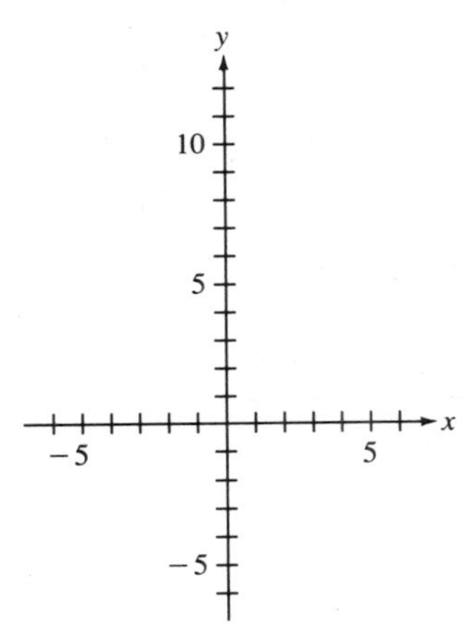

List the independent variable, the dependent variable, and the $f(x)$ in the following equation.

$$y = 7x + 6$$

**9.** independent variable = ______ **10.** dependent variable = ______ **11.** $f(x) =$ ______

Generate a table of values for the given equations. Use the values given in the table for the independent variable.

**12.** $y = 0.1x + 2$

| $x$ | $y$ |
|---|---|
| −2 | |
| −1 | |
| 0 | |
| 1 | |
| 2 | |

**13.** $a = b^3 + 1$

| $b$ | $a$ |
|---|---|
| −2 | |
| −1 | |
| 0 | |
| 1 | |
| 2 | |

Construct a table of values, and graph the equation.

**14.** $y = -3x + 4$

| $x$ | $y$ |
|---|---|
| −2 | |
| −1 | |
| 0 | |
| 1 | |

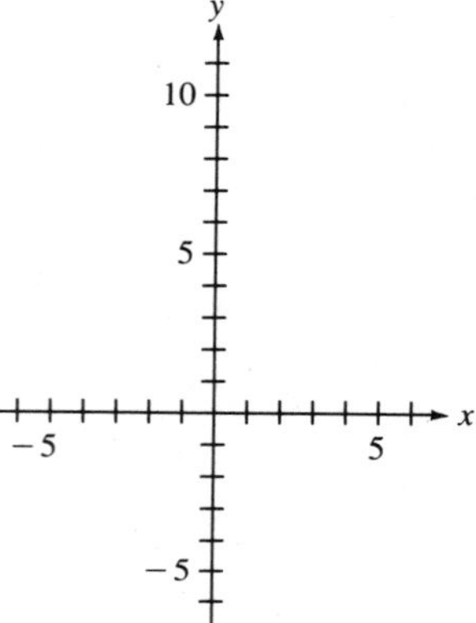


**15.** Construct a vertical bar graph using the data in the table.

| *Month* | *Number of computers sold* |
|---|---|
| Jan. | 110 |
| Feb. | 100 |
| Mar. | 140 |
| Apr. | 150 |

**16.** Construct a circle graph using the data in the table.

| *Plant* | *Percent of total production* |
|---|---|
| *A* | 50 |
| *B* | 20 |
| *C* | 30 |

# CHAPTER 11

# Trigonometry

Trigonometry is the branch of mathematics that deals with the relationships between sides and angles of triangles. Many practical problems can be solved using these relationships. This chapter will define the trigonometric functions and use them to solve various applied problems.

## Section 11-1 Trigonometric Ratios

**1**

The study of trigonometry is based on certain ratios in right triangles. These ratios are called *trigonometric functions.* Figure 11-1 shows three triangles: $AB_1C_1$, $AB_2C_2$, and $AB_3C_3$. The three right triangles all have the same angle $A$.

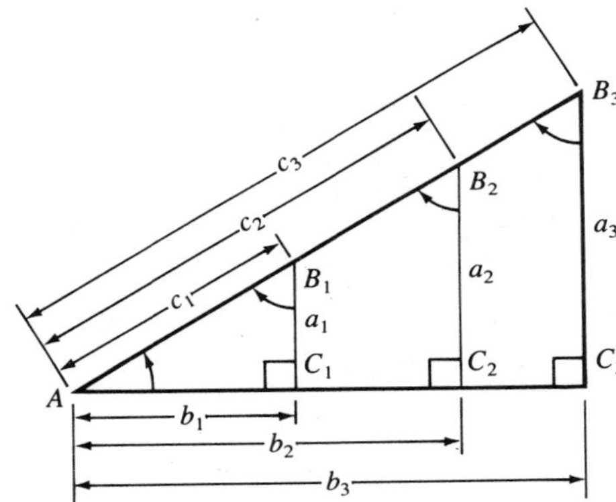


**Figure 11-1**

Corresponding sides for these triangles are proportional. That is,

$$\frac{\text{side } a_1}{\text{side } b_1} = \frac{\text{side } a_2}{\text{side } b_2} = \frac{\text{side } a_3}{\text{side } b_3} \qquad \text{and}$$

$$\frac{\text{side } a_1}{\text{side } c_1} = \frac{\text{side } a_2}{\text{side } c_2} = \frac{\text{side } a_3}{\text{side } c_3}$$

Name another set of proportional ratios.

$$\frac{\text{side } b_1}{\text{side } c_1} = \underline{\qquad} = \underline{\qquad}$$

**2**

The ratios will remain the same for sides of right triangles that have the same angle. When the angle changes, the ratios will change. These ratios are called trigonometric functions, or trig functions.

1. $\frac{\text{side } b_2}{\text{side } c_2} = \frac{\text{side } b_3}{\text{side } c_3}$

**3**

The six trig functions will be defined by referring to the right triangle in Figure 11-2.

Each of the six possible ratios in triangle $ABC$ will be defined relative to angle $A$. For the purpose of these definitions, side $a$ will be the side opposite angle $A$. Side $b$ will be the side adjacent to angle $A$. Side $c$ will be the hypotenuse.

Ratio 1, the sine (sin) of angle $A$:

$$\sin A = \frac{\text{side opposite angle } A}{\text{hypotenuse}} = \frac{a}{c}$$

Ratio 2, the cosine (cos) of angle $A$:

$$\cos A = \frac{\text{side adjacent angle } A}{\text{hypotenuse}} = \frac{b}{c}$$

Ratio 3, the tangent (tan) of angle $A$:

$$\tan A = \frac{\text{side opposite angle } A}{\text{side adjacent angle } A} = \frac{a}{b}$$

Ratio 4, the cotangent (cot) of angle $A$:

$$\cot A = \frac{\text{side adjacent angle } A}{\text{side opposite angle } A} = \frac{b}{a}$$

Ratio 5, the secant (sec) of angle $A$:

$$\sec A = \frac{\text{hypotenuse}}{\text{side adjacent angle } A} = \frac{c}{b}$$

Ratio 6, the cosecant (csc) of angle $A$:

$$\csc A = \frac{\text{hypotenuse}}{\text{side opposite angle } A} = \frac{c}{a}$$

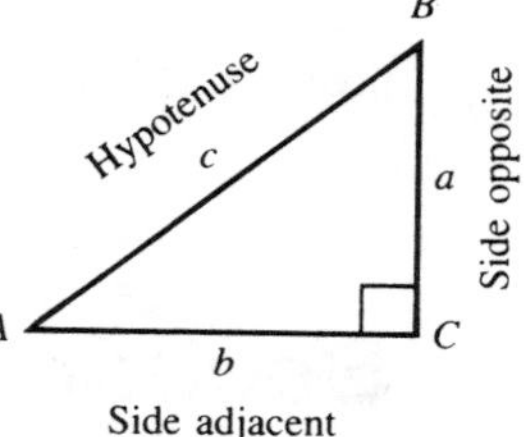


**Figure 11-2**

**4**

For each of the six trig functions mentioned in the previous frame, there is another function that contains the same sides.

Functions containing the side opposite and the hypotenuse:

$\sin A = \frac{a}{c}$ $\qquad$ $\csc A = \frac{c}{a}$

Functions containing the side adjacent and the hypotenuse:

$\cos A = \frac{b}{c}$ $\qquad$ $\sec A = \frac{c}{b}$

Functions containing the side opposite and the side adjacent:

$\tan A = \frac{a}{b}$ $\qquad$ $\cot A = \frac{b}{a}$

To solve most applied problems, only one set of trig functions are needed. In this text the sin, cos, and tan will be used.

**5**

The sine, cosine, and tangent ratios must be memorized for use in this text. Complete the following, referring to Figure 11-2.

**a.** $\sin A = \frac{\underline{\qquad}}{\text{hypotenuse}}$

**b.** $\tan A = \frac{\text{side opposite } \angle A}{\underline{\qquad}}$

**c.** $\cos A = \frac{\text{side adjacent } \angle A}{\underline{\qquad}}$

**d.** $\sin A = \frac{-}{c}$

**e.** $\cos A = \frac{b}{-}$

**f.** $\tan A = \frac{-}{a}$

**g.** $\frac{\text{side adjacent } \angle A}{\text{hypotenuse}} =$

**h.** $\frac{\text{side opposite } \angle A}{\text{side adjacent } \angle A} =$

## 6

The numerical value of sin $A$ can be found by performing the division $a/c$.

sine of $A$ when $a = 4$ and $c = 5$ is

$\sin A = 4/5$

$\sin A = 0.8000$

(In this text, trig functions will be calculated to four decimal places.)

**a.** Find sin $A$ when $a = 6.0$ and $c = 7.5$.

sin $A$ = ______

**b.** Find sin $A$ when $a = 8.0$ and $c = 10.0$.

sin $A$ = ______

Notice that sin $A$ is the same for each of the triangles above. That is because angle $A$ is approximately 53° in each case.

5. **a.** side opposite $\angle A$
**b.** side adjacent $\angle A$
**c.** hypotenuse
**d.** $a$
**e.** $c$
**f.** $b$
**g.** cos $A$
**h.** tan $A$

## 7

**a.** Find sin $A$ when $a = 7.1$ and $c = 7.8$.

sin $A$ = ______

**b.** Find sin $A$ when $a = 15$ and $c = 17$.

sin $A$ = ______

Notice that sin $A$ is different for each of the triangles above. That is because angle $A$ is different in each case.

6. **a.** 0.8000
**b.** 0.8000

## 8

The numerical values of cos $A$ and tan $A$ can be found by performing the divisions $b/c$ and $a/b$, respectively. For example, the cosine of $A$ when $b = 3$ and $c = 5$ is

$\cos A = 3/5$

$\cos A = 0.6000$

**a.** Find cos $A$ when $b = 7.2$ and $c = 11.0$.

cos $A$ = ______

**b.** Find tan $A$ when $a = 4.0$ and $b = 5.0$.

tan $A$ = ______

**c.** Find tan $A$ when $a = 14.0$ and $b = 14.0$.

tan $A$ = ______

**d.** Find cos $A$ when $b = 40.0$ and $c = 42.0$.

cos $A$ = ______

7. **a.** 0.9103
**b.** 0.8824

## 9

Answers to frame 8.

**a.** 0.6545 **b.** 0.8000 **c.** 1.000 **d.** 0.9524

# Section 11-2 Reading the Trigonometric Tables and Using the Calculator

**1**

The numerical values for the trigonometric functions have been calculated for all angles in degrees and subdivisions of degrees. This text will only be concerned with trigonometric functions of degrees. The table of trigonometric functions in the appendix (Table 6) lists the sine, cosine, and tangent for each degree from 0° to 90°. A section of that table is shown below.

| *Angle* | *Sine* | *Cosine* | *Tangent* |
|---|---|---|---|
| 26° | 0.4384 | 0.8988 | 0.4877 |
| 27° | 0.4540 | 0.8910 | 0.5095 |
| 28° | 0.4695 | 0.8829 | 0.5317 |
| 29° | 0.4848 | 0.8746 | 0.5543 |
| 30° | 0.5000 | 0.8660 | 0.5774 |
| 31° | 0.5150 | 0.8572 | 0.6009 |

This table can be used to find trigonometric functions of angles.

**Example** Find the cosine of 27°.

*Step 1.* Locate 27° in the column at the left.
*Step 2.* Move over two columns to the column under cosine.
*Step 3.* The cosine of 27° is 0.8910. This can be written as cos 27° = 0.8910.

Use the table to find the following trigonometric functions.

**a.** sin 31° = **b.** tan 28° = **c.** cos 30° =

**d.** sin 26° = **e.** cos 29° = **f.** tan 27° =

**1. a.** 0.5150 **b.** 0.5317 **c.** 0.8660 **d.** 0.4384 **e.** 0.8746 **f.** 0.5095

**2**

Use Table 6 in the appendix to find the following functions.

**a.** sin 42° = **b.** tan 58° = **c.** cos 89° =

**d.** tan 14° = **e.** cos 51° = **f.** sin 34° =

**2. a.** 0.6691 **b.** 1.6003 **c.** 0.0175 **d.** 0.2493 **e.** 0.6293 **f.** 0.5592

**3**

The trigonometric functions can be found by using a calculator with sine, cosine, and tangent keys. To find sin 40°, use the procedure below.

| *Enter* | *Press* | *Display* |
|---|---|---|
| 40 | SIN | 0.6427876 |

Round to four decimal places: 0.6428.

Use the calculator to find the following.

**a.** sin 42° = **b.** tan 58° = **c.** cos 89° =

**d.** tan 14° = **e.** cos 51° = **f.** sin 34° =

*Note:* Be sure that your calculator is in the degree mode before performing any trigonometric calculations.

## 4

Compare your calculator answers (frame 3) with the trigonometric functions found from the tables (frame 2). Are they the same?

**3. a.** 0.6691306 (0.6691)
**b.** 1.6003345 (1.6003)
**c.** 0.0174524 (0.0175)
**d.** 0.249328 (0.2493)
**e.** 0.6293204 (0.6293)
**f.** 0.5591929 (0.5592)

## 5

The table is adequate for finding trigonometric functions from 0° to 90°. Special conversion formulas are needed to find trigonometric functions greater than 90° when using a table. The calculator can find the trigonometric functions of any angle without conversion formulas.

Find the following trigonometric functions using a calculator.

**a.** sin 95° =   **b.** tan 241° =   **c.** cos 305° =

**d.** tan 182° =   **e.** cos 278° =   **f.** sin 140° =

**4.** yes

## 6

Trigonometric functions can be calculated for angles greater than 360°. One complete revolution is 360°, and machines often rotate through many revolutions. To use the trigonometric table, a degree greater than 360° must be converted to an equivalent degree less than 360°. This is done by dividing the angle greater than 360° by 360°. The remainder will be the equivalent degree.

**Example** Find the sine of 412° using the trig table.

*Step 1.* Find the equivalent degree of 412°.

$$\begin{array}{r|l} & 1 \\ \hline 360° & 412° \\ & 360° \\ \hline & 52° \end{array}$$

(remainder = equivalent degree)

*Step 2.* Find the sine of 52° from the table.

sin 52° = 0.7880

sin 412° = 0.7880

Since 412° and 52° are equivalent degrees, they will have the same sine.

Find the following using the trig table.

**a.** sin 385°   **b.** cos 445°   **c.** tan 432°   **d.** cos 378°

**5. a.** 0.9961947 (0.9962)
**b.** 1.8040478 (1.8040)
**c.** 0.5735764 (0.5736)
**d.** 0.0349208 (0.0349)
**e.** 0.1391731 (0.1392)
**f.** 0.6427876 (0.6428)

## 7

The number of revolutions a machine has made can be calculated by dividing the angle by 360°. The quotient is the number of full revolutions.

6. a. 0.4226
b. 0.0872
c. 3.0777
d. 0.9511

**Example** A machine rotates through 2,055°. How many full revolutions has it made?

```
         5      (quotient)
360° )2,055°
      1 800
       255°     (remainder)
```

The quotient is 5, so the machine has made five full revolutions.

How many full revolutions were made by machines rotating through the following angles?

**a.** 718° **b.** 950° **c.** 2,000° **d.** 3,205°

## 8

The trigonometric functions of large angles are given automatically on your calculator. Use your calculator to find the following.

**a.** sin 718° **b.** cos 2,000° **c.** tan 859° **d.** tan 1,300°

7. a. 1
b. 2
c. 5
d. 8

## 9

When the trigonometric function is known, the angle representing that function can be found. The "arc" notation is used to indicate this operation.

Arcsine 0.1736 means the angle whose sine is 0.1736.

8. a. −0.0349
b. −0.9397
c. −0.8693
d. 0.8391

Use the "arc" notation to indicate the angle of the following trigonometric functions.

**a.** The angle whose cosine is 0.7771 ________

**b.** The angle whose tangent is 0.9657 ________

**c.** The angle whose sine is 0.7660 ________

## 10

The calculator or the trig table can be used to find the angle when the trigonometric function is known.

9. a. arcosine 0.7771
b. arctangent 0.9657
c. arcsine 0.7660

When using the calculator the [INV] or [ARC] key is used to find the angle.

Most calculators have an [INV] key, and its use is illustrated.

Find arccos 0.7771.

| *Enter* | *Press* | *Display* |
|---|---|---|
| 0.7771 | [INV] [COS] | 39.004184 |

arccos 0.7771 = 39.00°

Use the calculator to find the following, rounding to two decimal places.

**a.** arcsin 0.6561 = ________ **b.** arccos 0.7193 = ________

**c.** arctan 12.0000 = ________ **d.** arctan 30.0000 = ________

**e.** arcsin 0.4500 = ________ **f.** arccos 0.8660 = ________

**11**

Answers to frame 10.

**a.** 41.00° **b.** 44.00° **c.** 85.24° **d.** 88.09° **e.** 26.74° **f.** 30.00°

# Exercise Set, Sections 11-1–11-2

## Trigonometric Ratios

Find the following, using the trigonometric table.

**1.** sin 19° = ______ **2.** cos 47° = ______ **3.** tan 125° = ______

**4.** tan 73° = ______ **5.** sin 52° = ______ **6.** cos 81° = ______

**7.** arccos 0.9659 = ______ **8.** arctan 0.6009 = ______ **9.** arcsin 0.9994 = ______

## Reading the Trigonometric Tables and Using the Calculator

Find the following, using a calculator.

**10.** cos 73° = ______ **11.** sin 15° = ______ **12.** tan 125° = ______

**13.** sin 510° = ______ **14.** tan 900° = ______ **15.** cos 41° = ______

**16.** arctan 10.0000 = ______ **17.** arcsin 0.5592 = ______ **18.** arccos 0.8387 = ______

Find an equivalent angle less than 360° for each of the following.

**19.** 475° = ______ **20.** 2,048° = ______

# Supplementary Exercise Set, Sections 11-1–11-2

Find the following, using the trigonometric table.

**1.** tan 21° = ______ **2.** sin 15° = ______ **3.** cos 42° = ______

**4.** tan 83° = ______ **5.** sin 63° = ______ **6.** cos 87° = ______

**7.** arccos 0.9903 = ______ **8.** arctan 0.8693 = ______ **9.** arcsin 0.3420 = ______

Find the following, using a calculator.

**10.** cos 78° = ______ **11.** sin 219° = ______ **12.** tan 482° = ______

**13.** arctan 0.1051 = ______ **14.** arcsin 0.6820 = ______ **15.** arccos 0.8418 = ______

Find an equivalent angle less than 360° for each of the following.

**16.** 675° = ______ **17.** 2,025° = ______ **18.** 760° = ______

# Section 11-3 Solving Right Triangles

**1**

The trigonometric ratios and the Pythagorean theorem can be used to find missing values in right triangles. When either of the two angles that are *not* a right angle and any one side is known, the other two sides can be found using the appropriate trigonometric ratios. The third angle can be found by subtracting the two known angles from 180°.

We can illustrate this procedure by using Figure 11-3 to find side $b$. Angle $A$ (a non-right angle) and side $a$ (5.2 cm) are given. The tangent of $A$ is used to solve for side $b$, because the tangent of $A$ equals the side opposite ($a$) over the side adjacent ($b$).

$$\tan A = \frac{a}{b}$$

$$\tan 40° = \frac{5.2 \text{ cm}}{b} \quad \text{(solve the equation for } b\text{)}$$

$$b = \frac{5.2 \text{ cm}}{\tan 40°}$$

$$b = \frac{5.2 \text{ cm}}{0.8391} \quad \text{or} \quad b = 6.2 \text{ cm}$$

The sin $A$ is used to solve for side $c$, because the sine of $A$ equals the side opposite ($a$) over the hypotenuse ($c$). Use Figure 11-3 to solve for side $c$.

$$\sin A = \frac{a}{c}$$

**a.** Solve for side $c$.

**b.** Angle $A = 40°$; angle $C = 90°$; what is angle $B$?

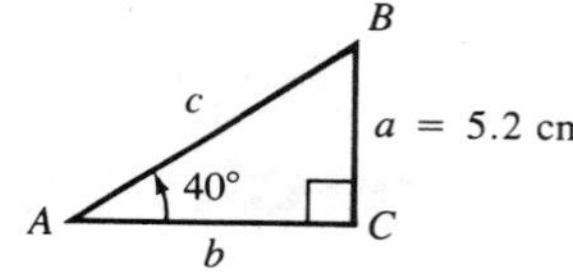


**Figure 11-3**

**1. a.** $c = 8.1$ cm
**b.** $\angle B = 180° - 40° - 90°$
$\angle B = 50°$

**2**

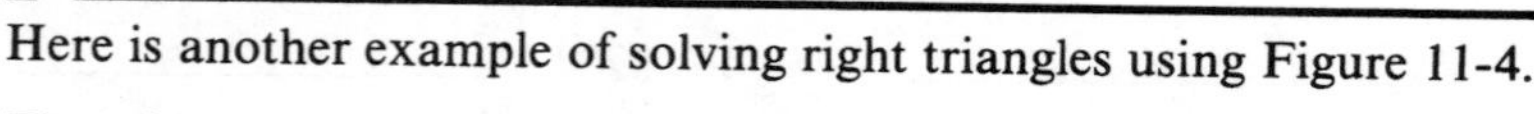

Here is another example of solving right triangles using Figure 11-4.

Find side $a$ using a trig ratio that contains angle $A$ and sides $a$ and $b$.

$$\tan A = \frac{a}{b}$$

$$\tan 35° = \frac{a}{17 \text{ m}} \quad \text{(solve the equation for } a\text{)}$$

$$a = (17 \text{ m})(\tan 35°)$$

$$a = (17 \text{ m})(0.7002)$$

$$a = 11.9\text{m} \quad \text{or} \quad 12 \text{ m}$$

Find side $c$ using $\cos A = \frac{b}{c}$.

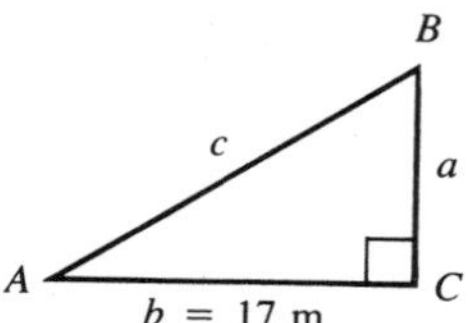


**Figure 11-4**

2. $c = 6.8$ m

## 3

A missing angle can be found when two sides of a right triangle are known.

Use Figure 11-5 to find angle $A$. Tan $A$ is used, because $a$ and $b$ are given.

$$\tan A = \frac{a}{b}$$

$$\tan A = \frac{3.0\text{ m}}{8.2\text{ m}}$$

$$\tan A = 0.3659$$

$$\angle A = \arctan 0.3659$$

$$\angle A = 20°$$

**Figure 11-5**

Use Figure 11-6 to solve the following.

**a.** Find angle $A$ when $b = 48$ cm and $c = 52$ cm.

**b.** Find angle $B$ when $b = 15$ ft and $a = 18$ ft.

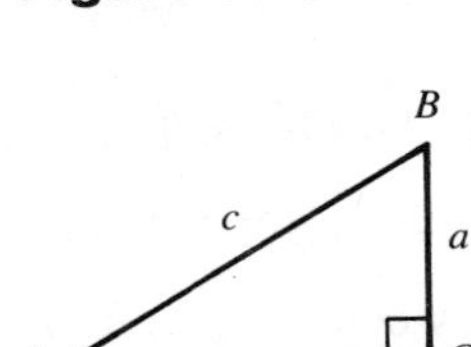


**Figure 11-6**

## 4

Answers to frame 3.

**a.** $\cos A = \frac{b}{c} = \frac{48\text{ cm}}{52\text{ cm}}$

$\cos A = 0.9231$
$\angle A = \text{arccosine } 0.9231$
$\angle A = 23°$

**b.** $\tan B = \frac{b}{a} = \frac{15\text{ ft}}{18\text{ ft}}$

$\tan B = 0.8333$
$\angle B = \text{arctangent } 0.8333$
$\angle B = 40°$

## 5

Right triangles can be in any position. All of the trigonometric relationships apply no matter what position the right triangle is in. Use Figure 11-7 to find the following.

**a.** Find side $a$ when $\angle A = 32°$ and side $c = 75$ ft.

**b.** Find side $a$ when $\angle B = 19°$ and side $b = 48.2$ cm.

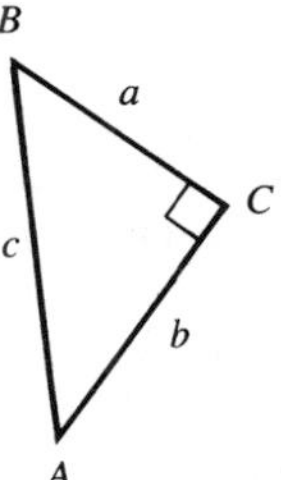


**Figure 11-7**

## 6

Answers to frame 5.

**a.** $\sin A = \frac{a}{c}$

$\sin 32° = \frac{a}{75\text{ ft}}$

$a = (75\text{ ft})(0.5299)$
$a = 40$ ft

**b.** $\tan B = \frac{b}{a}$

$\tan 19° = \frac{48.2\text{ cm}}{a}$

$a = \frac{48.2\text{ cm}}{0.3443}$

$a = 140$ cm

**7**

Letters other than $A$, $B$, and $C$ can be used to label triangles. Identify the following, using Figure 11-8.

**a.** Hypotenuse = ________

**b.** Side opposite $\angle K$ = ________

**c.** Side adjacent $\angle H$ = ________

**d.** $\cos K$ = ________  **e.** $\cos H$ = ________

**f.** $\tan H$ = ________  **g.** $\tan K$ = ________

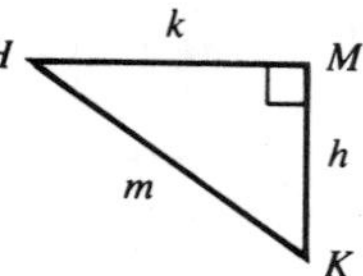


**Figure 11-8**

**8**

Use Figure 11-8 in the last frame to solve the following.

**a.** Find $k$ when $h = 9.9$ m and $\angle H = 61°$.

**b.** Find $k$ when $m = 72$ ft and $\angle K = 53°$.

**7. a.** $m$
**b.** $k$
**c.** $k$
**d.** $\frac{h}{m}$
**e.** $\frac{k}{m}$
**f.** $\frac{h}{k}$
**g.** $\frac{k}{h}$

**9**

Answers to frame 8.

**a.** $\tan 61° = \frac{9.9 \text{ m}}{k}$

$k = 5.5$ m

**b.** $\sin 53° = \frac{k}{72 \text{ ft}}$

$k = 58$ ft

**10**

The calculator and Figure 11-9 are used to solve the following problems.

**a.** Find side $a$ when $c = 35.2$ in. and $\angle A = 58.0°$.

$$\sin 58.0° = \frac{a}{35.2 \text{ in.}} \quad \text{or} \quad a = (35.2 \text{ in.})(\sin 58.0°)$$

| *Enter* | *Press* | *Display* |
|---|---|---|
| 58.0 | SIN × | 0.8480481 |
| 35.2 | = | 29.851293 |

Side $a = 29.9$ in.

**b.** Find side $b$ when side $a = 17$ m and $\angle A = 83°$.

$$\tan 83° = \frac{17 \text{ m}}{b} \quad \text{or} \quad b = \frac{17 \text{ m}}{\tan 83°}$$

| *Enter* | *Press* | *Display* |
|---|---|---|
| 17 | ÷ | 17 |
| 83 | TAN | 8.1443464 |
| | = | 2.0873375 |

Side $b = 2.1$ m.

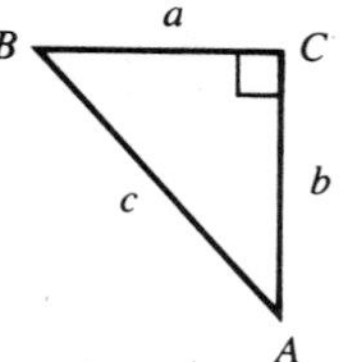


**Figure 11-9**

Using Figure 11-9 and the calculator, find the following.

**a.** Find side $a$ when side $b = 245$ ft and angle $A = 41.2°$.

**b.** Find side $b$ when side $c = 21$ ft and angle $B = 11°$.

**c.** Find side $c$ when side $a = 105$ ft and angle $A = 25.3°$.

**d.** Find side $a$ when side $b = 45.8$ m and angle $B = 73°$.

## 11

Answers to frame 10.

**a.** $\tan 41.2° = \frac{a}{245\text{ ft}}$

$a = 214$ ft

**b.** $\sin 11° = \frac{b}{21\text{ ft}}$

$b = 4.0$ ft

**c.** $\sin 25.3° = \frac{105}{c}$

$c = 246$ ft

**d.** $\tan 73° = \frac{45.8\text{ m}}{a}$

$a = 14.0$ m

## 12

Figure 11-10 shows a geometric figure with certain dimensions missing. The principles of trigonometry can be used to find the missing dimensions.

Find the missing dimension $v$.

The trig ratio used must contain the known angle; side $h$, which is also known; and side $v$.

$$\tan 40° = \frac{v}{h}$$

$$\tan 40° = \frac{v}{11\text{ cm}}$$

$$v = (11\text{ cm})(0.8391)$$

$$v = 9.2\text{ cm}$$

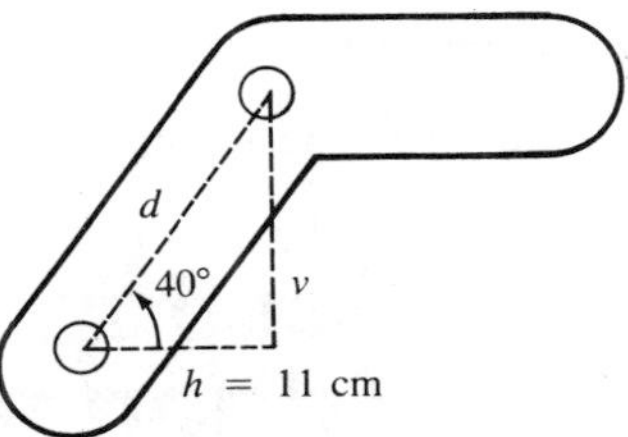


**Figure 11-10**

Find dimension $d$.

## 13

Answer to frame 12.

$\cos 40° = \frac{h}{d}$; $d = \frac{11}{\cos 40°}$; $d = 14$ cm

# Exercise Set, Section 11-3

## Solving Right Triangles

Find the missing value in each of the right triangles described below.

1. If side $c = 5.0$ in. and side $a = 3.1$ in., then $\angle A =$ __________.
2. If side $a = 7.1$ in. and side $b = 5.2$ in., then $\angle B =$ __________.
3. If side $b = 50$ ft and side $c = 75$ ft, then $\angle A =$ __________.
4. If side $a = 4.0$ m and side $c = 5.2$ m, then $\angle B =$ __________.
5. If side $a = 7.2$ in. and side $b = 7.2$ in., then $\angle A =$ __________.
6. If side $b = 35$ yd and side $c = 65$ yd, then $\angle B =$ __________.

Given the triangle in Figure 11-11, find the following.

7. Find $\angle P$ when side $p$ is 6.0 m and side $q$ is 4.2 m. $\angle P =$ __________
8. Find $\angle P$ when side $q$ is 5 mi and side $r$ is 7 mi. $\angle P =$ __________
9. Find $\angle Q$ when side $q$ is 357 mm and side $p$ is 420 mm. $\angle Q =$ __________
10. Find $\angle Q$ when side $q$ is 42 m and side $r$ is 66 m. $\angle Q =$ __________

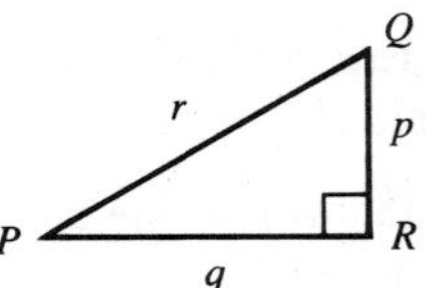

**Figure 11-11**

# Supplementary Exercise Set, Section 11-3

Find the missing value in each right triangle described below.

1. Side $c = 6.0$ ft and side $a = 4.1$ ft. $\angle A =$ __________
2. Side $c = 7.1$ cm and side $a = 5.2$ cm. $\angle A =$ __________
3. Side $a = 45$ mi and side $b = 30$ mi. $\angle B =$ __________
4. Side $b = 0.75$ cm and side $a = 0.75$ cm. $\angle A =$ __________
5. Side $a = 0.75$ cm and side $b = 0.75$ cm. $\angle B =$ __________
6. Side $a = 32$ yd and side $b = 11$ yd. $\angle A =$ __________
7. Side $a = 11$ yd and side $b = 32$ yd. $\angle B =$ __________
8. Side $a = 251.0$ cm and side $c = 345.5$ cm. $\angle B =$ __________
9. Side $a = 95.0$ m and side $b = 190.0$ m. $\angle A =$ __________
10. Side $b = 185$ yd and side $c = 250$ yd. $\angle B =$ __________

# Section 11-4 Solving Applied Trigonometry Problems

## 1

The previously demonstrated five-step problem-solving procedure and some principles of trigonometry are used to solve the following problem.

The pitch of a roof can be expressed as the angle the roof makes with the horizontal (see Figure 11-12). If a roof has a vertical rise of 5 ft and a horizontal distance from the center line of 16 ft, the pitch can be found using the following steps.

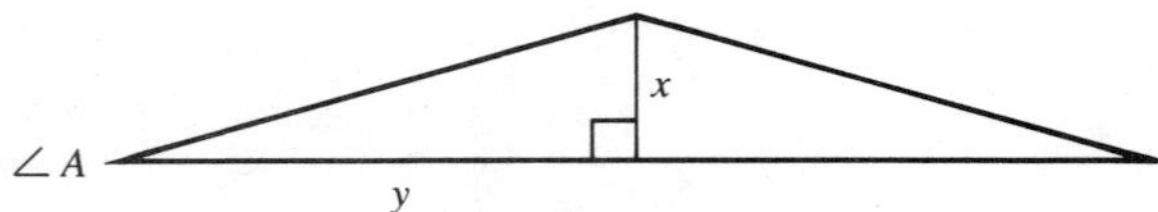

**Figure 11-12**

$x = 5$ ft (vertical rise)

$y = 16$ ft (horizontal distance)

pitch $= \angle A$

*Step 1.* What is being asked for?
pitch ($\angle A$) = ?

*Step 2.* What information is known?
$x = 5$ ft (vertical rise)
$y = 16$ ft (horizontal distance)

*Step 3.* Find a mathematical model that will solve the problem.
When the side opposite and the side adjacent to the angle are known, the following formula is used to solve for the angle.

$$\tan A = \frac{x}{y}$$

*Step 4.* Substitute the known data into the formula.

$$\tan A = \frac{5}{16}$$

*Step 5.* Do the calculation.

$$\tan A = 0.3125$$

$$\angle A = \arctan 0.3125$$

$$\text{pitch } (\angle A) = 17.35°$$

## 2

Use Figure 11-13 to solve for the phase angle ($\alpha$) of an alternating current circuit.

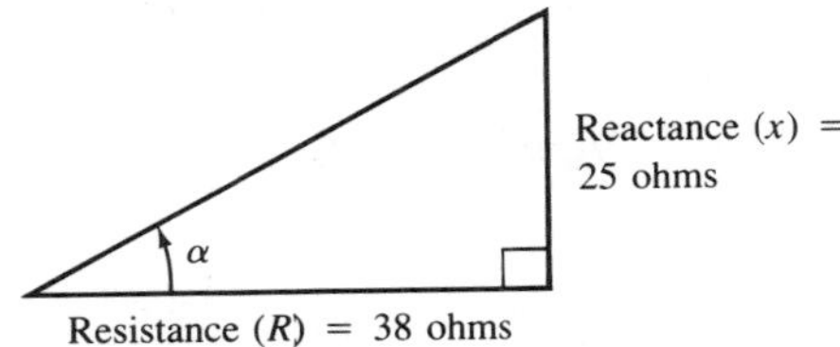


**Figure 11-13**

*Step 1.*

*Step 2.*

*Step 3.*

*Step 4.*

*Step 5.*

## 3

In the previous problems, diagrams were used to help solve the problems. If a diagram is not given, it is helpful to draw one. Solve the following problem by first drawing and labeling a diagram.

A surveyor sights the top of a building at an angle of elevation of 42° from a point 85 ft from the building. How high is the building?

**2. (1)** $\angle\alpha = ?$

**(2)** $x = 25$ ohms
$R = 38$ ohms

**(3)** $\tan = \frac{x}{R}$

**(4)** $\tan = \frac{25}{38}$

**(5)** arctan = 0.6579
$\angle\alpha = 33.34°$

## 4

Answer to frame 3.

$H = (\tan 42°)(85 \text{ ft})$
$H = 77 \text{ ft}$

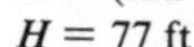

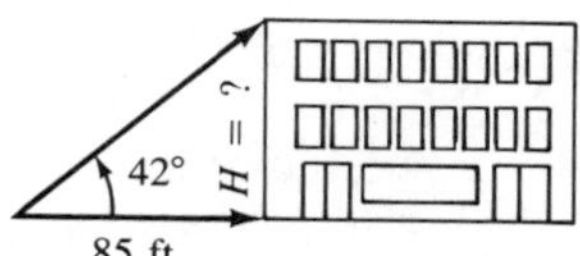


## 5

Figure 11-14 shows a triangular metal plate that will have holes drilled at points *A, B,* and *C*. Use the information in this diagram to find the distance from hole *A* to hole *B*.

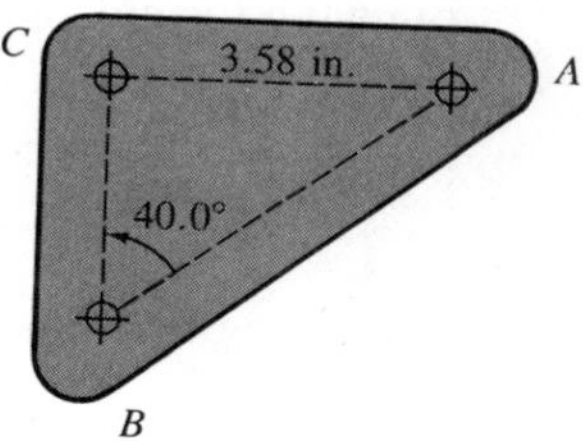


**Figure 11-14**

## 6

In some instances it is necessary to perform several operations to find a missing dimension. This procedure can be illustrated by finding $\angle A$ in Figure 11-15.

*Step 1.* Determine what is being asked for.

$$\angle A = ?$$

*Step 2.* Determine what information is already known.

Base = 3.8 in.

Short side = 0.8 in.

Long side = 2.5 in.

*Step 3.* Find a mathematical model that describes the relationship. (*Note:* Here it is necessary to draw additional lines and perform some preliminary calculations.)

$$\text{Tan } A = \frac{x}{3.8 \text{ in.}}$$

$$x = 2.5 \text{ in.} - 0.8 \text{ in.}$$

$$x = 1.7 \text{ in.}$$

*Step 4.* Substitute the data into the model.

$$\text{Tan } A = \frac{1.7 \text{ in.}}{3.8 \text{ in.}}$$

*Step 5.* Do the calculations.

Tan $A$ = __________

$\angle A$ = __________

**5.** $A \text{ to } B = \dfrac{3.58 \text{ in.}}{\sin 40°}$

$A \text{ to } B = 5.57 \text{ in.}$

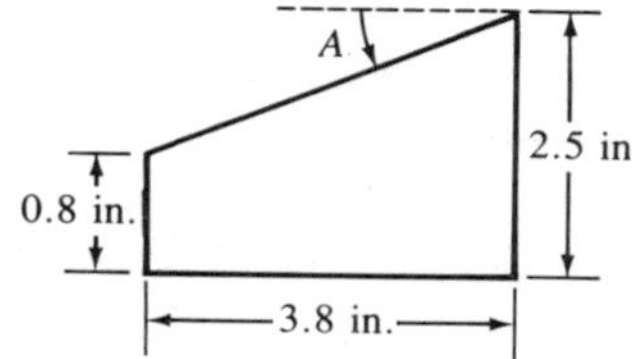


**Figure 11-15**

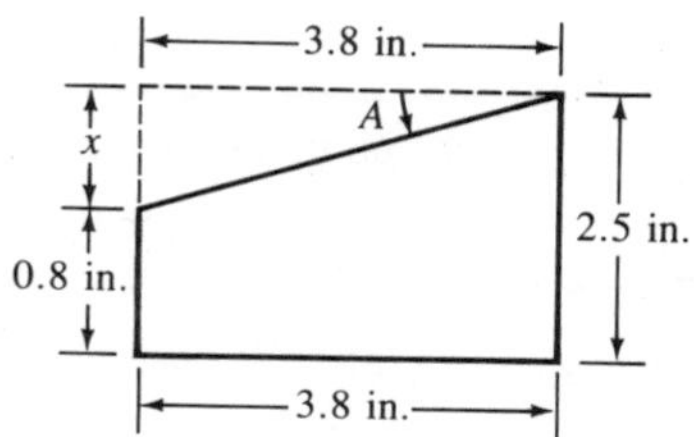

**7**

Find $\angle A$ in Figure 11-16.

**6.** Tan $A = 0.4474$
$\angle A = 24°$

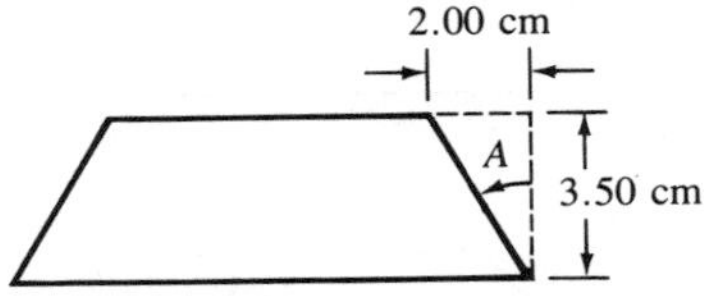


**Figure 11-16**

**8**

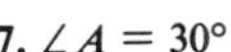

**7.** $\angle A = 30°$

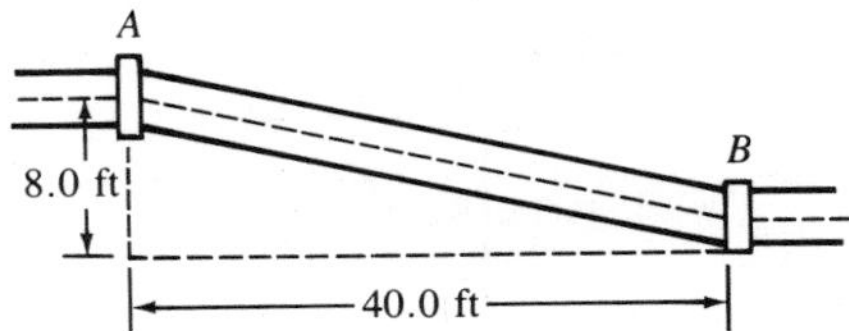


Use the five-step procedure to determine the length of the pipe from $A$ to $B$.

**9**

Answer to frame 8.

Length $A$ to $B = \sqrt{(8.0)^2 - (40.0)^2}$; length $A$ to $B = 40.8$ ft.

# Exercise Set, Section 11-4

## Solving Applied Trigonometry Problems

Solve the following problems using the five-step procedure.

1. If the roof of a building has a vertical rise of 4 ft and a horizontal distance of 15 ft, what angle does the roof make with the horizontal?

2. An alternating current has a reactance of 46 ohms and a resistance of 32 ohms. Find the phase angle $\alpha$.

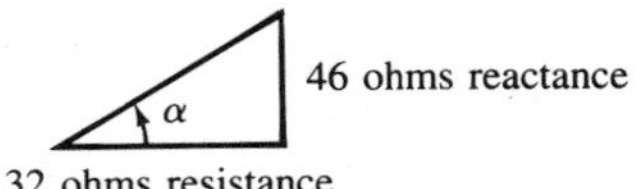


3. A surveyor is standing 65 ft from the foot of a building. The angle of elevation to the top of the building is 56°. How high is the building?

4. A 40-ft-tall tree casts a shadow 55 ft long. What is the angle of elevation of the sun?

5. Find angle $A$ in the diagram.

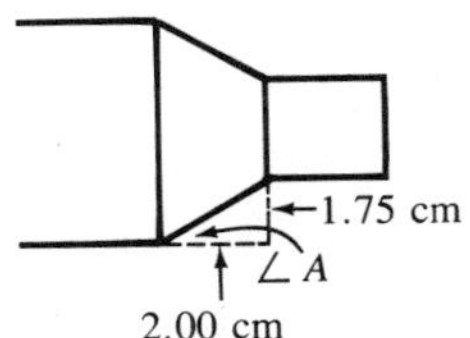

6. Find the length of $BC$ in the diagram.

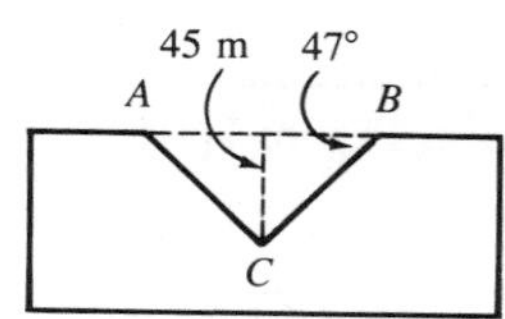


7. Find the value of $x$ in the drawing.

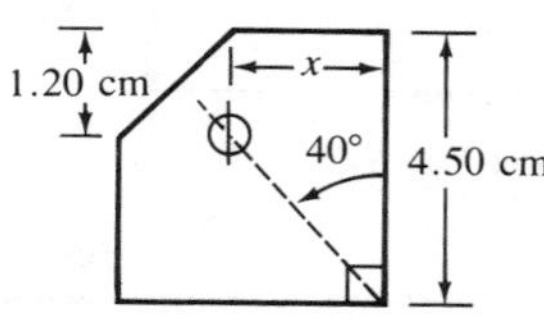


8. Find the value of angle $\theta$ in the drawing.

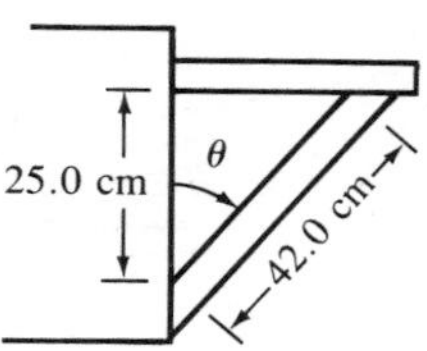


9. A 30-ft ladder leans against a building at an angle of 65°. How far is the foot of the ladder from the building?

10. Find the length of a trench that must be dug to connect a sewer 45 ft north and 15 ft east of an outlet.

## Supplementary Exercise Set, Section 11-4

1. Find the length of pipe $AB$ in the diagram.

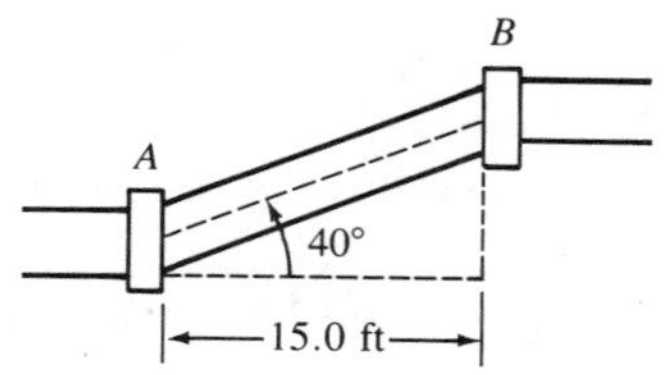


2. Use the diagram to find the distance between holes $A$ and $B$.

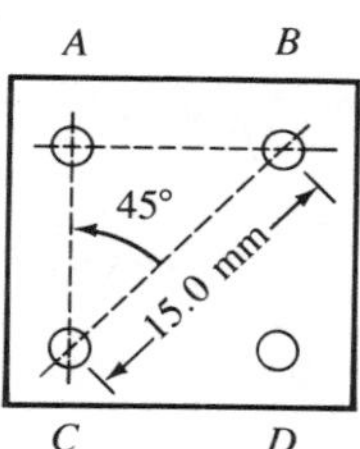


3. An alternating current has an inductive reactance of 72 ohms and a resistance of 60 ohms. What is the phase angle $\alpha$?

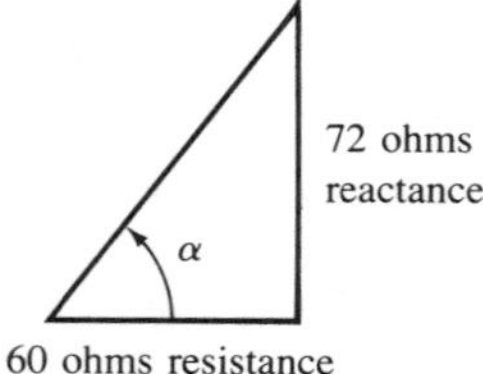

4. A bridge must span a river 70 ft wide. It must connect a road on the east bank with a road 25 feet upstream on the west bank. How long must the bridge be?

5. A loading platform is 4.5 ft above ground. A 20-ft ramp is used to get to the platform. What angle does the ramp make with the ground?

6. The roof of a building has a vertical rise of 5 ft and a horizontal distance of 10 ft. What angle does the roof make with the horizontal?

7. Find the distance from $B$ to $C$ in the diagram.

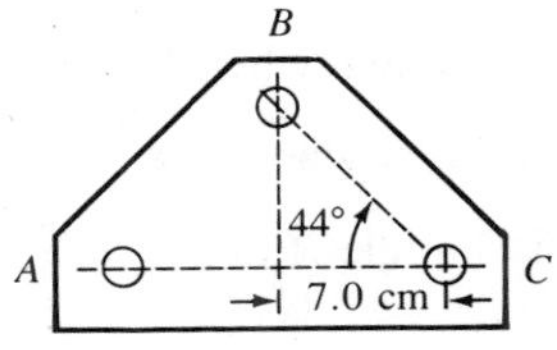


8. Find the depth of the thread in the diagram of a screw.

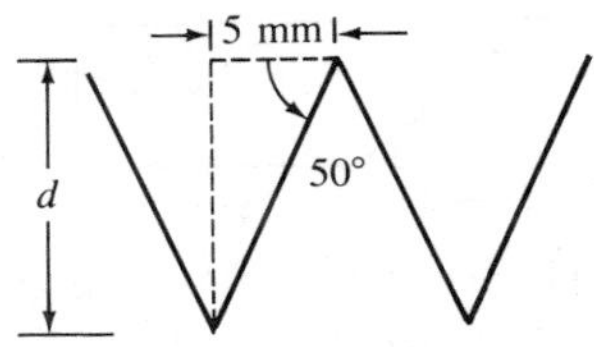


9. A road has a 10° incline. After traveling 5,000 ft, how much higher is the road?

10. What is the angle of elevation of the sun when a 75-ft-tall smokestack casts a shadow 100 ft long?

# Summary

1. The three most important trigonometric functions are

$$\sin A = \frac{\text{side opposite angle } A}{\text{hypotenuse}}$$

$$\cos A = \frac{\text{side adjacent angle } A}{\text{hypotenuse}}$$

$$\tan A = \frac{\text{side opposite angle } A}{\text{side adjacent angle } A}$$

2. Equivalent degrees are found by dividing the angle greater than 360° by 360°. The remainder is the equivalent angle.

# Chapter 11 Self-Test

## Trigonometry

1. Find $\angle A$ when $\angle B = 100°$ and $\angle C = 47°$.
2. Find the area of a triangle with a base of 2.86 ft and an altitude of 3.19 ft.

Use Figure 11-17 to solve the following problems.

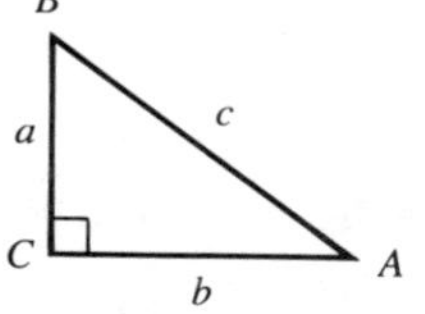


**Figure 11-17**

3. Find the hypotenuse when $a = 6.32$ ft and $b = 7.25$ ft.
4. Find side $a$ when the hypotenuse $= 7.78$ in. and side $b = 6.43$ in.
5. Find side $a$ when $b = 11$ ft and $\angle A = 41°$.
6. Find $\angle A$ when side $b = 18$ m and side $c = 27$ m.
7. Find angle $A$ in the diagram at the right.

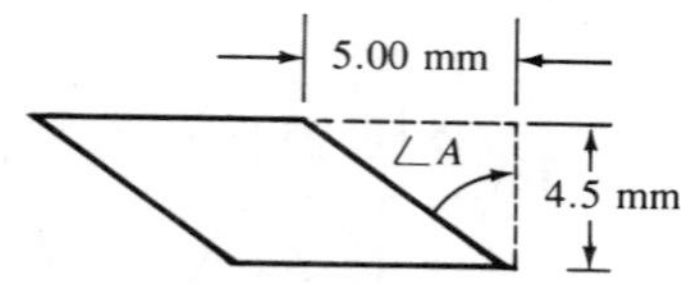


8. Determine the phase angle in an alternating current circuit when the inductive reactance is 54 ohms and the resistance is 42 ohms.

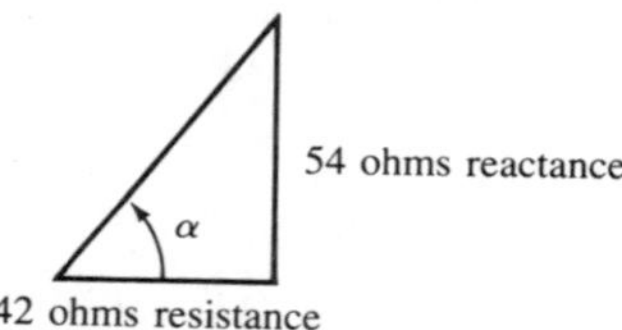


9. Find angle $H$ in the diagram below.

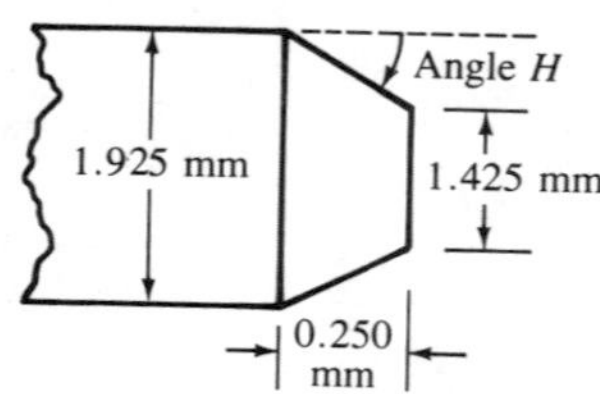


10. Find angle $K$ in the diagram below.

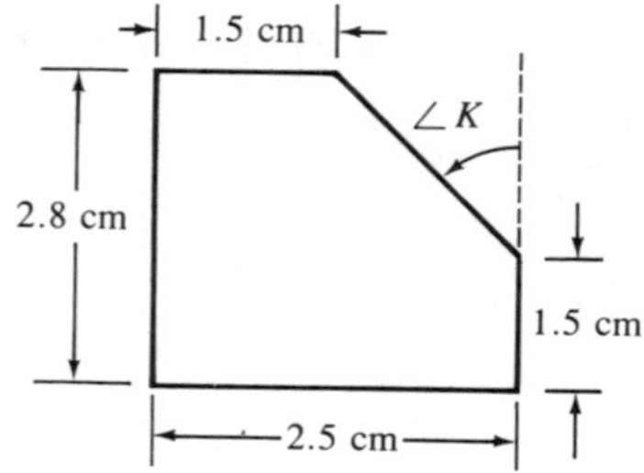

# APPENDIX

**Table 1** Units of the metric system

| *Prefix* | *Subdivision or multiple* |
|---|---|
| micro | 1,000,000 in a standard unit |
| milli | 1,000 in a standard unit |
| centi | 100 in a standard unit |
| deci | 10 in a standard unit |
| deka | 10 times the standard unit |
| hecto | 100 times the standard unit |
| kilo | 1,000 times the standard unit |
| mega | 1,000,000 times the standard unit |

**Table 2** Metric measures

| *Units of length* | *Subdivision or multiple* | *Units of volume* | *Subdivision or multiple* | *Units of weight* | *Subdivision or multiple* |
|---|---|---|---|---|---|
| 1 micrometer ($\mu$m) | 1,000,000 in a meter | 1 microliter ($\mu l$) | 1,000,000 in a liter | 1 microgram ($\mu$g) | 1,000,000 in a gram |
| 1 millimeter (mm) | 1,000 in a meter | 1 milliliter (m$l$) | 1,000 in a liter | 1 milligram (mg) | 1,000 in a gram |
| 1 centimeter (cm) | 100 in a meter | 1 centiliter (c$l$) | 100 in a liter | 1 centigram (cg) | 100 in a gram |
| 1 decimeter (dm) | 10 in a meter | 1 deciliter (d$l$) | 10 in a liter | 1 decigram (dg) | 10 in a gram |
| 1 meter (m) | standard | 1 liter ($l$) | standard | 1 gram (g) | most used unit |
| 1 dekameter (dkm) | 10 meters | 1 dekaliter (dk$l$) | 10 liters | 1 dekagram (dkg) | 10 grams |
| 1 hectometer (hm) | 100 meters | 1 hectoliter (h$l$) | 100 liters | 1 hectogram (hg) | 100 grams |
| 1 kilometer (km) | 1,000 meters | 1 kiloliter (K$l$) | 1,000 liters | 1 kilogram (kg) | 1,000 grams |
| 1 megameter (Mm) | 1,000,000 meters | 1 megaliter (M$l$) | 1,000,000 liters | 1 megagram (Mg) | 1,000,000 grams |

**Table 3** English measures

| *Units of length* | *Units of weight* | *Units of volume* |
|---|---|---|
| 1 foot (ft) = 12 inches (in.) | 1 pound (lb) = 16 ounces (oz) | 1 pint (pt) = 16 fluid ounces (fl oz) |
| 1 yard (yd) = 3 feet (ft) | 1 ton (tn) = 2,000 pounds (lb) | 1 quart (qt) = 2 pints (pt) |
| 1 mile (mi) = 5,280 feet (ft) | | 1 gallon (gal) = 4 quarts (qt) |
| 1 mile (mi) = 1,760 yards (yd) | | |

**Table 4** English–metric conversions

| *Units of length* | | *Units of volume* | | *Units of weight* | |
|---|---|---|---|---|---|
| | | *English* | *Metric* | *English* | *Metric* |
| 1 in. = 2.54 cm | 39.4 in. = 1 m | 1 gal | = 3.79 $l$ | 1 oz | = 28 g |
| 1 ft = 30.5 cm | 3.28 ft = 1 m | 1 qt | = 0.95 $l$ | 1 lb | = 454 g |
| 1 yd = 91.4 cm | 1.09 yd = 1 m | 0.26 gal | = 1 $l$ | 1 lb | = 0.45 kg |
| 1 mi = 1.6 km | 0.62 mi = 1 km | 1.05 qt | = 1 $l$ | 2.2 lb | = 1 kg |
| 0.39 in. = 1 cm | | | | | |

**Table 5** Common logarithms (mantissas of numbers)

| *N* | 0 | 1 | 2 | 3 | 4 | 5 | 6 | 7 | 8 | 9 |
|---|---|---|---|---|---|---|---|---|---|---|
| 10 | 0000 | 0043 | 0086 | 0128 | 0170 | 0212 | 0253 | 0294 | 0334 | 0374 |
| 11 | 0414 | 0453 | 0492 | 0531 | 0569 | 0607 | 0645 | 0682 | 0719 | 0755 |
| 12 | 0792 | 0828 | 0864 | 0899 | 0934 | 0969 | 1004 | 1038 | 1072 | 1106 |
| 13 | 1139 | 1173 | 1206 | 1239 | 1271 | 1303 | 1335 | 1367 | 1399 | 1430 |
| 14 | 1461 | 1492 | 1523 | 1553 | 1584 | 1614 | 1644 | 1673 | 1703 | 1732 |
| 15 | 1761 | 1790 | 1818 | 1847 | 1875 | 1903 | 1931 | 1959 | 1987 | 2014 |
| 16 | 2041 | 2068 | 2095 | 2122 | 2148 | 2175 | 2201 | 2227 | 2253 | 2279 |
| 17 | 2304 | 2330 | 2355 | 2380 | 2405 | 2430 | 2455 | 2480 | 2504 | 2529 |
| 18 | 2553 | 2577 | 2601 | 2625 | 2648 | 2672 | 2695 | 2718 | 2742 | 2765 |
| 19 | 2788 | 2810 | 2833 | 2856 | 2878 | 2900 | 2923 | 2945 | 2967 | 2989 |
| 20 | 3010 | 3032 | 3054 | 3075 | 3096 | 3118 | 3139 | 3160 | 3181 | 3201 |
| 21 | 3222 | 3243 | 3263 | 3284 | 3304 | 3324 | 3345 | 3365 | 3385 | 3404 |
| 22 | 3424 | 3444 | 3464 | 3483 | 3502 | 3522 | 3541 | 3560 | 3579 | 3598 |
| 23 | 3617 | 3636 | 3655 | 3674 | 3692 | 3711 | 3729 | 3747 | 3766 | 3784 |
| 24 | 3802 | 3820 | 3838 | 3856 | 3874 | 3892 | 3909 | 3927 | 3945 | 3962 |
| 25 | 3979 | 3997 | 4014 | 4031 | 4048 | 4065 | 4082 | 4099 | 4116 | 4133 |
| 26 | 4150 | 4166 | 4183 | 4200 | 4216 | 4232 | 4249 | 4265 | 4281 | 4298 |
| 27 | 4314 | 4330 | 4346 | 4362 | 4378 | 4393 | 4409 | 4425 | 4440 | 4456 |
| 28 | 4472 | 4487 | 4502 | 4518 | 4533 | 4548 | 4564 | 4579 | 4594 | 4609 |
| 29 | 4624 | 4639 | 4654 | 4669 | 4683 | 4698 | 4713 | 4728 | 4742 | 4757 |
| 30 | 4771 | 4786 | 4800 | 4814 | 4829 | 4843 | 4857 | 4871 | 4886 | 4900 |
| 31 | 4914 | 4928 | 4942 | 4955 | 4969 | 4983 | 4997 | 5011 | 5024 | 5038 |
| 32 | 5051 | 5065 | 5079 | 5092 | 5105 | 5119 | 5132 | 5145 | 5159 | 5172 |
| 33 | 5185 | 5198 | 5211 | 5224 | 5237 | 5250 | 5263 | 5276 | 5289 | 5302 |
| 34 | 5315 | 5328 | 5340 | 5353 | 5366 | 5378 | 5391 | 5403 | 5416 | 5428 |
| 35 | 5441 | 5453 | 5465 | 5478 | 5490 | 5502 | 5514 | 5527 | 5539 | 5551 |
| 36 | 5563 | 5575 | 5587 | 5599 | 5611 | 5623 | 5635 | 5647 | 5658 | 5670 |
| 37 | 5682 | 5694 | 5705 | 5717 | 5729 | 5740 | 5752 | 5763 | 5775 | 5786 |
| 38 | 5798 | 5809 | 5821 | 5832 | 5843 | 5855 | 5866 | 5877 | 5888 | 5899 |
| 39 | 5911 | 5922 | 5933 | 5944 | 5955 | 5966 | 5977 | 5988 | 5999 | 6010 |
| 40 | 6021 | 6031 | 6042 | 6053 | 6064 | 6075 | 6085 | 6096 | 6107 | 6117 |
| 41 | 6128 | 6138 | 6149 | 6160 | 6170 | 6180 | 6191 | 6201 | 6212 | 6222 |
| 42 | 6232 | 6243 | 6253 | 6263 | 6274 | 6284 | 6294 | 6304 | 6314 | 6325 |
| 43 | 6335 | 6345 | 6355 | 6365 | 6375 | 6385 | 6395 | 6405 | 6415 | 6425 |
| 44 | 6435 | 6444 | 6454 | 6464 | 6474 | 6484 | 6493 | 6503 | 6513 | 6522 |
| 45 | 6532 | 6542 | 6551 | 6561 | 6571 | 6580 | 6590 | 6599 | 6609 | 6618 |
| 46 | 6628 | 6637 | 6646 | 6656 | 6665 | 6675 | 6684 | 6693 | 6702 | 6712 |
| 47 | 6721 | 6730 | 6739 | 6749 | 6758 | 6767 | 6776 | 6785 | 6794 | 6803 |
| 48 | 6812 | 6821 | 6830 | 6839 | 6848 | 6857 | 6866 | 6875 | 6884 | 6893 |
| 49 | 6902 | 6911 | 6920 | 6928 | 6937 | 6946 | 6955 | 6964 | 6972 | 6981 |
| 50 | 6990 | 6998 | 7007 | 7016 | 7024 | 7033 | 7042 | 7050 | 7059 | 7067 |
| 51 | 7076 | 7084 | 7093 | 7101 | 7110 | 7118 | 7126 | 7135 | 7143 | 7152 |
| 52 | 7160 | 7168 | 7177 | 7185 | 7193 | 7202 | 7210 | 7218 | 7226 | 7235 |
| 53 | 7243 | 7251 | 7259 | 7267 | 7275 | 7284 | 7292 | 7300 | 7308 | 7316 |
| 54 | 7324 | 7332 | 7340 | 7348 | 7356 | 7364 | 7372 | 7380 | 7388 | 7396 |
| *N* | 0 | 1 | 2 | 3 | 4 | 5 | 6 | 7 | 8 | 9 |

**Table 5** continued

| *N* | 0 | 1 | 2 | 3 | 4 | 5 | 6 | 7 | 8 | 9 |
|---|---|---|---|---|---|---|---|---|---|---|
| 55 | 7404 | 7412 | 7419 | 7427 | 7435 | 7443 | 7451 | 7459 | 7466 | 7474 |
| 56 | 7482 | 7490 | 7497 | 7505 | 7513 | 7520 | 7528 | 7536 | 7543 | 7551 |
| 57 | 7559 | 7566 | 7574 | 7582 | 7589 | 7597 | 7604 | 7612 | 7619 | 7627 |
| 58 | 7634 | 7642 | 7649 | 7657 | 7664 | 7672 | 7679 | 7686 | 7694 | 7701 |
| 59 | 7709 | 7716 | 7723 | 7731 | 7738 | 7745 | 7752 | 7760 | 7767 | 7774 |
| 60 | 7782 | 7789 | 7796 | 7803 | 7810 | 7818 | 7825 | 7832 | 7839 | 7846 |
| 61 | 7853 | 7860 | 7868 | 7875 | 7882 | 7889 | 7896 | 7903 | 7910 | 7917 |
| 62 | 7924 | 7931 | 7938 | 7945 | 7952 | 7959 | 7966 | 7973 | 7980 | 7987 |
| 63 | 7993 | 8000 | 8007 | 8014 | 8021 | 8028 | 8035 | 8041 | 8048 | 8055 |
| 64 | 8062 | 8069 | 8075 | 8082 | 8089 | 8096 | 8102 | 8109 | 8116 | 8122 |
| 65 | 8129 | 8136 | 8142 | 8149 | 8156 | 8162 | 8169 | 8176 | 8182 | 8189 |
| 66 | 8195 | 8202 | 8209 | 8215 | 8222 | 8228 | 8235 | 8241 | 8248 | 8254 |
| 67 | 8261 | 8267 | 8274 | 8280 | 8287 | 8293 | 8299 | 8306 | 8312 | 8319 |
| 68 | 8325 | 8331 | 8338 | 8344 | 8351 | 8357 | 8363 | 8370 | 8376 | 8382 |
| 69 | 8388 | 8395 | 8401 | 8407 | 8414 | 8420 | 8426 | 8432 | 8439 | 8445 |
| 70 | 8451 | 8457 | 8463 | 8470 | 8476 | 8482 | 8488 | 8494 | 8500 | 8506 |
| 71 | 8513 | 8519 | 8525 | 8531 | 8537 | 8543 | 8549 | 8555 | 8561 | 8567 |
| 72 | 8573 | 8579 | 8585 | 8591 | 8597 | 8603 | 8609 | 8615 | 8621 | 8627 |
| 73 | 8633 | 8639 | 8645 | 8651 | 8657 | 8663 | 8669 | 8675 | 8681 | 8686 |
| 74 | 8692 | 8698 | 8704 | 8710 | 8716 | 8722 | 8727 | 8733 | 8739 | 8745 |
| 75 | 8751 | 8756 | 8762 | 8768 | 8774 | 8779 | 8785 | 8791 | 8797 | 8802 |
| 76 | 8808 | 8814 | 8820 | 8825 | 8831 | 8837 | 8842 | 8848 | 8854 | 8859 |
| 77 | 8865 | 8871 | 8876 | 8882 | 8887 | 8893 | 8899 | 8904 | 8910 | 8915 |
| 78 | 8921 | 8927 | 8932 | 8938 | 8943 | 8949 | 8954 | 8960 | 8965 | 8971 |
| 79 | 8976 | 8982 | 8987 | 8993 | 8998 | 9004 | 9009 | 9015 | 9020 | 9025 |
| 80 | 9031 | 9036 | 9042 | 9047 | 9053 | 9058 | 9063 | 9069 | 9074 | 9079 |
| 81 | 9085 | 9090 | 9096 | 9101 | 9106 | 9112 | 9117 | 9122 | 9128 | 9133 |
| 82 | 9138 | 9143 | 9149 | 9154 | 9159 | 9165 | 9170 | 9175 | 9180 | 9186 |
| 83 | 9191 | 9196 | 9201 | 9206 | 9212 | 9217 | 9222 | 9227 | 9232 | 9238 |
| 84 | 9243 | 9248 | 9253 | 9258 | 9263 | 9269 | 9274 | 9279 | 9284 | 9289 |
| 85 | 9294 | 9299 | 9304 | 9309 | 9315 | 9320 | 9325 | 9330 | 9335 | 9340 |
| 86 | 9345 | 9350 | 9355 | 9360 | 9365 | 9370 | 9375 | 9380 | 9385 | 9390 |
| 87 | 9395 | 9400 | 9405 | 9410 | 9415 | 9420 | 9425 | 9430 | 9435 | 9440 |
| 88 | 9445 | 9450 | 9455 | 9460 | 9465 | 9469 | 9474 | 9479 | 9484 | 9489 |
| 89 | 9494 | 9499 | 9504 | 9509 | 9513 | 9518 | 9523 | 9528 | 9533 | 9538 |
| 90 | 9542 | 9547 | 9552 | 9557 | 9562 | 9566 | 9571 | 9576 | 9581 | 9586 |
| 91 | 9590 | 9595 | 9600 | 9605 | 9609 | 9614 | 9619 | 9624 | 9628 | 9633 |
| 92 | 9638 | 9643 | 9647 | 9652 | 9657 | 9661 | 9666 | 9671 | 9675 | 9680 |
| 93 | 9685 | 9689 | 9694 | 9699 | 9703 | 9708 | 9713 | 9717 | 9722 | 9727 |
| 94 | 9731 | 9736 | 9741 | 9745 | 9750 | 9754 | 9759 | 9763 | 9768 | 9773 |
| 95 | 9777 | 9782 | 9786 | 9791 | 9795 | 9800 | 9805 | 9809 | 9814 | 9818 |
| 96 | 9823 | 9827 | 9832 | 9836 | 9841 | 9845 | 9850 | 9854 | 9859 | 9863 |
| 97 | 9868 | 9872 | 9877 | 9881 | 9886 | 9890 | 9894 | 9899 | 9903 | 9908 |
| 98 | 9912 | 9917 | 9921 | 9926 | 9930 | 9934 | 9939 | 9943 | 9948 | 9952 |
| 99 | 9956 | 9961 | 9965 | 9969 | 9974 | 9978 | 9983 | 9987 | 9991 | 9996 |
| *N* | 0 | 1 | 2 | 3 | 4 | 5 | 6 | 7 | 8 | 9 |

**Table 6** Four-place values of trigonometric functions
To find the value of a trigonometric function of an angle between 0° and 45°, read down the left side of the table and across the column heads at the top of the table. To find the value of a trigonometric function of an angle between 45° and 90°, read up the right side of the table and across the column heads at the bottom of the table.

| *angle* | *sin* | *cos* | *tan* | *cot* | |
|---|---|---|---|---|---|
| **0°** | .0000 | 1.0000 | .0000 | ∞ | **90°** |
| **1** | .0175 | .9998 | .0175 | 57.2900 | **89** |
| **2** | .0349 | .9994 | .0349 | 28.6363 | **88** |
| **3** | .0523 | .9986 | .0524 | 19.0811 | **87** |
| **4** | .0698 | .9976 | .0699 | 14.3007 | **86** |
| **5°** | .0872 | .9962 | .0875 | 11.4301 | **85°** |
| **6** | .1045 | .9945 | .1051 | 9.5144 | **84** |
| **7** | .1219 | .9925 | .1228 | 8.1443 | **83** |
| **8** | .1392 | .9903 | .1405 | 7.1154 | **82** |
| **9** | .1564 | .9877 | .1584 | 6.3138 | **81** |
| **10°** | .1736 | .9848 | .1763 | 5.6713 | **80°** |
| **11** | .1908 | .9816 | .1944 | 5.1446 | **79** |
| **12** | .2079 | .9781 | .2126 | 4.7046 | **78** |
| **13** | .2250 | .9744 | .2309 | 4.3315 | **77** |
| **14** | .2419 | .9703 | .2493 | 4.0108 | **76** |
| **15°** | .2588 | .9659 | .2679 | 3.7321 | **75°** |
| **16** | .2756 | .9613 | .2867 | 3.4874 | **74** |
| **17** | .2924 | .9563 | .3057 | 3.2709 | **73** |
| **18** | .3090 | .9511 | .3249 | 3.0777 | **72** |
| **19** | .3256 | .9455 | .3443 | 2.9042 | **71** |
| **20°** | .3420 | .9397 | .3640 | 2.7475 | **70°** |
| **21** | .3584 | .9336 | .3839 | 2.6051 | **69** |
| **22** | .3746 | .9272 | .4040 | 2.4751 | **68** |
| **23** | .3907 | .9205 | .4245 | 2.3559 | **67** |
| **24** | .4067 | .9135 | .4452 | 2.2460 | **66** |
| **25°** | .4226 | .9063 | .4663 | 2.1445 | **65°** |
| **26** | .4384 | .8988 | .4877 | 2.0503 | **64** |
| **27** | .4540 | .8910 | .5095 | 1.9626 | **63** |
| **28** | .4695 | .8829 | .5317 | 1.8807 | **62** |
| **29** | .4848 | .8746 | .5543 | 1.8040 | **61** |
| **30°** | .5000 | .8660 | .5774 | 1.7321 | **60°** |
| **31** | .5150 | .8572 | .6009 | 1.6643 | **59** |
| **32** | .5299 | .8480 | .6249 | 1.6003 | **58** |
| **33** | .5446 | .8387 | .6494 | 1.5399 | **57** |
| **34** | .5592 | .8290 | .6745 | 1.4826 | **56** |
| **35°** | .5736 | .8192 | .7002 | 1.4281 | **55°** |
| **36** | .5878 | .8090 | .7265 | 1.3764 | **54** |
| **37** | .6018 | .7986 | .7536 | 1.3270 | **53** |
| **38** | .6157 | .7880 | .7813 | 1.2799 | **52** |
| **39** | .6293 | .7771 | .8098 | 1.2349 | **51** |
| **40°** | .6428 | .7660 | .8391 | 1.1918 | **50°** |
| **41** | .6561 | .7547 | .8693 | 1.1504 | **49** |
| **42** | .6691 | .7431 | .9004 | 1.1106 | **48** |
| **43** | .6820 | .7314 | .9325 | 1.0724 | **47** |
| **44** | .6947 | .7193 | .9657 | 1.0355 | **46** |
| **45°** | .7071 | .7071 | 1.0000 | 1.0000 | **45°** |
| | *cos* | *sin* | *cot* | *tan* | *angle* |

**Table 7** Inch to millimeter conversions

| *Inches to millimeters* | | *Millimeters to inches* | |
|---|---|---|---|
| *in.* | *mm* | *mm* | *in.* |
| 0.001 | 0.025 | 1 | 0.039 |
| 0.002 | 0.051 | 2 | 0.079 |
| 0.003 | 0.076 | 3 | 0.118 |
| 0.004 | 0.102 | 4 | 0.158 |
| 0.005 | 0.127 | 5 | 0.197 |
| 0.006 | 0.152 | 6 | 0.236 |
| 0.007 | 0.178 | 7 | 0.276 |
| 0.008 | 0.203 | 8 | 0.315 |
| 0.009 | 0.229 | 10 | 0.394 |
| 0.010 | 0.254 | 12 | 0.472 |
| 0.020 | 0.508 | 16 | 0.630 |
| 0.030 | 0.762 | 20 | 0.787 |
| 0.040 | 1.016 | 25 | 0.984 |
| 0.050 | 1.270 | 30 | 1.181 |
| 0.060 | 1.524 | 35 | 1.378 |
| 0.070 | 1.778 | 40 | 1.575 |
| 0.080 | 2.032 | 45 | 1.772 |
| 0.090 | 2.286 | 50 | 1.968 |
| 0.100 | 2.540 | 55 | 2.165 |
| 0.200 | 5.080 | 60 | 2.362 |
| 0.300 | 7.620 | 65 | 2.559 |
| 0.400 | 10.160 | 70 | 2.756 |
| 0.500 | 12.700 | 75 | 2.953 |
| 0.600 | 15.240 | 80 | 3.150 |
| 0.700 | 17.780 | 85 | 3.346 |
| 0.800 | 20.320 | 90 | 3.543 |
| 0.900 | 22.860 | 95 | 3.740 |
| 1.000 | 25.400 | 100 | 3.937 |

**Table 8** Fractional inch to millimeter conversions

| *in.* | *mm* | *in.* | *mm* | *in.* | *mm* | *in.* | *mm* |
|---|---|---|---|---|---|---|---|
| 1/64 | 0.397 | 17/64 | 6.747 | 33/64 | 13.097 | 49/64 | 19.447 |
| 1/32 | 0.794 | 9/32 | 7.144 | 17/32 | 13.494 | 25/32 | 19.844 |
| 3/64 | 1.191 | 19/64 | 7.541 | 35/64 | 13.890 | 51/64 | 20.240 |
| 1/16 | 1.587 | 5/16 | 7.937 | 9/16 | 14.287 | 13/16 | 20.637 |
| 5/64 | 1.984 | 21/64 | 8.334 | 37/64 | 14.684 | 53/64 | 21.034 |
| 3/32 | 2.381 | 11/32 | 8.731 | 19/32 | 15.081 | 27/32 | 21.431 |
| 7/64 | 2.778 | 23/64 | 9.128 | 39/64 | 15.478 | 55/64 | 21.828 |
| 1/8 | 3.175 | 3/8 | 9.525 | 5/8 | 15.875 | 7/8 | 22.225 |
| 9/64 | 3.572 | 25/64 | 9.922 | 41/64 | 16.272 | 57/64 | 22.622 |
| 5/32 | 3.969 | 13/32 | 10.319 | 21/32 | 16.669 | 29/32 | 23.019 |
| 11/64 | 4.366 | 27/64 | 10.716 | 43/64 | 17.065 | 59/64 | 23.415 |
| 3/16 | 4.762 | 7/16 | 11.113 | 11/16 | 17.462 | 15/16 | 23.812 |
| 13/64 | 5.159 | 29/64 | 11.509 | 45/64 | 17.859 | 61/64 | 24.209 |
| 7/32 | 5.556 | 15/32 | 11.906 | 23/32 | 18.256 | 31/32 | 24.606 |
| 15/64 | 5.953 | 31/64 | 12.303 | 47/64 | 18.653 | 63/64 | 25.003 |
| 1/4 | 6.350 | 1/2 | 12.700 | 3/4 | 19.050 | 1 | 25.400 |

# ANSWERS TO EXERCISE SETS

## Exercise Set, Sections 1-1–1-2

**1.** hundreds **2.** hundred thousands **3.** billions **4.** tenths **5.** thousandths **6.** ones **7.** 4,025,012 **8.** 6,083,002,065 **9.** 7,021 **10.** 602 **11.** 4,003 **12.** 15,037 **13.** 0.9 **14.** 0.0308 **15.** 7.021 **16.** 0.37 **17.** 0.0016 **18.** 4.09 **19.** 12.9 **20.** 507.0112

## Exercise Set, Sections 1-3–1-5

**1.** 43 **2.** 77 **3.** 7,852 **4.** 338 **5.** 55,886 **6.** 5.25 **7.** 43.79 **8.** 97.16 **9.** 15.5 **10.** 32.852 **11.** 33,193 **12.** 21.99 **13.** 196.187 **14.** 105.37 **15.** 67.07 **16.** 41.69 **17.** 1.43 **18.** 8.45 **19.** 38.04 **20.** 11 **21.** 183 **22.** 44 **23.** 8,489 **24.** 25 **25.** 93.55 **26.** 2.1 **27.** 89.9 **28.** 76.13 **29.** 3.01 **30.** 16.02 **31.** 58.41 **32.** 52.013 **33.** 8.79 **34.** 0.707 **35.** 1.001 **36.** 0.96

## Exercise Set, Section 1-6

**1.** 159 **2.** 288 **3.** 520 **4.** 679 **5.** 1.86 **6.** 1.68 **7.** 1.8 **8.** 3.04 **9.** 819 **10.** 5,850 **11.** 16.45 **12.** 0.4365 **13.** 4,230 **14.** 40,588 **15.** 2.6524 **16.** 415.294 **17.** 644 **18.** 558 **19.** 27.6 **20.** 15.6 **21.** 3.32 **22.** 8.73 **23.** 2.3 **24.** 0.91 **25.** 1.84 **26.** 7.52 **27.** 12,662 **28.** 52,688 **29.** 15.9159 **30.** 6.2376 **31.** 67,000 **32.** 470 **33.** 4,830 **34.** 732,000 **35.** 52 **36.** 2,780 **37.** 14 **38.** 4,121 **39.** 18.3 **40.** 1,010

## Exercise Set, Section 1-7

**1.** 51 **2.** 22.08 **3.** 27.22 **4.** 52 **5.** 13.33 **6.** 9 **7.** 7 **8.** 29.1 **9.** 27 **10.** 12.71 **11.** 59.11 **12.** 9 **13.** 0.56 **14.** 5.24 **15.** 1.01 **16.** 9.9 **17.** 0.74 **18.** 7.06 **19.** 7 **20.** 13.25 **21.** 24.1 **22.** 290.83 **23.** 62.88 **24.** 15.45 **25.** 7.18 **26.** 11.74 **27.** 18.13 **28.** 105.21 **29.** 169.89 **30.** 91.58 **31.** 12.25 **32.** 3.22 **33.** 1.78 **34.** 25.5 **35.** 0.0372 **36.** 0.045 **37.** 700 **38.** 0.0956 **39.** 0.01784

## Exercise Set, Sections 1-8–1-10

**1.** hundredths **2.** tenths **3.** thousandths **4.** hundredths **5.** 5.9 **6.** 3.61 **7.** 31.85 **8.** two **9.** three **10.** four **11.** three **12.** three **13.** five **14.** two **15.** three **16.** 11.4 **17.** 115 **18.** 2.83 **19.** 6.6 **20.** 64.2 **21.** 2,288 **22.** 10 m*ℓ* 23. 3 *ℓ* **24.** 20 *ℓ* **25.** 2,000 mg **26.** 4,000 g **27.** 50 *K* **28.** 321.67 grams **29.** 0.13 inches **30.** 252 miles **31.** 1.380 inches **32.** 12.90 gallons **33.** 5.9 volts **34.** $48.38 **35.** $0.18 **36.** 1.31 feet **37.** 200 feet **38.** 114 **39.** 102.5 inches **40. a.** 667; 568; 757; 667 **b.** 508; 521; 550; 543; 537 **c.** 2659

## Exercise Set, Section 2-1

**1.** 0.44 **2.** 0.72 **3.** 0.45 **4.** 0.67 **5.** 0.40 **6.** 0.33 **7.** 0.75 **8.** 0.94 **9.** 16/32 **10.** 24/32 **11.** 28/32 **12.** 6/32 **13.** 12/48 **14.** 36/48 **15.** 3/48 **16.** 10/48 **17.** 7/9 **18.** 1/2 **19.** 1/3 **20.** 1/8 **21.** 3/5 **22.** 7/8 **23.** 1/9 **24.** 1/3 **25.** 1 4/5 **26.** 2 1/3 **27.** 6 1/4 **28.** 5 **29.** 3 1/6 **30.** 5 3/7 **31.** 6 3/5 **32.** 8 1/3 **33.** 5/2 **34.** 9/8 **35.** 65/3 **36.** 119/8 **37.** 11/6 **38.** 16/7 **39.** 38/3 **40.** 205/9

## Exercise Set, Section 2-2

**1.** 3/32 **2.** 7/40 **3.** 1/2 **4.** 1/11 **5.** 1/18 **6.** 1/3 **7.** 1/12 **8.** 1/16 **9.** 1/9 **10.** 3/16 **11.** 1/9 **12.** 1/12 **13.** 1/15 **14.** 1/6 **15.** 1/18 **16.** 21/8 **17.** 45/8 **18.** 10/3 **19.** 1 **20.** 6 **21.** 12 **22.** 3/4 **23.** 12/7 **24.** 9/2 **25.** 25/64 **26.** 1 **27.** 1/8 **28.** 5/2 **29.** 1/4 **30.** 7/6 **31.** 1/6 **32.** 1/8 **33.** 5/24 **34.** 1/32 **35.** 20 **36.** 45 **37.** 50 **38.** 36 **39.** 384

## Exercise Set, Section 2-3

**1.** 4/7 **2.** 11/16 **3.** 7/15 **4.** 11/12 **5.** 4/9 **6.** 11/14 **7.** 3/4 **8.** 1/2 **9.** 1/2 **10.** 7/8 **11.** 7/9 **12.** 1/2 **13.** 8/21 **14.** 7/12 **15.** 4/7 **16.** 1/8 **17.** 1/3 **18.** 1/8 **19.** 1/4 **20.** 1/2 **21.** 1/2 **22.** 1/8 **23.** 7/24 **24.** 3/8 **25.** 27/32 **26.** 5/8 **27.** 0.29 **28.** 0.56 **29.** 0.77 **30.** 0.38

## Exercise Set, Section 2-4

**1.** 72 **2.** 24 **3.** 8 **4.** 6 **5.** 35 **6.** 18 **7.** 14/15 **8.** 5/8 **9.** 11/18 **10.** 31/45 **11.** 1/16 **12.** 23/16 **13.** (3)(5)(5) **14.** (2)(2)(7) **15.** (3)(31) **16.** (1)(29) **17.** (5)(5)(5) **18.** (3)(2)(2)(2)(5)(5)(5) **19.** (3)(3)(5) **20.** (3)(3)(3)(3)

## Exercise Set, Section 2-5

**1.** 15 7/8 in. **2.** 4 11/16 ft **3.** 1 1/12 oz **4.** 4 3/16 ft **5.** 1.08 oz **6.** 12 units **7.** 15 grams **8.** 29 1/16 in. **9.** 3/16 in. **10.** 3/8 in. **11.** 6 1/8 $ft^2$ **12.** 187 1/2 ft **13.** 10 1/2 ft **14.** 37 1/2 in. **15.** 10 1/8 in.

## Exercise Set, Sections 3-1–3-2

**1.** 8 **2.** 2 **3.** 7 **4.** −6 **5.** 15 **6.** 8 **7.** 2 **8.** 74 **9.** 7 **10.** 18 **11.** 14 **12.** 30 **13.** −17 **14.** −17 **15.** −35 **16.** −52 **17.** 4 **18.** 0 **19.** −10 **20.** −1 **21.** 4 **22.** 4 **23.** −11 **24.** −3 **25.** −9 **26.** −16 **27.** −12 **28.** −9 **29.** −6 **30.** 2 **31.** −15 **32.** 4 **33.** 0 **34.** 0 **35.** −9 **36.** 66 **37.** −62 **38.** −14 **39.** 1 **40.** −2

## Exercise Set, Sections 3-3–3-4

**1.** −27 **2.** −80 **3.** 42 **4.** 81 **5.** 72 **6.** −100 **7.** 28 **8.** −36 **9.** 120 **10.** 6 **11.** −12 **12.** −24 **13.** 31 **14.** 7 **15.** −3 **16.** −6 **17.** 21 **18.** −6 **19.** −12 **20.** 4 **21.** −20 **22.** 120 **23.** −82 **24.** 50 **25.** −84 **26.** 4,050 **27.** 74,980 **28.** −71 **29.** −8 **30.** 7 **31.** −104 **32.** 133 **33.** 36 **34.** −140 **35.** −5 **36.** −33 **37.** 42 **38.** −7.5 **39.** −5.5 **40.** 17

## Exercise Set, Sections 3-5–3-6

**1.** 2 **2.** 5 **3.** 3 **4.** 10 **5.** 5 **6.** 5 **7.** 5 **8.** 17 **9.** 7 **10.** 12 **11.** 7 **12.** 8 **13.** 5/3 **14.** 6 **15.** 10 **16.** $5x + 19$ **17.** $11x + 35$ **18.** $3x + 18$ **19.** $9x + 15$ **20.** $6x + 16$ **21.** $2x + 8$ **22.** $5x + 23$ **23.** $16h + 6$ **24.** $20x + 30$ **25.** $21x + 9$ **26.** $10x - 36$ **27.** $-4x + 8$ **28.** $-x + 1$ **29.** $9x - 6$ **30.** $10x - 21$ **31.** $2x + 21$

## Exercise Set, Sections 3-7–3-9

**1.** 13 **2.** −2 **3.** 18 **4.** −5 **5.** 18 **6.** 1 **7.** −11 **8.** −4 **9.** 3 **10.** 5 **11.** 7 **12.** 6 **13.** 6 **14.** −10 **15.** 0 **16.** 7 **17.** 18 **18.** 3 **19.** −27 **20.** 27 **21.** 1 **22.** −10 **23.** 4.2 **24.** 5.2 **25.** 5 **26.** −2 **27.** 5 **28.** −7 **29.** 5 **30.** 2 **31.** −3 **32.** 4

## Exercise Set, Section 3-10

**1.** $TW = W_1 + 4W_2$ **2.** $V = L \times W \times H$ **3.** $A = 1/4T$ **4.** $T = P_1 + 1.2P_1$ **5.** $C_1 = 2/3T$ **6.** $A = (G_1 + G_2 + G_3)/3$ **7.** km = 1.6 mi **8.** Total = $4C + 4tax$ **9.** $V_a = V_1 + V_2 + V_3$ **10.** $A = lw$; $l = 3w$; $A = (3w)(w)$ or $3w^2$

## Exercise Set, Section 4-1

**1.** 15 **2.** 22 **3.** −32 **4.** 12 **5.** 4.68 **6.** −4.16 **7.** 15.12 **8.** 25 **9.** 15 **10.** 14 **11.** 8 **12.** 9 **13.** 2.5 **14.** 1.43 **15.** 6 **16.** 3.18 **17.** 4.57 **18.** 2.92 **19.** 1.11 **20.** 2.8 **21.** 0.5 **22.** 2 **23.** 0.89 **24.** 1 **25.** 18 **26.** 0.04 **27.** 26 **28.** 229 **29.** 0.05 **30.** 81.3 **31.** 51 **32.** 1.76

## Exercise Set, Sections 4-2–4-3

**1.** 313 K **2.** 68°F **3.** 76.1°F **4.** 4.1°F **5.** 5.2 **6.** −2.7 **7.** 20°C **8.** −10°C **9.** 10°C **10.** −30°C **11.** 80 **12.** 5 **13.** 180 **14.** 150 **15.** 833 **16.** 18.8 **17.** 5.0 **18.** 43.7 **19.** 648 **20.** 15.9 **21.** 9.5 **22.** 1329 **23.** 73.9 **24.** 15.0 **25.** 356 **26.** 102 **27.** 35.00 **28.** 0.43 **29.** 13 **30.** 5.1

## Exercise Set, Section 4-4

**1.** $P_1 = P_2V_2/V_1$ **2.** $V_2 = P_1V_1/P_2$ **3.** $V_2 = T_2V_1/T_1$ **4.** $T_1 = T_2V_1/V_2$ **5.** $P_2 = P_1V_1T_2/T_1V_2$ **6.** $T_1 = T_2P_1V_1/P_2V_2$ **7.** $n = PV/RT$ **8.** $T = PV/nR$ **9.** $x = (Y - b)/m$ **10.** $h = (m + r)/k$ **11.** $f = (h + k)/(b - a)$ **12.** $x = (z - y)/(h - k)$ **13.** $h = kb/47$ **14.** $R = nw/k$ **15.** $x = sz + m$

# Exercise Set, Section 4-5

**1.** $R = P/I^2$ **2.** $T_2 = nT_1 + T_1$ **3.** $g = Wv^2/F_cr$ **4.** $m = KT/4\pi^2$ **5.** $h = A - \pi r^2/2\pi r$ **6.** $K = Fd_2/Q_1Q_2$ **7.** $G = 550Pt/W$ **8.** $\lambda = v/f$ **9.** $i = 2\pi Bd/m$ **10.** $W = P - 2L/2$

# Exercise Set, Section 4-6

**1.** 300 ml/600 K = $V_2$/400 K; $V_2$ = 200 ml **2.** 35 l/245 K = $V_2$/300 K; $V_2$ = 42.9 l **3.** 500 ml/273 K = $V_2$/173 K; $V_2$ = 317 K **4.** 10 l/273 K = $V_2$/546 K; $V_2$ = 20 l **5.** 600 ml/900 K = $V_2$/600 K; $V_2$ = 400 ml **6.** 3.50 l/500 K = $V_2$/750 K; $V_2$ = 5.25 l **7.** $P$ = 625 lb/15.5 in.$^2$; $P$ = 21.0 psi **8.** $A$ = 15.0 in. × 18.0 in. = 270 in.$^2$; $P$ = (2500 lb/in.$^2$)(270 in.$^2$); $P$ = 338 tons **9.** $\Delta L$ = (0.12 in.)(220 A)(40 in.)/(100,000 A)(0.50 in.); $\Delta L$ = 0.021 in. **10.** $P$ = 150 lb/3.0 in.$^2$ = 50 psi

# Exercise Set, Sections 5-1–5-2

**1.** 1:5 **2.** 2:11 **3.** 2:9 **4.** 15:23 **5.** 1:9 **6.** 5:1 **7.** 6:1 **8.** 8:3 **9.** 1:3 **10.** 1:10 **11.** 6:1 **12.** 3.17:1 **13.** 0.44:1 **14.** 0.75:1 **15.** 0.29:1 **16.** $3/x = 6/7$ **17.** $5/8 = N/10$ **18.** $K/8 = 1/4$ **19.** $7/9 = 14/N$ **20.** $2/3 = x/9$ **21.** $7/N = 14/28$ **22.** 3 **23.** 5 **24.** 9 **25.** 15 **26.** 3 **27.** 16 **28.** 1.25 **29.** 200 **30.** 100 **31.** 30 **32.** 4.5 **33.** 0.2 **34.** 9 **35.** 3.2 **36.** 14 oz **37.** 39 g **38.** 3.4 **39.** 3.4 g **40.** 13.4 g

# Exercise Set, Sections 5-3–5-4

**1.** 0.375 **2.** 3:1 **3.** 10.1:1 **4.** 3:1 **5.** 0.03/1 = $x$/2000; $x$ = 60 hp **6.** 20/50 = $x$/745; $x$ = 298 ft **7.** 3/8 = $x$/24; $x$ = 9 ft **8.** 650/800 = 310/$T_2$; $T_2$ = 382 K **9.** 6/1 = $x$/14; $x$ = 84 capsules **10.** 200 ml/$v$ = 300 K/400 K; $v$ = 267 ml **11.** 1/0.25 = $x$/6; $x$ = 24 **12.** 1/6.50 = 144/$x$; $x$ = $936 **13.** 0.072 **14.** 3.48 **15.** 5.0 **16.** 0.10 **17.** 0.408 **18.** 0.45 **19.** 0.042 **20.** 0.076 **21.** $V_2$ = (5.00)(250)/7.00; $V_2$ = 179 ml **22.** 16/8 = $R$/20; $R$ = 40 rev **23.** 4/2 = $x$/17; $x$ = 34 houses **24.** 0.25/1 = $x$/8; $x$ = 2 in. **25.** 25.5 gal/510 mi = $x$/765 mi; $x$ = 38.3 gal

# Exercise Set, Section 6-1

**1.** 1/4 **2.** 51/100 **3.** 1/3 **4.** 7/10 **5.** 5/8 **6.** 2/5 **7.** 41/50 **8.** 17/200 **9.** 3 **10.** 2 1/2 **11.** 1 3/4 **12.** 15 **13.** 6 2/3 **14.** 1 2/5 **15.** 12 1/2 **16.** 0.85 **17.** 0.52 **18.** 0.09 **19.** 0.41 **20.** 0.15 **21.** 0.01 **22.** 0.91 **23.** 0.38 **24.** 0.42 **25.** 0.11 **26.** 0.0065 **27.** 0.0003 **28.** 0.0075 **29.** 0.0099 **30.** 0.024 **31.** 0.07 **32.** 0.0106 **33.** 0.0901

# Exercise Set, Section 6-2

**1.** 1% **2.** 28% **3.** 25% **4.** 100% **5.** 66 2/3% **6.** 10% **7.** 75% **8.** 33 1/3% **9.** 87 1/2% **10.** 37% **11.** 40% **12.** 78% **13.** 97.2% **14.** 19.91% **15.** 4% **16.** 18% **17.** 1% **18.** 23% **19.** 75.3% **20.** 10.3% **21.** 83.5% **22.** 13.5% **23.** 6.1% **24.** 275% **25.** 27.7% **26.** 0.04% **27.** 100.6% **28.** 590.8% **29.** 92.2% **30.** 0.9%

# Exercise Set, Section 6-3

**1.** 69 **2.** 84 **3.** 44 **4.** 94 **5.** 2.5 **6.** 141 **7.** $x$/100 = 3/150; $x$ = 2% **8.** $x$/100 = 40/60; $x$ = 67% **9.** 11% **10.** 150% **11.** 150% **12.** 133% **13.** 70/100 = 58/$x$, $x$ = 83% **14.** 40.0/100 = 217/$x$, $x$ = 542 **15.** 5.0 **16.** 1.40 **17.** 2.8 **18.** 125 **19.** 20% **20.** 40

# Exercise Set, Section 6-4

**1.** 59% **2.** 0.45 lb **3.** 6.3 kg **4.** 8 parts **5.** 60% **6.** 90 workers **7.** 25% **8.** 1.3 hr **9.** 42/100 = 21/$x$, $x$ = 50 **10.** 144 hr **11.** 5% **12.** 11/100 = 534/$x$, $x$ = 4,855 parts **13.** 83% **14.** 75% **15.** 350 kg **16.** 11%

# Exercise Set, Sections 7-1–7-3

**1.** 100 **2.** 1,000 **3.** 10 **4.** 10 **5.** 1,000 **6.** 100 **7.** 1,000 **8.** 1,000 **9.** 100 **10.** 10 **11.** 1,930 **12.** 0.87 **13.** 0.132 **14.** 86 **15.** 670 **16.** 9.76 **17.** 1.280 **18.** 18 **19.** 8 **20.** 1.4 **21.** 16 **22.** 8.3 **23.** 100 **24.** 4.41 **25.** 8.29

# Exercise Set, Sections 7-4–7-6

**1.** 1,000 **2.** 10 **3.** 1 **4.** 1,000 **5.** 0.697 **6.** 2,600 **7.** 0.827 **8.** 840 **9.** 1.285 **10.** 1,600 **11.** 1.657 **12.** 250 **13.** 1,000 **14.** 100 **15.** 1,000 **16.** 10 **17.** 2,500 **18.** 0.415 **19.** 150 **20.** 2.16 **21.** 8,900 **22.** 0.089 **23.** 2,870 **24.** 0.487 **25.** 0.95 **26.** 39.4 **27.** 2.2 **28.** 454 **29.** 28 **30.** 30 **31.** 1.6 **32.** 2.54 **33.** 0.66 **34.** 2.27 **35.** 128 **36.** 236 **37.** 10.2 **38.** 3.16 **39.** 840 **40.** 20 **41.** 2 **42.** 9.5 **43.** 25 **44.** 10 **45.** 4,540 **46.** 1.26 **47.** 1.68 **48.** 1,900 **49.** 179 **50.** 1.2

# Exercise Set, Sections 7-7–7-8

**1.** 261 **2.** 473 **3.** 89.6 **4.** −12.2 **5.** −23.3 **6.** 5 **7.** −40 **8.** 20 **9.** 7.2 **10.** 50 **11.** 9.53 **12.** 19.1 **13.** 15.9 **14.** 7.94

# Exercise Set, Section 7-9

**1.** 13/16 in. **2.** 1 5/8 in. **3.** 0.7 cm **4.** 4.2 cm **5.** 1,700 rpm **6.** 4,600 rpm **7.** 6.33 **8.** 4.07 **9.** 33.21 **10.** 22.28

# Exercise Set, Sections 8-1–8-2

**1.** 184.96 **2.** 0.49 **3.** 6.25 **4.** 0.176 **5.** 6.554 **6.** 0.008 **7.** 7.626 **8.** 65.944 **9.** 11.45 **10.** 26.46 **11.** 1.273 **12.** 0.53 **13.** 7.21 **14.** 4.65 **15.** 0.5

# Exercise Set, Sections 8-3–8-4

**1.** $10^3$ **2.** $10^1$ **3.** $10^0$ **4.** $5.87 \times 10^3$ **5.** $1.83 \times 10^5$ **6.** $4.02 \times 10^5$ **7.** $2 \times 10^{-2}$ **8.** $4.3 \times 10^{-3}$ **9.** $9.7 \times 10^{-5}$ **10.** $10^7$ **11.** $10^{-2}$ **12.** $10^1$ **13.** $10^{-5}$ **14.** $6 \times 10^{-2}$ **15.** $8 \times 10^3$ **16.** $1.2 \times 10^5$ **17.** $6.4 \times 10^{-4}$ **18.** $1.296 \times 10^4$ **19.** $1.95 \times 10^{-4}$ **20.** $3.485 \times 10^6$ **21.** $10^3$ **22.** $3.5 \times 10^{-6}$ **23.** $4 \times 10^{-2}$ **24.** $3 \times 10^2$

# Exercise Set, Sections 9-1–9-2

**1.** 50° **2.** 80° **3.** 50° **4.** 50° **5.** 80° **6.** 130° **7.** 30° **8.** 80° **9.** 50° **10.** 9.06 m **11.** 4.83 $m^2$ **12.** 36.8 cm **13.** 85 $cm^2$ **14.** 60 $cm^2$ **15.** 24.60 cm **16.** 32.3 $cm^2$ **17.** 60.8 m **18.** 231 $m^2$ **19.** 77 $cm^2$ **20.** 40 cm **21.** 3.13 $in.^2$

# Exercise Set, Sections 9-3–9-4

**1.** 120°; 135°; 108° **2.** 27 ft **3.** 105 mm **4.** 342 cm **5.** 390 ft **6.** 32.8 $cm^3$ **7.** 829 $cm^3$ **8.** 0.36 $mm^3$ **9.** 4,000 $ft^3$ **10.** 25 g **11.** 340 g **12.** 10.8 g **13.** 28.6 g **14.** 160 g **15.** 121 g

# Exercise Set, Section 9-5

**1.** 30° **2.** 69.4° **3.** 1.09 $in.^2$ **4.** 4.10 $m^2$ **5.** 52.1 $ft^2$ **6.** 19.8 $cm^2$ **7.** $c = \sqrt{a^2 + b^2}$ **8.** $a = \sqrt{c^2 - b^2}$ **9.** 9.28 ft **10.** 19.26 mm **11.** 6.88 in. **12.** 1.84 cm

# Exercise Set, Sections 9-6–9-7

**1.** 4.00 cm **2.** 12.6 cm **3.** 12.6 $cm^2$ **4.** 56.2 $in.^2$ **5.** 54.0 cm **6.** Area of circle = 67 $in.^2$; Area of sector = 5.77 $in.^2$ **7.** 315 $cm^3$ **8.** 224 $cm^2$ **9.** 53 $in.^2$ **10.** 35 $cm^2$

# Exercise Set, Section 10-1

1.

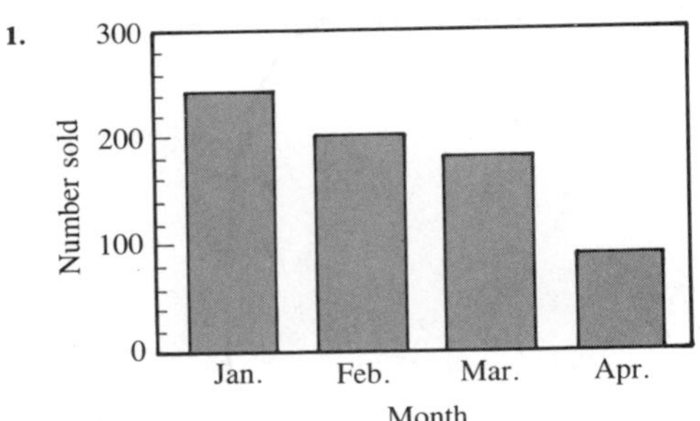


2.

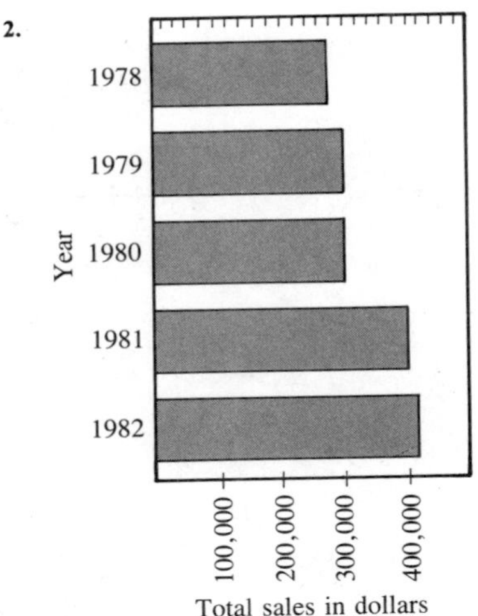


3.

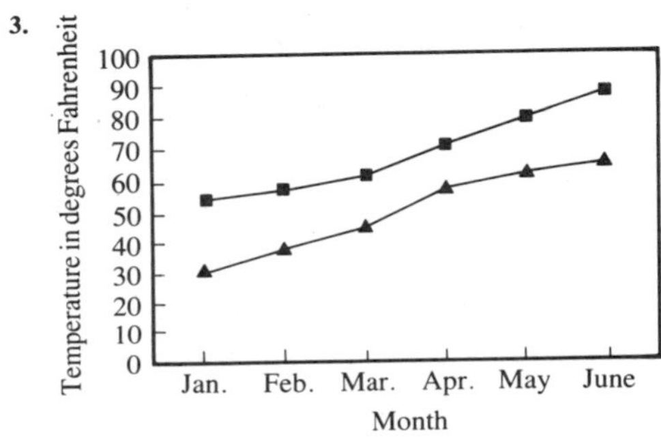


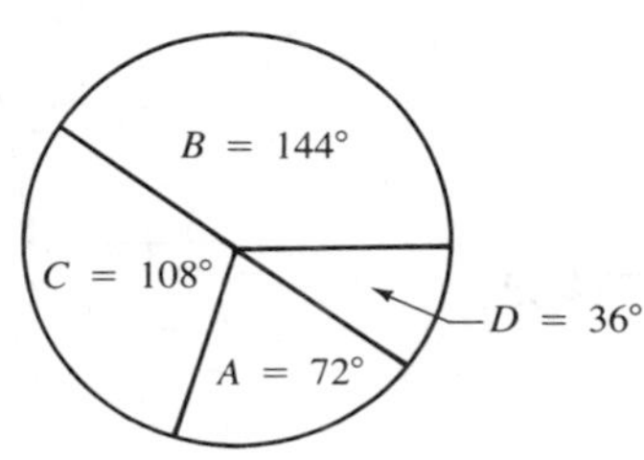


5.

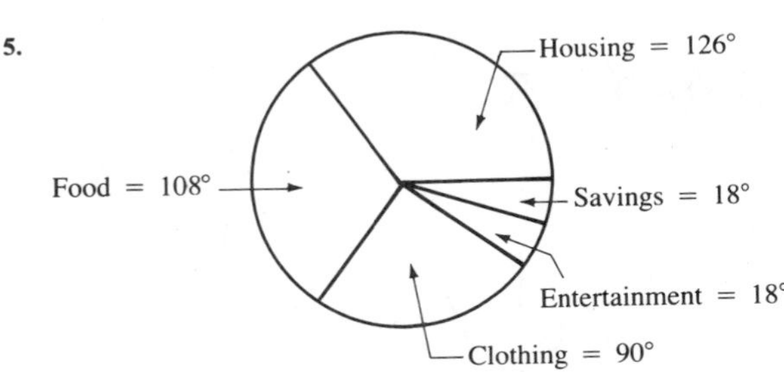


# Exercise Set, Sections 10-2–10-3

**1.** 4 **2.** 4 **3.** −2 **4.** 3 **5.** −4 **6.** −2 **7.** 2 **8.** −4 Answers **9.** to **12.**

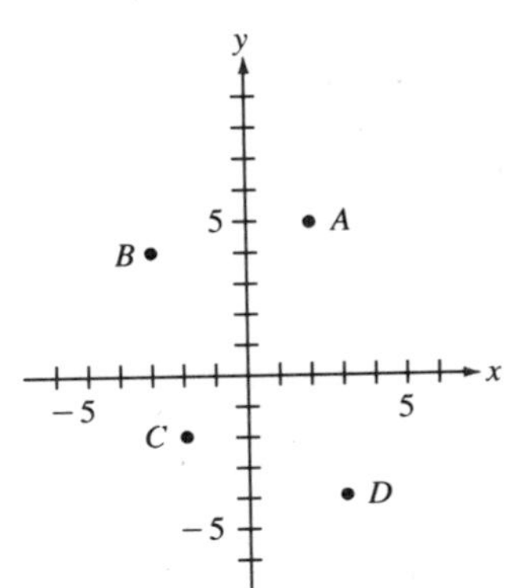

**13.**

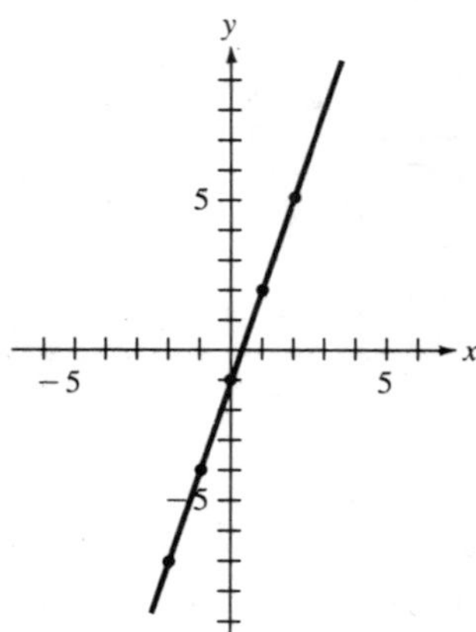


**14.**

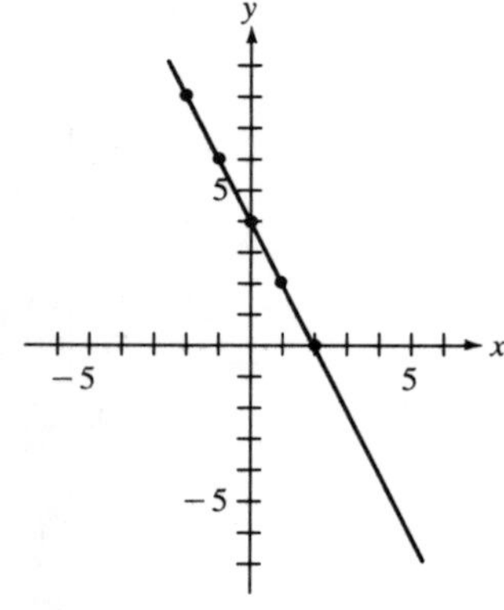


**15.**

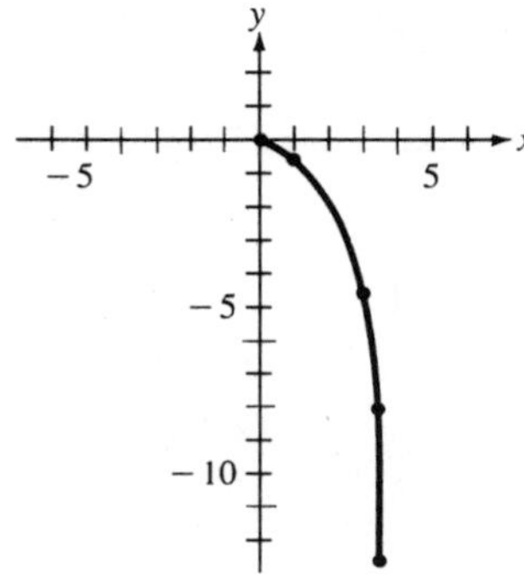


**16.**

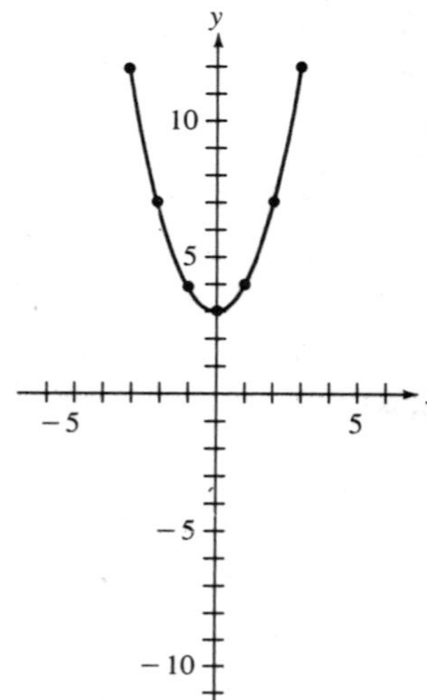


**17.**

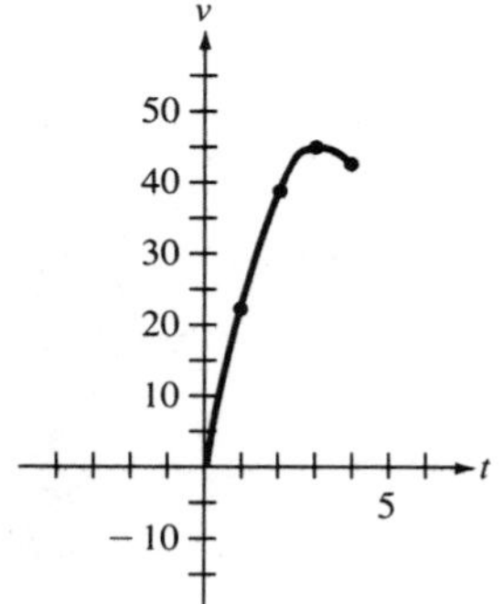


**18.**

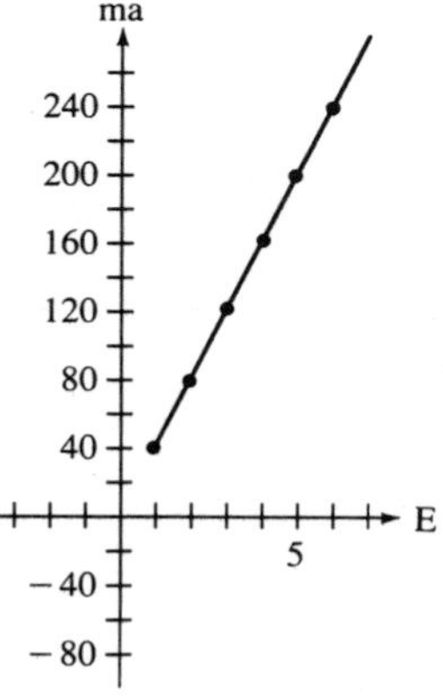


# Exercise Set, Sections 10-4–10-5

**1.** $t$ **2.** $V$ **3.** $x$ **4.** $y$ **5.** $2x^3 + 4$ **6.** $x^2 - 2x$

**7.** $y = 0.5x + 6$

| $x$ | $y$ |
|---|---|
| 0 | 6 |
| 1 | 6.5 |
| 2 | 7 |
| 3 | 7.5 |
| 4 | 8 |
| 5 | 8.5 |

**8.** $a = b^2 + 2$

| $b$ | $a$ |
|---|---|
| −3 | 11 |
| −2 | 6 |
| −1 | 3 |
| 0 | 2 |
| 1 | 3 |
| 2 | 6 |
| 3 | 11 |

**9.** $y = -x + 4$

| $x$ | $y$ |
|---|---|
| −3 | 7 |
| −2 | 6 |
| −1 | 5 |
| 0 | 4 |
| 1 | 3 |
| 2 | 2 |
| 3 | 1 |

**10.** $k = m^3 - 2$

| $m$ | $k$ |
|---|---|
| −2 | −10 |
| −1 | −3 |
| 0 | −2 |
| 1 | −1 |
| 2 | 6 |

**11.** $y = 2x + 6$

| $x$ | $y$ |
|---|---|
| −3 | 0 |
| −2 | 2 |
| 0 | 6 |
| 1 | 8 |
| 2 | 10 |
| 3 | 12 |

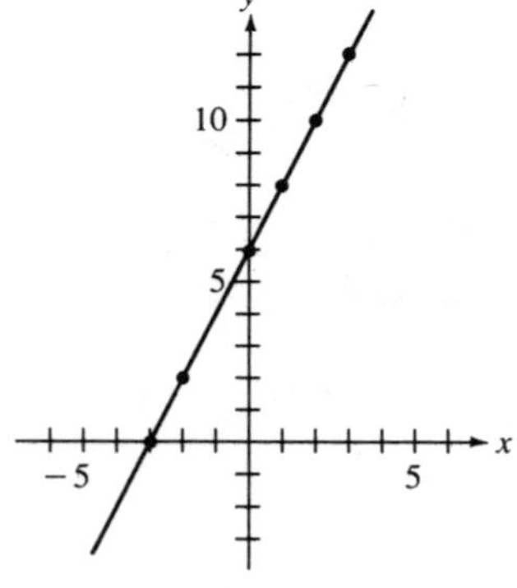

**12.** $y = -2x + 3$

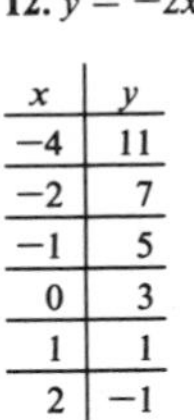

| $x$ | $y$ |
|---|---|
| −4 | 11 |
| −2 | 7 |
| −1 | 5 |
| 0 | 3 |
| 1 | 1 |
| 2 | −1 |

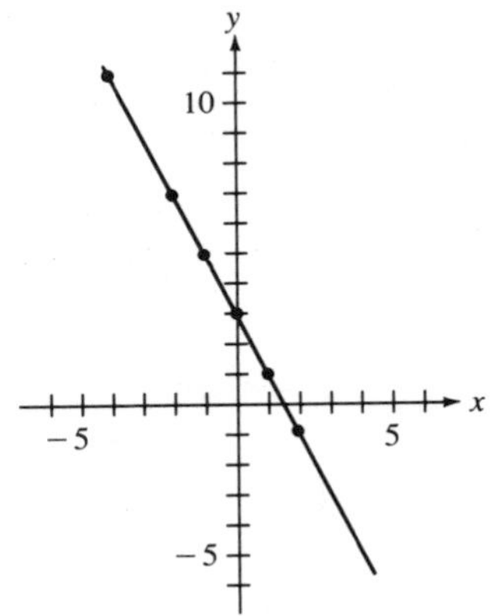


**13.** $y = 2x^2 - 1$

| $x$ | $y$ |
|---|---|
| 0 | −1 |
| 1 | 1 |
| 2 | 7 |
| 3 | 17 |



**14.** $a = b^3 + 1$

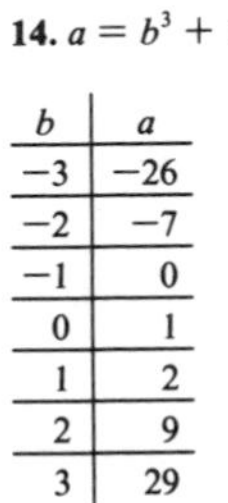

| $b$ | $a$ |
|---|---|
| −3 | −26 |
| −2 | −7 |
| −1 | 0 |
| 0 | 1 |
| 1 | 2 |
| 2 | 9 |
| 3 | 29 |

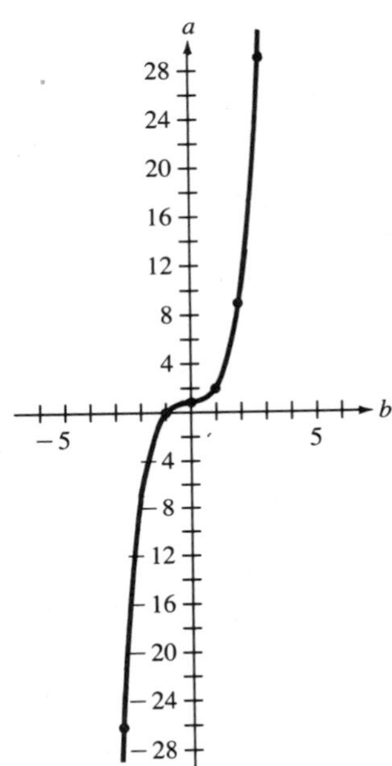


# Exercise Set, Sections 11-1–11-2

**1.** 0.3256 **2.** 0.6820 **3.** −1.4281 **4.** 3.2708 **5.** 0.7880 **6.** 0.1564 **7.** 15.01° **8.** 31.00° **9.** 88.02° **10.** 0.2924 **11.** 0.2588 **12.** −1.4281 **13.** 0.5000 **14.** 0.0000 **15.** 0.7547 **16.** 84.29° **17.** 34.00° **18.** 33.00° **19.** 115° **20.** 248°

# Exercise Set, Section 11-3

**1.** 38.32° **2.** 36.22° **3.** 48.19° **4.** 39.72° **5.** 45° **6.** 32.58° **7.** 55.01° **8.** 44.42° **9.** 40.36° **10.** 39.52°

# Exercise Set, Section 11-4

**1.** tan of angle = 4/15; angle = 15° **2.** tan of angle = 46/32; angle = 55° **3.** tan 56° = height/65; height = 96 ft **4.** tan of angle = 40/55; angle = 36° **5.** tan $A$ = 1.75/2.00; $\angle A$ = 41.2° **6.** sin 47° = 45 m/$BC$; $BC$ = 62 m **7.** tan 40° = $x$/3.30 cm; $x$ = 2.77 cm **8.** cos $\theta$ = 25.0/42.0; $\theta$ = 53.5 **9.** distance = (30 ft)(cos 65°); distance = 13 ft **10.** length = $\sqrt{45^2 + 15^2}$ ; length = 47 ft

# ANSWERS TO CHAPTER SELF-TESTS

## Chapter 1 Self-Test

**1.** ten-thousands **2.** hundredths **3.** 2,600,015 **4.** 0.0404 **5.** 32,349 **6.** 82.25 **7.** 23.7 **8.** 40.842 **9.** 265 **10.** 102.65 **11.** 92.9 **12.** 29.01 **13.** 29.4 **14.** 0.0987 **15.** 8.5854 **16.** 830,000 **17.** 86.50 **18.** 104.33 **19.** 35 **20.** 0.2 **21.** hundredths **22.** 4 **23.** 2.58 **24.** 62.64 **25.** 0.37 **26.** 2.482 **27.** 96 in. **28.** 100 mg **29.** 12.1 min **30.** 13.5 quarts **31.** 166 pieces **32.** 32.5 gallons **33.** 1.76 in.

## Chapter 2 Self-Test

**1.** 0.28 **2.** 0.37 **3.** 0.46 **4.** 0.38 **5.** 6/30 **6.** 25/30 **7.** 9/30 **8.** 4/30 **9.** 7/8 **10.** 2/5 **11.** 1/4 **12.** 1/4 **13.** 2 1/4 **14.** 2 2/3 **15.** 5 3/4 **16.** 7/3 **17.** 11/8 **18.** 53/3 **19.** 4/5 **20.** 5/6 **21.** 1/4 **22.** 13/24 **23.** 0.13 **24.** 0.65 **25.** 1/32 **26.** 1/9 **27.** 1/2 **28.** 1 **29.** 8 **30.** 10 **31.** 1/4 **32.** 3/2 **33.** 4/3 **34.** 25 **35.** 1/16 **36.** 3/40 **37.** 3 7/8 ft **38.** 9 1/8 watts **39.** 28 samples **40.** 18 in.

## Chapter 3 Self-Test

**1.** 0 **2.** 7 **3.** 36 **4.** 18 **5.** 10 **6.** −12 **7.** −4 **8.** −15 **9.** 32 **10.** 9 **11.** −18 **12.** −56 **13.** −32 **14.** 33 **15.** 36 **16.** −12 **17.** −12 **18.** 9.75 **19.** −18 **20.** −58 **21.** −6 **22.** 3150 **23.** 8 **24.** 6 **25.** 5 **26.** 6 **27.** $4x + 18$ **28.** $5x + 10$ **29.** $5x + 14$ **30.** $19h + 6$ **31.** 14 **32.** −2 **33.** 7 **34.** −22 **35.** 22 **36.** 7 **37.** $x = 4$ **38.** $2x = 10; x = 5$ **39.** $V = IR$ **40.** $A = T - B - C$

## Chapter 4 Self-Test

**1.** 28 **2.** 25 **3.** 40.5 **4.** 1.11 **5.** 12 **6.** 1.62 **7.** 5.73 **8.** 1.4 **9.** 190 **10.** 301° **11.** −193° **12.** 0.64 **13.** 20° **14.** 40 **15.** 2.2 **16.** 24 **17.** 331 **18.** 10.4 **19.** 19.8 **20.** 856 **21.** 1,620 **22.** $V_2 = T_2V_1/T_1$ **23.** $P_2 = P_1V_1T_2/T_1V_2$ **24.** $h = 5/K$ **25.** $x = Y - b/m$ **26.** $xy - by = -h - k; y = (h + k)/(b - x)$ **27.** $h = zm/k$ **28.** $a = Ps + m$ **29.** $K = wz/24$ **30.** $h = 2\pi^2 m/T$ **31.** $L = P - 2w/2$ **32.** 200 mℓ/300 K = $V_2$/450 mℓ; $V_2$ = 300 mℓ **33.** 100 mℓ/200 K = 300 mℓ/$T_2$; $T_2$ = 600 K **34.** $P$ = (25 psi)($A$)/20 lb; $P$ = (5 psi)($A$)/4 lb **35.** 445 psi

## Chapter 5 Self-Test

**1.** 1:8 **2.** 2:7 **3.** 4:1 **4.** 4:1 **5.** 4:1 **6.** 2.91:1 **7.** 0.8:1 **8.** 30:9 **9.** $4/x = 8/12$ **10.** $5/9 = Y/10$ **11.** 6 **12.** 1.67 **13.** 1.75 **14.** 32 g **15.** $3/10 = x/150$; $x$ = 45 cubic ft **16.** 1,200/8,200 = $x$/27,800; $x$ = $4,068.29 **17.** 0.48 **18.** 0.074 **19.** 41 **20.** $V_2 = P_1V_1/P_2$; (5 atm)(200 mℓ)/9 atm; $V_2$ = 111 mℓ

## Chapter 6 Self-Test

**1.** 21/50 **2.** 2/3 **3.** 65/100 or 13/20 **4.** 4 **5.** 1 3/4 **6.** 15 **7.** 0.95 **8.** 0.09 **9.** 0.0025 **10.** 2% **11.** 30% **12.** 33 1/3% **13.** 70% **14.** 46.7% **15.** 0.6% **16.** 58.8% **17.** 7.8% **18.** 8.6% **19.** 0.05% **20.** 45 **21.** 96 **22.** 11% **23.** 150% **24.** 44/100 = 36/$x$, $x$ = 82 **25.** 110/100 = 42/$x$, $x$ = 38% **26.** 80.0/100 = $N$/210, $N$ = 168 kg **27.** 18/42 = $N$/100, $N$ = 43% **28.** 2.0/100 = $N$/85.0, $N$ = 83 in. **29.** $N$/100 = 18/32, $N$ = 56% **30.** 6/100 = $N$/220, $N$ = 13 pieces **31.** (460 − 500)/500 = −8% **32.** (10 − 9.2)/10 = 8%

## Chapter 7 Self-Test

**1.** 1,000 **2.** 100 **3.** 10 **4.** 1,000 **5.** 100 **6.** 10 **7.** 1,000 **8.** 1,000 **9.** 2,830 **10.** 0.92 **11.** 0.484 **12.** 56 **13.** 0.681 **14.** 4,800 **15.** 4,500 **16.** 0.082 **17.** 1.6 **18.** 28 **19.** 454 **20.** 0.95 **21.** 30 **22.** 2.54 **23.** 6.32 **24.** 1.1 **25.** 4.29 **26.** 1,800 **27.** 15,000 **28.** 255 **29.** 20 **30.** 64.4 **31.** −27.8 **32.** 23.8 **33.** 8.73 **34.** 4.18 **35.** 10 + 1.5 + 0.07 = 11.57

## Chapter 8 Self-Test

**1.** 201.64 **2.** 0.036 **3.** 0.672 **4.** 16 **5.** 0.004 **6.** 14.2 **7.** 19.672 **8.** 0.992 **9.** 1.367 **10.** 8.602 **11.** 0.255 **12.** $10^1$ **13.** $4.26 \times 10^3$ **14.** $8.1 \times 10^{-4}$ **15.** $10^8$ **16.** $10^{-7}$ **17.** $6 \times 10^{-7}$ **18.** $3.72 \times 10^3$ **19.** $5.6 \times 10^6$ **20.** $1.827 \times 10^{-1}$ **21.** $10^4$ **22.** $3 \times 10^{-10}$ **23.** $4 \times 10^{-2}$ **24.** $3 \times 10^2$ **25.** $2 \times 10^{-9}$

# Chapter 9 Self-Test

**1.** 30° **2.** 50° **3.** 100° **4.** 22.82 cm **5.** 27.7 $cm^2$ **6.** 12.4 m **7.** 9.6 $m^2$ **8.** 19.9 $cm^3$ **9.** 490 $cm^3$ **10.** 1.58 $in.^3$ **11.** 105 g **12.** 72.6 $cm^2$ **13.** 29.1 in. **14.** area of circle = 314 $in.^2$; $40°/360° = x/314$ $in.^2$; $x = 35$ $in.^2$ **15.** 51 $ft^2$

# Chapter 10 Self-Test

**1.** 3 **2.** 4 **3.** −3 **4.** −4 Answers to **5.** and **6.**

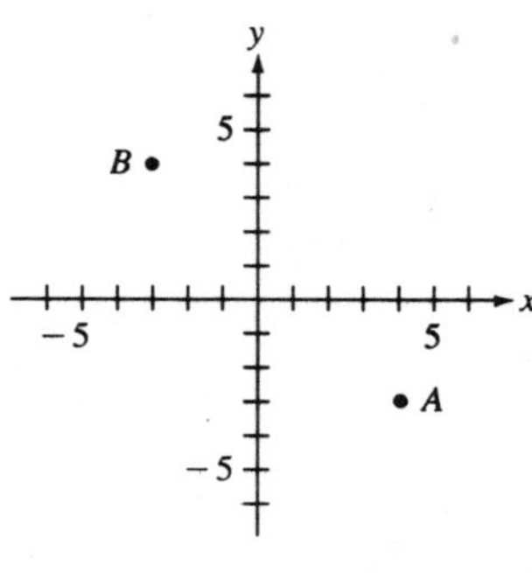


**7.**

**8.**

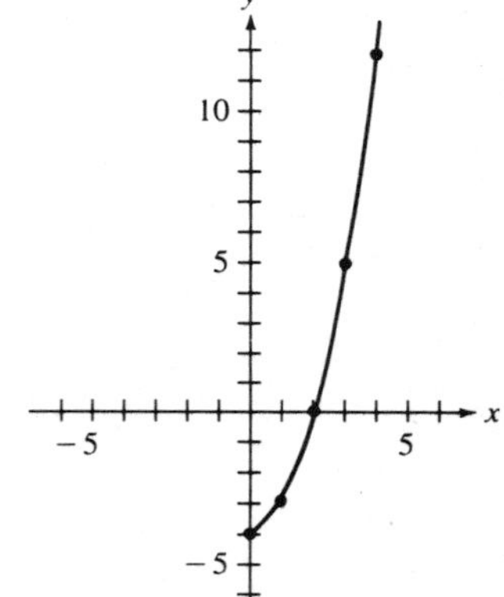


**9.** $x$ **10.** $y$ **11.** $7x + 6$ **12.** $y = 0.1x + 2$

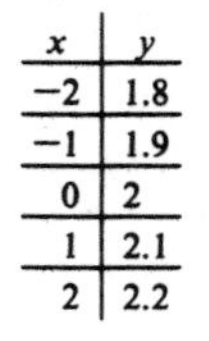

| $x$ | $y$ |
|---|---|
| −2 | 1.8 |
| −1 | 1.9 |
| 0 | 2 |
| 1 | 2.1 |
| 2 | 2.2 |

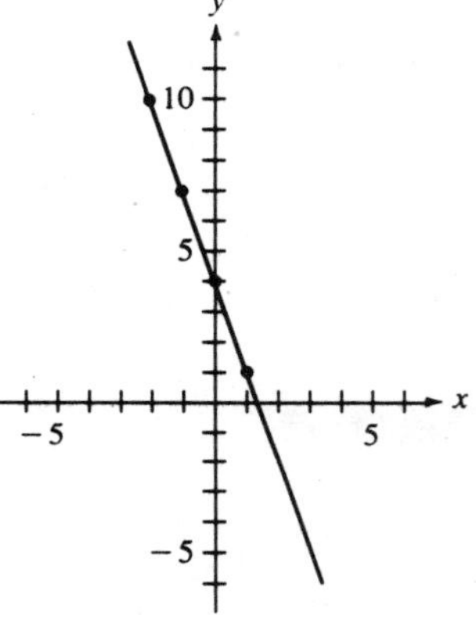


**13.** $a = b^3 + 1$

| $b$ | $a$ |
|---|---|
| −2 | −7 |
| −1 | 0 |
| 0 | 1 |
| 1 | 2 |
| 2 | 9 |

**14.** $y = -3x + 4$

| $x$ | $y$ |
|---|---|
| −2 | 10 |
| −1 | 7 |
| 0 | 4 |
| 1 | 1 |

**15.**

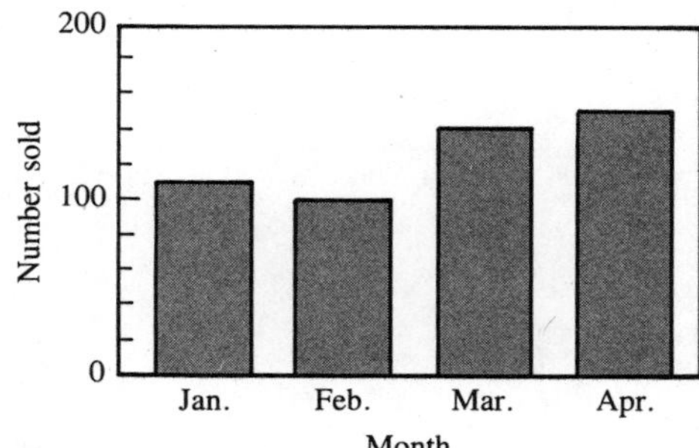


**16.**

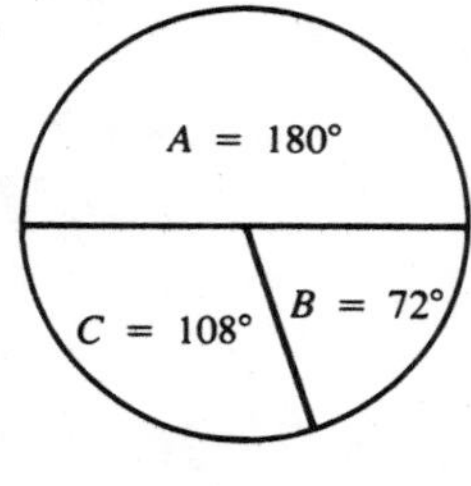


# Chapter 11 Self-Test

**1.** 33° **2.** 4.56 $ft^2$ **3.** $c = \sqrt{a^2 + b^2}$, $c = 9.62$ ft **4.** $a = \sqrt{c^2 - b^2}$, $a = 4.38$ in. **5.** $a = (\tan 41°)(11 \text{ ft})$, $a = 9.6$ ft **6.** $\cos A = 18/27$, $A = 48°$ **7.** $\tan A = 5.00/4.5$, $A = 48°$ **8.** $\tan \alpha = 54/42$, $\alpha = 52°$ **9.** $\tan H = 0.250/0.250$, $H = 45°$ **10.** $\tan K = 1.0/1.3$, $K = 38°$

# INDEX